Prokaryotic diversity: mechanisms and significance

The true extent of prokaryote diversity, encompassing the spectrum of variability among bacteria, remains unknown. Early discussions on prokaryote diversity were frequently devoted to sterile arguments about 'how much?' or 'how many?'. Increasingly, however, the focus is turning towards trying to understand why prokaryote diversification occurs, its underlying mechanisms, and its likely impact. The significance of such studies has a broad appeal, and the popular scientific press frequently highlights such topics as the emergence of new diseases, the attribution of existing diseases to hitherto unrealized actions of prokaryotes, and the activities of prokaryotes in key environmental processes. The dynamic nature of the prokaryotic world, and continuing advances in the technological tools available to this field of study, ensure that the latest story illustrating prokaryote diversity is never far away. This book will appeal to a wide variety of microbiologists. Its coverage ranges from studies of prokaryotes in specialized environmental niches to broad examinations of prokaryote evolution and diversity, and the mechanisms underlying them. Topics include: bacteria of the gastrointestinal tract, unculturable organisms in the mouth and in the soil, the question of a link between chlamydia and heart disease, organisms from extreme environments, the diversity of archaea and their phages, comparative genomics and the emergence of pathogens, spread of genomic islands between clinical and environmental organisms, core genes, minimal genomes needed for life, horizontal gene transfer, genomic islands and the evolution of catabolic pathways, phenotypic innovation, and patterns and extent of biodiversity.

Niall A. Logan is Professor of Systematic Bacteriology in the School of Life Sciences, Glasgow Caledonian University, UK.

Hilary M. Lappin-Scott is Professor of Environmental Microbiology in the School of Biosciences, University of Exeter, UK.

Petra C. F. Oyston is Team Leader in Molecular Bacteriology at Dstl Porton Down, Salisbury, UK.

Symposia of the Society for General Microbiology

Managing Editor: Dr Melanie Scourfield, SGM, Reading, UK
Volumes currently available:

43	Transposition
45	Control of virus diseases
47	Prokaryotic structure and function – a new perspective
52	Population genetics of bacteria
53	Fifty years of antimicrobials: past perspectives and future trends
54	Evolution of microbial life
55	Molecular aspects of host–pathogen interactions
56	Microbial responses to light and time
57	Microbial signalling and communication
58	Transport of molecules across microbial membranes
59	Community structure and co-operation in biofilms
60	New challenges to health: the threat of virus infection
61	Signals, switches, regulons and cascades: control of bacterial gene expression
62	Microbial subversion of host cells
63	Microbe–vector interactions in vector-borne diseases
64	Molecular pathogenesis of virus infections
65	Micro-organisms and Earth systems – advances in geomicrobiology

SIXTY-SIXTH SYMPOSIUM OF THE
SOCIETY FOR GENERAL MICROBIOLOGY
HELD AT THE UNIVERSITY OF WARWICK APRIL 2006

Edited by
N. A. Logan, H. M. Lappin-Scott & P. C. F. Oyston

prokaryotic diversity: mechanisms and significance

Published for the Society for General Microbiology

CAMBRIDGE UNIVERSITY PRESS
Cambridge, New York, Melbourne, Madrid, Cape Town,
Singapore, São Paulo

Cambridge University Press
The Edinburgh Building, Cambridge CB2 2RU, UK

Published in the United States of America by
Cambridge University Press, New York

www.cambridge.org
Information on this title: www.cambridge.org/9780521869355

First published 2006

Printed in the United Kingdom at the University Press, Cambridge

A catalogue record for this publication is available from the British Library

ISBN 13   978 0 521 86935 5 hardback
ISBN 10   0 521 86935 8 hardback

Typeface   Sabon (Adobe) 10·5/13·5 pt   *System*   QuarkXPress™   [SGM]

Front cover illustration: Coloured scanning electron micrograph of a mixture of saprophytic cocci and rods covering the surface of a piece of decomposing leaf on a compost heap. Eye of Science / Science Photo Library

CONTENTS

Contributors vii

Editors' Preface xi

R. R. Colwell
Microbial diversity in the era of genomics 1

M. C. Horner-Devine, J. Green and B. J. M. Bohannan
Patterns in prokaryotic biodiversity 19

E. V. Koonin, K. S. Makarova, N. V. Grishin and Y. I. Wolf
A putative RNA-interference-based immune system in prokaryotes: the epitome of prokaryotic genomic diversity 39

H. J. Flint
The significance of prokaryote diversity in the human gastrointestinal tract 65

H. J. E. Beaumont, S. M. Gehrig, R. Kassen, C. G. Knight, J. Malone, A. J. Spiers and P. B. Rainey
The genetics of phenotypic innovation 91

R. Gil, V. Pérez-Brocal, A. Latorre and A. Moya
Minimal genomes required for life 105

V. Daubin and E. Lerat
Evolution of the core of genes 123

K. M. Stedman, A. Clore and Y. Combet-Blanc
Biogeographical diversity of archaeal viruses 131

L. A. Campbell and C.-C. Kuo
Is there a link between *Chlamydia* and heart disease? 145

W. G. Wade
Unculturable oral bacteria 163

R. W. Titball and M. Duffield
Comparative genomics – what do such studies tell us about the emergence and spread of key pathogens? 175

J. Klockgether, O. N. Reva and B. Tümmler
Spread of genomic islands between clinical and environmental strains 187

A. Morningstar, W. H. Gaze, S. Tolba and E. M. H. Wellington
Evolving gene clusters in soil bacteria 201

A. Ventosa
Unusual micro-organisms from unusual habitats: hypersaline environments 223

S. Lacour, M. Gaillard and J. R. van der Meer
Genomic islands and evolution of catabolic pathways 255

A. M. Osborn
Horizontal gene transfer and its role in the emergence of new phenotypes 275

Index 295

CONTRIBUTORS

Beaumont, H. J. E.
School of Biological Sciences, University of Auckland, Private Bag 92019, Auckland, New Zealand

Bohannan, B. J. M.
Department of Biological Sciences, Stanford University, Stanford, CA 94305, USA

Campbell, L. A.
Department of Pathobiology, University of Washington, Seattle, WA 98195, USA

Clore, A.
Department of Biology, Center for Life in Extreme Environments, Portland State University, PO Box 751, Portland, OR 97207-0751, USA

Colwell, R. R.
University of Maryland, College Park, MD 20742, USA, and Johns Hopkins Bloomberg School of Public Health, 615 North Wolfe Street, Baltimore, MD 21205, USA

Combet-Blanc, Y.
Laboratoire de Microbiologie IRD, Université de Provence, CESB/ESIL case 925, 163 avenue de Luminy, F-13288 Marseille Cedex 9, France

Daubin, V.
Laboratoire de Biométrie et Biologie Evolutive, 43 Bld du 11 Novembre 1918, Université Lyon 1, 69622 Villeurbanne cedex, France

Duffield, M.
Defence Science and Technology Laboratory, Porton Down, Salisbury, Wiltshire SP4 0JQ, UK

Flint, H. J.
Microbial Ecology Group, Rowett Research Institute, Greenburn Road, Bucksburn, Aberdeen AB21 9SB, UK

Gaillard, M.
Department of Fundamental Microbiology, University of Lausanne, CH-1015 Lausanne, Switzerland

Gaze, W. H.
Department of Biological Sciences, University of Warwick, Coventry CV4 7AL, UK

Gehrig, S. M.
Department of Plant Sciences, University of Oxford, South Parks Road, Oxford OX1 3RB, UK

Gil, R.
Insituto Cavanilles de Biodiversidad y Biología Evolutiva, Universitat de València, Apartado Postal 22085, 46071 València, Spain

Green, J.
School of Natural Sciences, University of California, Merced, CA 95344, USA

Grishin, N. V.
Department of Biochemistry, University of Texas Southwestern Medical Center, 5323 Harry Hines Blvd, Dallas, TX 75390-9050, USA

Horner-Devine, M. C.
School of Aquatic and Fishery Sciences, University of Washington, Seattle, WA 98195, USA

Kassen, R.
Department of Biology and Centre for Advanced Research in Environmental Genomics, University of Ottawa, 150 Louis Pasteur, Ottawa, ON, K1N 6N5, Canada

Klockgether, J.
Klinische Forschergruppe, OE 6711, Medizinische Hochschule Hannover, D-30625 Hannover, Germany

Knight, C. G.
School of Chemistry, University of Manchester, Faraday Building, Box 88, Sackville St, Manchester M60 1QD, UK

Koonin, E. V.
National Center for Biotechnology Information, National Library of Medicine, National Institutes of Health, Bethesda, MD 20894, USA

Kuo, C.-C.
Department of Pathobiology, University of Washington, Seattle, WA 98195, USA

Lacour, S.
Department of Fundamental Microbiology, University of Lausanne, CH-1015 Lausanne, Switzerland

Latorre, A.
Insituto Cavanilles de Biodiversidad y Biología Evolutiva, Universitat de València, Apartado Postal 22085, 46071 València, Spain

Lerat, E.
Laboratoire de Biométrie et Biologie Evolutive, 43 Bld du 11 Novembre 1918, Université Lyon 1, 69622 Villeurbanne cedex, France

Makarova, K. S.
National Center for Biotechnology Information, National Library of Medicine, National Institutes of Health, Bethesda, MD 20894, USA

Malone, J.
Division of Molecular Microbiology, Biozentrum, University of Basel, Klingelbergstrasse 70, CH-4056 Basel, Switzerland

Morningstar, A.
Department of Biological Sciences, University of Warwick, Coventry CV4 7AL, UK

Moya, A.
Insituto Cavanilles de Biodiversidad y Biología Evolutiva, Universitat de València, Apartado Postal 22085, 46071 València, Spain

Osborn, A. M.
Department of Animal and Plant Sciences, The University of Sheffield, Western Bank, Sheffield S10 2TN, UK

Pérez-Brocal, V.
Insituto Cavanilles de Biodiversidad y Biología Evolutiva, Universitat de València, Apartado Postal 22085, 46071 València, Spain

Rainey, P. B.
School of Biological Sciences, University of Auckland, Private Bag 92019, Auckland, New Zealand, and Department of Plant Sciences, University of Oxford, South Parks Road, Oxford OX1 3RB, UK

Reva, O. N.
Klinische Forschergruppe, OE 6711, Medizinische Hochschule Hannover, D-30625 Hannover, Germany

Spiers, A. J.
Department of Plant Sciences, University of Oxford, South Parks Road, Oxford OX1 3RB, UK

Stedman, K. M.
Department of Biology, Center for Life in Extreme Environments, Portland State University, PO Box 751, Portland, OR 97207-0751, USA

Titball, R. W.
Defence Science and Technology Laboratory, Porton Down, Salisbury, Wiltshire SP4 0JQ, UK

Tolba, S.
Department of Biological Sciences, University of Warwick, Coventry CV4 7AL, UK

Tümmler, B.
Klinische Forschergruppe, OE 6711, Medizinische Hochschule Hannover, D-30625 Hannover, Germany

van der Meer, J. R.
Department of Fundamental Microbiology, University of Lausanne, CH-1015 Lausanne, Switzerland

Ventosa, A.
Department of Microbiology and Parasitology, Faculty of Pharmacy, University of Seville, 41012 Seville, Spain

Wade, W. G.
King's College London Dental Institute at Guy's, King's College and St Thomas' Hospitals, Infection Research Group, London SE1 9RT, UK

Wellington, E. M. H.
Department of Biological Sciences, University of Warwick, Coventry CV4 7AL, UK

Wolf, Y. I.
National Center for Biotechnology Information, National Library of Medicine, National Institutes of Health, Bethesda, MD 20894, USA

EDITORS' PREFACE

There have been many important recent developments in our knowledge of the breadth of prokaryote diversity, our understanding of the driving forces behind that diversity, and of its significance for our lives and for fundamental processes upon Earth. It has become clear that the microbes we know about are actually just the tip of a biological iceberg. In fact, the majority of microbes are unculturable on laboratory media at present. Much of our attention has been focused on pathogens, understanding their interaction with the host and how to prevent disease. However, there is a growing appreciation that without microbes fundamental ecological processes would not be balanced. For example, microbes in the ocean have a direct influence on the composition of the atmosphere we breathe.

A major advance in allowing us to understand the extent and nature of microbial diversity has been the development of genome sequencing. In parallel, there has been the development of tools to allow whole-genome comparisons. This has facilitated the study of microbial diversity and evolution, such as allowing the tracking of unculturable organisms, the study of organisms from extreme environments, and of medical and environmental bacteria and interactions between them. It has given us insights into the exchanges of genes between organisms, resulting in an understanding of the emergence of pathogens, a process which involves both gene acquisition and gene loss. Genomic comparison has helped to identify core genes, to the point where we can predict a minimal genome needed for life, which can be supplemented by the horizontal transfer of genomic islands, phenotypic innovation and catabolic pathway evolution. This has all culminated in the development of broader concepts such as the extent of prokaryote diversity and the patterns discernible within it.

This symposium volume aims to review these themes. Some reviews will discuss the extent of microbial diversity, from extreme environments to unculturables, while others will discuss how this wonderful diversity emerged. We extend our thanks to the authors for their contributions, despite their already heavy workloads, to Melanie Scourfield for her help in assembling this volume, and to the staff at SGM, particularly Josiane Dunn and Janet Hurst, for their help in organizing the Symposium held at Warwick in 2006.

Niall A. Logan
Hilary M. Lappin-Scott
Petra C. F. Oyston

Microbial diversity in the era of genomics

Rita R. Colwell[1,2]

[1]University of Maryland, College Park, MD 20742, USA

[2]Johns Hopkins Bloomberg School of Public Health, 615 North Wolfe Street, Baltimore, MD 21205, USA

THE CONTINUING SAGA OF THE MICROBIAL SPECIES

Fundamental to the assessment of global microbial diversity is understanding the unit of measurement of diversity, namely the microbial species. Historically, definition of a microbial species has been accomplished by employing relatively crude methods based on laboratory culture. Limitations of both the methods and the philosophical construct underpinning the species definition for micro-organisms have been elaborately detailed by Woese (2004), who pointed out the fallacies of the 'concept of a bacterium' that led to the inclusion of all bacteria (as then understood) within the single grouping of 'prokaryotes', with a shared 'prokaryotic' organization evolved ultimately from a common prokaryotic ancestor (Stanier & van Niel, 1962; Stanier *et al.*, 1957, 1963). Woese (2004) tracks the history of the 'prokaryote–eukaryote' dichotomy to the protozoologist Edouard Chatton (1938), whose reasoning was that, just as nucleated cells represented a monolithic grouping structurally and phylogenetically, non-nucleated cells (bacteria) also do so. Woese (2004) provides an eloquent explanation for the lack of philosophical understanding impelled by, on the one hand, technological advance and, on the other, 'fundamentalist reductionism' (the reductionism of 19th century classical physics), namely the view that living systems can be completely understood in terms of the properties of their constituent parts. The unfortunate aspect of fundamentalist reductionism is that it ignores the notion of emergent properties. The definition of a 'micro-organism' is, in itself, elusive. The term generally denotes members of the domains *Bacteria* and *Archaea*, as well as microscopic members of the domain *Eukarya*, i.e. the unicellular algae, some fungi and protists (Hughes Martiny *et al.*, 2006). A more concise definition has been offered, namely that a micro-organism has a

SGM symposium 66: Prokaryotic diversity: mechanisms and significance.
Editors N. A. Logan, H. M. Lappin-Scott & P. C. F. Oyston. Cambridge University Press. ISBN 0 521 86935 8

mass of less than 10^{-5} g and a length of less than 500 μm (Hughes Martiny *et al.*, 2006).

The emergence of higher levels of organization with attendant, qualitatively new properties in the form of complex cells has nevertheless left a trail in existing molecular sequence data. Woese (2004) drew upon the canonical pattern found in sequence comparisons, pointing out that, for nearly all of the proteins involved in transcription and translation, the archaeal and bacterial versions of each, although homologous, are clearly dissimilar (Woese, 1987; Woese *et al.*, 2000). Canonical pattern, then, represents the molecular 'fossil remains' of an evolutionary saltation (Woese, 1987; Woese *et al.*, 2000) and provides the traces of microbial evolutionary history.

The dilemma for microbial systematicists is that bacteria, generally speaking, are characterized by extensive intraspecies variation. Sequence variation is found in all species, as is the presence or absence of whole genes or clusters of genes. It is necessary to obtain the entire genomic sequence for all the DNA relevant for a species to have the actual 'species' genome, as distinct from an individual genome (Lan & Reeves, 2000). Bacteria can carry plasmids or lysogenic bacteriophages and these are usually present in only some strains of a species. Furthermore, groups of genes can occur in the chromosome of some strains and not others, the best known of such groupings being the pathogenicity islands (PAIs), carrying genes that confer aspects of pathogenicity. Lan & Reeves (2000) provide a useful review of genome comparison methods, i.e. macrorestriction mapping or pulsed-field gel electrophoresis, analysis of genome size variation, horizontal gene transfer and niche adaptation and gene loss during bacterial evolution. One conclusion drawn from comparative analyses is that, based on presently available data, up to 20 % of the DNA in any given strain can be absent in another, related strain.

Nevertheless, if we accept the concept of a species genome comprising all genes found in the species, two components are critical: a core set of genes and auxiliary genes. That is, genes found in most of the individuals of a species, the core set of genes, are those genes that determine properties characteristic of all members of the species. Each strain may have, in addition, some auxiliary genes that determine properties found in some, but not all members of the species. Genes found in 95% or more of isolates form a core set and genes found in 1–95 % of isolates comprise the auxiliary set of genes, while those present in < 1 % represent foreign genes or genes being lost from the species (Lan & Reeves, 2000).

The first attempt to use quantitative genome data to establish a genome-based species definition for prokaryotes was made by Konstantinidis & Tiedje (2005a, b), who compared the gene content of 70 closely related and fully sequenced strains to determine whether species boundaries exist and also to determine whether the ecology of an organism affected shared gene content. The mean nucleotide identity of shared genes between two strains was concluded to be a robust method for comparing genetic relatedness. Their conclusion was that approximately 94 % mean nucleotide identity corresponded to the 70 % DNA–DNA relatedness standard of the current species definition (Wayne *et al.*, 1987) that a bacterial species is essentially a collection of strains characterized by at least one diagnostic phenotypic trait and whose purified DNA shows at least 70 % reassociation, i.e. DNA–DNA relatedness (Brenner *et al.*, 2001). Konstantinidis & Tiedje (2005a, b) concluded that a more stringent and natural definition for prokaryotic species was needed to accommodate ecological distinctiveness of bacteria, i.e. would be more predictive of phenotype and ecological potential of species. They opined that only strains showing higher than 99 % mean nucleotide identity, or less at the nucleotide level but sharing an overlapping ecological niche, should belong to the same species, since such strains show few gene differences, e.g. <5 % difference.

Almost three decades ago, Woese & Fox (1977a, b) introduced rRNA as a taxonomic marker tool and thereby extended the tree paradigm, namely bifurcating trees in which evolutionary lineages split and evolve independently from each other, to microorganisms. The tree of life is the standard imagery employed to depict species evolution, employing a common root of all life on Earth and a bifurcating evolutionary process. Reticulate models of evolutionary history that incorporate gene transfer have proven valuable for microbiology. As pointed out by Gogarten & Townsend (2005), organismal mutualisms and lineage reticulation are supplemented by horizontal gene transfer, as processes that lead to the network-like histories of living organisms. Essentially, all functional categories of genes are susceptible to horizontal gene transfer, even rRNA operons (Gogarten *et al.*, 2002) and those genes associated with phylum-defining characteristics. The main difference, then, between prokaryotes and multicellular eukaryotes is the 'fuzzy' species boundaries of prokaryotes (Lawrence, 2002; Hanage *et al.*, 2005).

A perspective recently offered by Gevers *et al.* (2005) is that there is no widely accepted concept of species for prokaryotes and that assignment of isolates to species depends on phenotypic or genome similarity. In a study by Tettelin *et al.* (2005), the results of which tend to support the view of Gevers *et al.* (2005), six strains of *Streptococcus agalactiae* (group B streptococcus) were sequenced and their sequences were compared with each of the other strains and two previously sequenced strains. The eight genomes covered all five of the major disease-causing serotypes. Each of the eight strains contained unique genes, with about 80 % of the group B streptococcus genome conserved in all eight strains. The other 20 % was either partially shared or specific to only one strain. The authors calculated how many new genes appeared with each additional strain sequenced, considering all possible permutations. Their observation was that each genome sequence yielded a significant number of new genes. Previous work with *Escherichia coli* has also shown very high diversity between strains, most likely arising from horizontal gene transfer. Clearly, as Gevers *et al.* (2005) state, current methods for defining prokaryotic species are inadequate and incapable of keeping pace with the levels of diversity that are being discovered in nature.

It becomes clear, then, that a sequence-based approach to microbial systematics, in which comparisons between strains are carried out by computer, is the future of microbial systematics and taxonomy. It is generally conceded that it simply may not be possible to delineate groups or clusters of strains within a continuous spectrum of genotypic variation, i.e. clustering may not occur in every instance. It may also be possible that a new group may not be recognized if only one or a few genotypes have been isolated. Even more difficult is the situation whereby an important phenotype is conferred by a plasmid and not by stable chromosomal loci.

Despite these shortcomings of contemporary microbial systematics, much has been gained from genomic and evolutionary studies of microbial species. For example, it is only through complete genome sequencing that it has been possible to detect gene loss. Because of the value of genomics to microbial systematics, Eisen & Fraser (2003) call for the integration of genome analysis and evolutionary analysis into a 'phylogenomics'. Their view is that not only has analysis of complete genomes contributed to better understanding of horizontal gene transfer, but they suggest that horizontal gene transfer should be carefully analysed to rule out unusual rates of evolution, strong selection or gene loss before attributing variation only to horizontal gene transfer in every instance (Eisen, 2000).

Undoubtedly, genomics has influenced key concepts in microbiology and will place microbial systematics on a much more sound footing (Ward & Fraser, 2005) and the extremes in genomic analyses provide new boundaries but, at the same time, reinforce the 'fuzziness' of the microbial species boundaries. For example, the large genome (400 nm) and unusual coding capacity (1·2 Mb) of mimivirus, a DNA virus that infects amoebae, is similar to that of a small bacterium (La Scola *et al.*, 2003). It exemplifies the shifting boundaries between the currently accepted microbial domains, *Bacteria*, *Archaea*, *Eukarya* and viruses. The genome and properties of the mimivirus (Benson *et al.*, 1999) suggest it to be an ancient form between a virus and a bacterium (Xiao *et al.*, 2005). The discovery of very large ('*Epulopiscium*') and very small ('*Nanoarchaeum*') bacterial cells is paralleled by a report of bacterial genome sizes ranging from smaller than that of the mimivirus to larger than some of the fungal genomes (Angert *et al.*, 1993; Janssen *et al.*, 1997; Huber *et al.*, 2002). In other areas, such as autotroph–heterotroph, aerobe–anaerobe, free living–symbiont and psychrophile–thermophile, diversity may be more appropriately viewed as spectra, rather than dichotomies (Ward & Fraser, 2005).

A fascinating new finding by Khayat *et al.* (2005) is the result of an analysis of an archaeal virus capsid protein. The Sulfolobus turreted icosahedral virus (STIV) infects *Sulfolobus solfataricus*, an acidophilic, hyperthermophilic bacterium that grows optimally at pH 2–4 and at >80 °C and has become a model for analysis of hyperthermophilic archaea and their viruses (Pfeifer *et al.*, 1994). The crystal structure of the major capsid protein of the *Sulfolobus* virus, an archaeal virus isolated from an acidic hot spring (pH 2–4, 72–92 °C) in Yellowstone National Park, was found to be nearly identical to the major capsid protein structures of the eukaryotic *Paramecium bursaria Chlorella virus* (PBCV) and the bacteriophage PRD1, and showed a common fold with the mammalian adenovirus (Khayat *et al.*, 2005).

Since STIV spends at least part of its life cycle in the acidic, high-temperature extracellular environment of the hot springs from which it was isolated, the capsid must tolerate high temperature and acidic conditions to maintain infectivity. The exceptional stability of the capsid is explained partly in that the major capsid protein subunit is a tightly packed protein with few cavities and a small total cavity volume. Although it is difficult to make a case for specific structural features that are responsible for the increased thermostability of thermophilic organisms, some factors can be considered influential.

In the study carried out by Khayat *et al.* (2005), structural analysis of the capsid architecture, determined by fitting the subunit into an electron cryomicroscopy reconstruction of the virus, led to the identification of a number of key interactions, akin to those observed in the mammalian adenovirus, bacteriophage PRD1 and green algal PBCV-1. The authors conclude that these viral capsids, or capsid proteins, and not the viruses themselves, originated and evolved from a common ancestor, following a proposal by Bamford (2003) who suggested that the fundamental structural aspects of a virus, e.g. the capsid architecture and perhaps the genome packaging machinery, are, in principle, the 'self' or 'soul' of the virus and inherited from a viral ancestor. Attributes like host recognition and adaptation are probably acquired from the host through horizontal gene transfer, rather than inherited from the ancestor (Bamford *et al.*, 2002). The net result is that structural homology can reveal phylogenetic relationships, with the argument that the mammalian adenovirus, bacteriophage PRD1 and green algal PBCV-1 share a common ancestor, not just their viral capsids (Khayat *et al.*, 2005).

As pointed out by Burnett (2006), virus origins have been difficult to determine because of their diversity and complexity, the way they were studied and their small size. Each virus family, until recently, was studied separately and, structurally, the viruses could be imaged only by low-resolution electron microscopy. Also, most viruses are too large to be analysed by crystallography. With the report by Khayat *et al.* (2005) and others, the viruses can now be considered to fall into lineages identifiable by features in common with ancient precursors (Bamford *et al.*, 2002). Thus, these findings in toto cast light on virus origins, leading Burnett (2006) to speculate that viruses are descendants of ancient self-reproducing systems that aided early cells to develop by shuttling genes between the protected environments in divisible lipid vesicles, with the evidence pointing to a primordial role for viruses, rather than a late appearance. Thus, the viruses may soon be arrayed in their own phylogenomics and tree of life.

Bacteria, perhaps, may have endless diversity and, from the cell to higher organizations of micro-organisms, metagenomics has emerged as a powerful tool for analysing microbial communities, regardless of the ability of member organisms to be cultured in the laboratory. Metagenomics is based on the genomic analysis of microbial DNA extracted directly from communities in environmental samples. Essentially, metagenomics provides genomics on a huge scale and enables surveys of various micro-organisms present in a specific environment, e.g. water, soil or air. Information concerning biological functions within the community can be integrated and the structure of microbial communities can be probed. Most importantly, the massive uncultured microbial diversity present in the environment finally can be investigated in detail that was not possible previously.

For decades, it has been axiomatic in microbiology that more than 99 % of micro-organisms present in natural environments are not readily culturable (MacLeod, 1965). In fact, most microbial species in the environment have yet to be described and, therefore, global microbial diversity is only beginning to be mapped. Approaches that have been used in the past to explore the diversity of microbial communities are necessarily biased because of the limitations of the culture methods used. Culture-independent methods, namely 16S rRNA gene sequence analysis and, more recently, direct sequencing, provide valuable information about taxa and species in the environment. Spiegelman *et al.* (2005) provide a useful survey of methods for characterization of microbial consortia and communities, including molecular biological, biochemical and microbiological methods. 'Slicing' complex microbial communities by application of these various methods permits construction of a comprehensive picture over and above culture, notably PCR amplification of 16S rRNA gene sequences, identification of phylogenetic anchors on metagenomics clones and analysis of random end sequences of metagenomics clones. The phylogenetic anchor approach involves sequencing all clones that contain a given phylogenetic marker, and collection of clones containing 16S rRNA phylogenetic information provides a basis for making inferences from the nucleotide sequence concerning functions carried out by micro-organisms from which the fragments of metagenomics DNA originated.

The Microbial Observatories Program, funded by the United States National Science Foundation (NSF), is proving a rich source of information concerning microbial diversity. Functional anchor, the converse of the phylogenetic anchor approach, is being used to describe microbial life in Alaskan soil at the Bonanza Creek Long Term Ecological Research Microbial Observatory (funded by NSF) (Handelsman, 2005; Schloss & Handelsman, 2005).

In essence, metagenomics attempts to analyse the complex genomes of microbial niches, and such analyses are initiated by the isolation of environmental DNAs, methods for which are reviewed by Streit & Schmitz (2004). By combining sequence analysis of large insert libraries of environmental DNA with genetic and functional analysis, one gains significant insight into the genomic potential and ecological roles of both the cultured and uncultured micro-organisms present in the sample.

Several examples are illustrative of the power of metagenomics. Recent analysis of genome fragments recovered directly from marine bacterioplankton allowed detection of a new bacterial rhodopsin, proteorhodopsin (Beja *et al.*, 2000; de la Torre *et al.*, 2003). Furthermore, widespread distribution of proteorhodopsin genes among divergent marine bacterial taxa (de la Torre *et al.*, 2003), large abundance and spectrum adaptation of proteorhodopsin proteins to different habitats, combined with genetic and biophysical data, indicate that the proteorhodopsin-based bacterial phototrophy is globally distributed and a very significant ocean microbial process worldwide (Streit & Schmitz, 2004).

A major sequencing approach used to analyse Sargasso Sea water samples revealed more DNA sequences than ever described before (Venter *et al.*, 2004). This accomplishment is expected to open up the possibility of investigating the microbial world and its evolution on a scale never before dreamed possible. However, new bioinformatic tools will be needed to reap full benefit from the torrent of information that will flow from its wide application in answering the question of 'what's out there'.

Examples of metagenomics applied to real-life problems include the study of the complex metagenome of biofilms growing on rubber-coated valves within drinking-water networks. Using culture-independent methods, Schmeisser *et al.* (2003) identified more than 100 kb of contiguous environmental DNA sequences in these films.

The field of metagenomics has moved so rapidly since 1991 (Schmidt *et al*., 1991) that molecular approaches to assessing microbial diversity are now considered an essential component of the research microbiologist's toolbox, perhaps comparable to the microscope (Oremland *et al*., 2005). These methods, based on genome sequencing, will be expanded, but will also spur development of novel culturing methods to capture a greater fraction of the as-yet uncultured microbial diversity being discovered. For example, diluted nutrient broth, long incubation times and other variations in culturing micro-organisms have enabled some successes in culturing isolates from bacterial groups poorly represented in culture collections. Microtitre plates, automated imaging and flow cytometry are among the tools used to enhance culture success, each employed with the intent of attempting to simulate and mimic the natural environment.

A combination of traditional and environmental genomic approaches to advance understanding of biogeochemical processes is described by Oremland *et al*. (2005). The anaerobic oxidation of methane was first documented by geochemists almost 40 years ago and, through environmental genomics, a link to sulfate reduction has been determined (Michaelis *et al*., 2002). The powerful advances offered by metagenomics, however, are constrained by the need to identify and annotate genes and proteins of unknown functions, as well as to test the validity of those already annotated in the available databases. Understanding the function of novel genes and proteins within microbial niches and their role in global cycles is the current challenge.

Although environmental genetic and genomic tools are making it possible to identify and characterize the vast reservoir of microbial life by cultivation-independent means, it remains important to know which cells are active and what activities micro-organisms mediate in various habitats. What makes it particularly difficult to obtain this information is the functional redundancy of the microbial world. Individual cells can carry out many different functions, but certain metabolic activities are common to many different taxa, e.g. glucose utilization. Methods that employ stable isotypes are now being developed to link genetic and genomic information with the information collected on the metabolic activities of microbial cells in environmental samples. Stable isotope probing (SIP; also referred to as secondary ion mass spectrometry or SIMS) employs a high-sensitivity ion microprobe designed for microscale isotopic and elemental analysis of complex minerals that also functions as an ion microscope by direct ion-imaging of samples. This allows precise location of small inclusions for isotopic analysis. When the method is coupled with fluorescent *in situ* hybridization (FISH-SIMS), individual cells in natural samples can be monitored.

SIP is proving to be an incredibly powerful tool for microbial ecologists because it, as FISH-SIMS, links taxonomic identity with function. Examples of applications are provided by Whitby *et al.* (2005) and include cycling of carbon and nitrogen and degradation of xenobiotics. It also allows a culture-independent method for determining how complex microbial communities respond to changes in temperature, pH and substrate concentrations. As with every method, there are limitations to SIP, e.g. the micro-organism being observed must be actively replicating when the isotope is added, which may not always be the case. This problem can be circumvented by RNA-SIP (Whitby *et al.*, 2005). No doubt we will see ample application of this approach to microbial ecology in the immediate future.

Clearly, environmental sequencing has contributed significantly to our understanding of microbial diversity and this is exemplified by the expansion of archaeal diversity, measured by the accumulation of archaeal rRNA sequences submitted to GenBank (Robertson *et al.*, 2005). The evidence now supports the *Archaea* as a cosmopolitan group, not restricted to 'extreme' environments. More comprehensive phylogenetic trees of the *Archaea* have been constructed. Also, representatives of the *Archaea* have been detected in ocean water (DeLong, 2005), ocean sediments (Knittel *et al.*, 2005), solid gas hydrates (Mills *et al.*, 2005), tidal-flat sediments (Kim *et al.*, 2005), freshwater lakes (Keough *et al.*, 2003), soil (Ochsenreiter *et al.*, 2003), plant roots (Simon *et al.*, 2000), peatlands (Galand *et al.*, 2005) and petroleum-contaminated aquifers (Kleikemper *et al.*, 2005) as well as the human mouth and gut (Lepp *et al.*, 2004). Furthermore, Robertson *et al.* (2005) point out that archaea may occupy a significant fraction of the total microbiota, usually about 10 % of the total rRNA phylotypes, with the rest usually being bacteria. Nevertheless, the archaea can dominate the microbial presence in some environments, notably at high temperature and low pH (Futterer *et al.*, 2004) or the deep-sea environment (DeLong *et al.*, 1994).

It is important to emphasize that, for years, microbiologists characterized the archaea as obligate extremophiles growing well in environments too harsh for other organisms. Because of their limited physiological diversity, based on archaea grown in the laboratory, it was suggested that these micro-organisms were metabolically constrained to only a few environmental niches, e.g. currently cultured crenarchaeota are sulfur-metabolizing thermophiles. By using cultivation-independent methods, vast numbers of crenarchaeota have been discovered in cold, oxic ocean waters. The numerical dominance of marine crenarchaeota, estimated at 10^{28} cells in the world's oceans, indicate that they play a major role in global biogeochemical cycles. Isotopic analyses of marine crenarchaeal lipids indicate that these planktonic archaea fix inorganic carbon and the crenarchaeota must now be considered significant players in global carbon and nitrogen cycles (Konneke *et al.*, 2005).

MICROBIAL BIOGEOGRAPHY – DOES IT EXIST?

For the past century, microbiologists have considered a basic principle of microbial ecology to be that micro-organisms are not subject to geographical boundaries, but instead are dispersed globally and thrive wherever they find a hospitable environment, such as freshwater ponds, open ocean, mangrove swamps or hot springs. The Dutch biologist Lourens Baas-Becking (1934) postulated that 'Everything is everywhere; the environment selects.' However, the information provided by methods that do not require culturing suggests that micro-organisms can no longer be believed to be infinitely mobile (Whitfield, 2005). A more nuanced view is compelling, with an embedded caution about what the molecular picture of microbial diversity shows and a questioning of whether diversity at the microbial level can be fully understood using the species-based view of plant and animal ecology. Some investigators insist that biogeography, namely the study of distribution of biodiversity over space and time, cannot be applied to anything smaller than 1 mm. However, a biogeography of microbial species can provide insights into those mechanisms that generate and maintain diversity, such as speciation, extinction, dispersal and species interactions [Brown & Lomolino, 1998; an updated, 3rd edition of this definitive textbook on the biogeography of macro-organisms has recently been published (Lomolino *et al.*, 2005)]. Thus, geographical distributions of micro-organisms are being studied extensively, since genetic methods have shown clearly that culture-based studies of the past almost completely overlooked the vast majority of microbial diversity (Ward *et al.*, 1990; Øvreås, 2000; Floyd *et al.*, 2005). As microbial diversity is sampled more deeply and widely, a far greater appreciation of microbial diversity has been gained and the basis for a microbial biogeography prepared (Schloss & Handelsman, 2004; Venter *et al.*, 2004).

An informative, albeit early, review of the biogeography of micro-organisms is provided by Hughes Martiny *et al.* (2006). Their analysis of published data suggests that microbial assemblages exhibit both environmental segregation and biogeographical provincialism, supporting the idea that free-living micro-organisms exhibit biogeographical patterns. The proposition posited by Baas-Becking (1934) that 'the environment selects' is, in part, responsible for spatial variation in microbial diversity. In contrast, the idea that 'everything is everywhere' does not appear to be valid. Instead, as Hughes Martiny *et al.* (2006) propose, the legacies of historical events have left lasting signatures on the distributions of microbial assemblages and they wisely recommend that new microbial biogeography studies should not be undertaken without systematically sampling and recording data from a variety of distances, habitats and environmental conditions to distinguish more clearly between contemporary versus historical factors. Furthermore, the body of evidence that is accumulating indicates that microbial composition affects many, if not all, ecosystem processes, including CO_2 respiration and decomposition (McGrady-Steed *et al.*, 1997) and nitrogen cycling (Horz *et al.*, 2004).

The claim that 'everything is everywhere' implies that micro-organisms have such enormous dispersal capabilities that they very quickly eliminate effects of past evolutionary and ecological events, such as genetic divergence arising from barriers to gene flow and differences in assemblages arising from colonization sequence. That is, the hypothesis predicts that variations in the distribution of micro-organisms and their assemblages reflect only the effect of contemporary environmental variation within a single global biogeographical province. Studies undertaken at the Northern Temperate Lakes Microbial Observatory have revealed that the relative influence of historical versus environmental factors is, in fact, related to the scale of sampling. Significant variation in community composition across multiple spatial scales was reported and the variation was linked to morphological features of the lake that reduced water exchange from different areas (Yannarell & Triplett, 2004, 2005).

Quantitative information regarding the environmental prevalence of individual genotypes that allows inference of their ecological importance or competitive success is lacking for natural bacterial populations. Thompson *et al.* (2005) analysed the diversity and annual dynamics of a group of coastal bacterioplankton (>99 % 16S rRNA gene sequence identity to *Vibrio splendidus*). The interesting discovery that they have reported is that *V. splendidus* consists of at least 1000 distinct genotypes, each occurring at extremely low environmental concentrations of less than one cell per millilitre. The coastal assemblage of *V. splendidus*, considered a phylogenetically discrete cluster based on nearly identical (<1 % divergent) 16S rRNA gene sequences in an analysis of bacterioplankton community structure, yielded ribotype clusters representing ecologically differentiated units, termed ecotypes or populations (Thompson *et al.*, 2005). The high degree of heterogeneity detected among the *V. splendidus* genomes suggests that the ribotype clusters within the population persist, most likely reflecting the differentiation of subpopulations specialized to particular environmental components. In a similar study, Louis *et al.* (2003) showed that *Vibrio cholerae* genomic variants were related to season, rather than geographical location, in Chesapeake Bay.

In a comparative analysis of the distribution of *V. cholerae* in Chesapeake Bay, the Bay of Bengal and coastal waters of Peru, the seasonal pattern in the incidence of *V. cholerae*, when analysed using both molecular methods and remote sensing via satellite observation (Colwell, 1996; Lobitz *et al.*, 2000), showed that the incidence of cholera in Bangladesh is linked to environmental factors, climate and seasonality (Huq *et al.*, 2005; Colwell *et al.*, 2003). Thus, remote sensing via satellite represents yet another tool that will be much more widely employed by microbial ecologists in the future. Clouds of desert dust, containing significant numbers of micro-organisms and transported from continent to continent through the atmosphere, can be monitored by remote sensing (Griffin, 2005).

HABITAT DIVERSITY

A wealth of information is available in the literature on microbial diversity in water, soil, air, marine, freshwater, polluted and unpolluted environments and other diverse environments in temperate, tropical and both low and high latitude areas of the world. One of the recent interesting and provocative discoveries is the 'champion bantamweight' of bacteria, '*Pelagibacter ubique*' (SAR-11), one of the smallest self-replicating cells, with only 1354 genes (Giovannoni *et al.*, 2005). The bacterium has no duplications, no viral genes, no introns and no non-coding sequences. The paring down of the genome of this bacterium appears to be a result of an evolutionary process for conserving genetic energy for an organism that is highly abundant (about 10^{28} cells on Earth) in an environment offering limited resources, the Sargasso Sea. It is present in all oceans and throughout nearly all of the water column in the world's oceans. It is estimated to have a combined weight exceeding that of all the fish in the world's oceans (Ashley, 2005; Giovannoni *et al.*, 2005).

Another example of the many new discoveries in microbial diversity is *Silicibacter pomeroyi*, a member of the *Roseobacter* clade found in abundance in coastal and oceanic mixed-layer bacterioplankton (Giovannoni & Rappé, 2000), the genome of which has been sequenced. The genome analysis indicates that this micro-organism relies on a lithoheterotrophic strategy using inorganic compounds (carbon monoxide and sulfide) to supplement heterotrophy (Moran *et al.*, 2004). It also possesses genes that assist in its association with plankton and suspended particles, namely the genes for uptake of algae-derived compounds, use of metabolites from reducing microzones, rapid growth and cell-density-dependent regulation. The diversity of the microbial world is only beginning to be documented in its fullest richness.

A significant biomass exists buried deep within the sediments of the world's oceans (Parkes *et al.*, 1994; Whitman *et al.*, 1998), and a previously unknown and diverse microbial community has been revealed within this biosphere by employing 16S rRNA gene sequence analysis. Many previously unidentified and uncultured groups of organisms, some without clear phylogenetic affiliation, have been reported as components of this habitat (Webster *et al.*, 2004). Novel sponge-associated bacteria (Ridley *et al.*, 2005), diverse micro-organisms in drinking water (Hill *et al.*, 2005) and an unexpectedly high bacterial diversity in Arctic tundra relative to boreal forest soils (Neufeld & Mohn, 2005), as well as novel symbionts in the midgut of mosquitoes (Lindh *et al.*, 2005), have also recently been reported. Even old data, re-evaluated and analysed, are providing new species, reported by Salzberg *et al.* (2005), who scanned a publicly available database and discovered the genomic sequence of an unknown bacterial species, probably related to the bacterium *Wolbachia*, embedded within the genome sequences of the fruit fly *Drosophila* (Salzberg *et al.*, 2005). Clearly, the 'captured diversity' in culture collections is significantly limited and not representative of the flora of the natural environment, especially since some phylum-level groups are significantly underrepresented (Floyd *et al.*, 2005).

CONCLUSIONS

In the era of genomics, the perspective of microbial ecologists has been expanded and the panorama that the world of microbial diversity comprises is only beginning to be known. In the decade ahead, we will see revealed a grand tapestry of life reflecting richness and abundance of the microbial world. With this new knowledge, there will be a powerful reinforcement of the biocomplexity of life on this planet: life that is complex, integrated and both genetically and phenetically (functionally) interwoven. Sustainability of the blue planet rests unmistakably and irrevocably on the vast and intricate microbial diversity that can be appreciated and, hopefully, understood.

ACKNOWLEDGEMENTS

The author gratefully acknowledges Matt Kane and Maryanna Henkart of the National Science Foundation for helpful discussions during preparation of the manuscript.

REFERENCES

Angert, E. R., Clements, K. D. & Pace, N. R. (1993). The largest bacterium. *Nature* **362**, 239–241.

Ashley, S. (2005). Lean gene machine. *Sci Am* **293**, 26–28.

Baas-Becking, L. G. M. (1934). *Geobiologie of Inleiding Tot de Milieukunde*. The Hague: Van Stockkum & Zoon (in Dutch).

Bamford, D. H. (2003). Do viruses form lineages across different domains of life? *Res Microbiol* **154**, 231–236.

Bamford, D. H., Burnett, R. M. & Stuart, D. I. (2002). Evolution of viral structure. *Theor Popul Biol* **61**, 461–470.

Beja, O., Aravind, L., Koonin, E. V. & 9 other authors (2000). Bacterial rhodopsin: evidence for a new type of phototrophy in the sea. *Science* **289**, 1902–1906.

Benson, S. D., Bamford, J. K., Bamford, D. H. & Burnett, R. M. (1999). Viral evolution revealed by bacteriophage PRD1 and human adenovirus coat protein structures. *Cell* **98**, 825–833.

Brenner, D. J., Staley, J. T. & Krieg, N. R. (2001). Classification of procaryotic organisms and the concept of bacterial speciation. In *Bergey's Manual of Systematic Bacteriology*, 2nd edn, vol. 1, pp. 27–31. Edited by D. R. Boone, R. W. Castenholz & G. M. Garrity. New York: Springer.

Brown, J. H. & Lomolino, M. V. (1998). *Biogeography*, 2nd edn. Sunderland, MA: Sinauer.

Burnett, R. M. (2006). More barrels from the viral tree of life. *Proc Natl Acad Sci U S A* **103**, 3–4.

Chatton, E. (1938). *Titres et Travaux Scientifiques (1906–1937)*. Sottano, Italy: Sète (in French).

Colwell, R. R. (1996). Global climate and infectious disease: the cholera paradigm. *Science* **274**, 2025–2031.

Colwell, R. R., Huq, A., Islam, M. S. & 9 other authors (2003). Reduction of cholera in Bangladeshi villages by simple filtration. *Proc Natl Acad Sci U S A* **100**, 1051–1055.

de la Torre, J. R., Christianson, L. M., Beja, O., Suzuki, M. T., Karl, D. M., Heidelberg, J. & DeLong, E. F. (2003). Proteorhodopsin genes are distributed among divergent marine bacterial taxa. *Proc Natl Acad Sci U S A* **100**, 12830–12835.

DeLong, E. F. (2005). Microbial community genomics in the ocean. *Nat Rev Microbiol* **3**, 459–469.

DeLong, E. F., Wu, K. Y., Prézelin, B. B. & Jovine, R. V. M. (1994). High abundance of archaea in Antarctic marine picoplankton. *Nature* **371**, 695–697.

Eisen, J. A. (2000). Horizontal gene transfer among microbial genomes: new insights from complete genome analysis. *Curr Opin Genet Dev* **10**, 606–611.

Eisen, J. A. & Fraser, C. M. (2003). Phylogenomics: intersection of evolution and genomics. *Science* **300**, 1706–1707.

Floyd, M. M., Tang, J., Kane, M. & Emerson, D. (2005). Captured diversity in a culture collection: case study of the geographic and habitat distributions of environmental isolates held at the American Type Culture Collection. *Appl Environ Microbiol* **71**, 2813–2823.

Futterer, O., Angelov, A., Liesegang, H., Gottschalk, G., Schleper, C., Schepers, B., Dock, C., Antranikian, G. & Liebl, W. (2004). Genome sequence of *Picrophilus torridus* and its implications for life around pH 0. *Proc Natl Acad Sci U S A* **101**, 9091–9096.

Galand, P. E., Fritze, H., Conrad, R. & Yrjala, K. (2005). Pathways for methanogenesis and diversity of methanogenic archaea in three boreal peatland ecosystems. *Appl Environ Microbiol* **71**, 2195–2198.

Gevers, D., Cohan, F. M., Lawrence, J. G. & 8 other authors (2005). Re-evaluating prokaryotic species. *Nature Rev Microbiol* **3**, 733–739.

Giovannoni, S. J. & Rappé, M. (2000). Evolution, diversity and molecular ecology of marine prokaryotes. In *Microbial Ecology of the Oceans*, pp. 47–48. Edited by D. Kirchman. New York: Wiley.

Giovannoni, S. J., Tripp, H. J., Givan, S. & 11 other authors (2005). Genome streamlining in a cosmopolitan oceanic bacterium. *Science* **309**, 1242–1245.

Gogarten, J. P. & Townsend, J. P. (2005). Horizontal gene transfer, genome innovation and evolution. *Nat Rev Microbiol* **3**, 679–687.

Gogarten, J. P., Doolittle, W. F. & Lawrence, J. G. (2002). Prokaryotic evolution in light of gene transfer. *Mol Biol Evol* **19**, 2226–2238.

Griffin, D. W. (2005). Clouds of desert dust and microbiology: a mechanism of global dispersion. *Microbiol Today* **32**, 180–182.

Hanage, W. P., Fraser, C. & Spratt, B. G. (2005). Fuzzy species in recombinogenic bacteria. *BMC Biol* **3**, 6.

Handelsman, J. (2005). Sorting out metagenomes. *Nat Biotechnol* **23**, 38–39.

Hill, V. R., Polaczyk, A. L., Hahn, D., Narayanan, J., Cromeans, T. L., Roberts, J. M. & Amburgey, J. E. (2005). Development of a rapid method for simultaneous recovery of diverse microbes in drinking water by ultrafiltration with sodium polyphosphate and surfactants. *Appl Environ Microbiol* **71**, 6878–6884.

Horz, H.-P., Barbrook, A., Field, C. B. & Bohannan, B. J. M. (2004). Ammonia-oxidizing bacteria respond to multifactorial global change. *Proc Natl Acad Sci U S A* **101**, 15136–15141.

Huber, H., Hohn, M. J., Rachel, R., Fuchs, T., Wimmer, V. C. & Stetter, K. O. (2002). A new phylum of *Archaea* represented by a nanosized hyperthermophilic symbiont. *Nature* **417**, 63–67.

Hughes Martiny, J. B., Bohannan, B. J. M., Brown, J. H. & 13 other authors (2006). Microbial biogeography: putting microorganisms on the map. *Nat Rev Microbiol* **4**, 102–112.

Huq, A., Sack, R. B., Nizam, A. & 10 other authors (2005). Critical factors influencing the occurrence of *Vibrio cholerae* in the environment of Bangladesh. *Appl Environ Microbiol* **71**, 4645–4654.

Janssen, P. H., Schuhmann, A., Mörschel, E. & Rainey, F. A. (1997). Novel anaerobic ultramicrobacteria belonging to the *Verrucomicrobiales* lineage of bacterial descent isolated by dilution culture from anoxic rice paddy soil. *Appl Environ Microbiol* **63**, 1382–1388.

Keough, B. P., Schmidt, T. M. & Hicks, R. E. (2003). Archaeal nucleic acids in picoplankton from great lakes on three continents. *Microb Ecol* **46**, 238–248.

Khayat, R., Tang, L., Larson, E. T., Lawrence, C. M., Young, M. & Johnson, J. E. (2005). Structure of an archaeal virus capsid protein reveals a common ancestry to eukaryotic and bacterial viruses. *Proc Natl Acad Sci U S A* **102**, 18944–18949.

Kim, B. S., Oh, H. M., Kang, H. & Chun, J. (2005). Archaeal diversity in tidal flat sediment as revealed by 16S rDNA analysis. *J Microbiol* **43**, 144–151.

Kleikemper, J., Pombo, S. A., Schroth, M. H., Sigler, W. V., Pesaro, M. & Zeyer, J. (2005). Activity and diversity of methanogens in a petroleum hydrocarbon-contaminated aquifer. *Appl Environ Microbiol* **71**, 149–158.

Knittel, K., Losekann, T., Boetius, A., Kort, R. & Amann, R. (2005). Diversity and distribution of methanotrophic archaea at cold seeps. *Appl Environ Microbiol* **71**, 467–479.

Konneke, M., Bernhard, A. E., de la Torre, J. R., Walker, C. B., Waterbury, J. B. & Stahl, D. A. (2005). Isolation of an autotrophic ammonia-oxidizing marine archaeon. *Nature* **437**, 543–546.

Konstantinidis, K. T. & Tiedje, J. M. (2005a). Genomic insights that advance the species definition for prokaryotes. *Proc Natl Acad Sci U S A* **102**, 2567–2572.

Konstantinidis, K. T. & Tiedje, J. M. (2005b). Towards a genome-based taxonomy for prokaryotes. *J Bacteriol* **187**, 6258–6264.

Lan, R. & Reeves, P. R. (2000). Intraspecies variation in bacterial genomes: the need for a species genome concept. *Trends Microbiol* **8**, 396–401.

La Scola, B., Audic, S., Robert, C., Jungang, L., de Lamballerie, X., Drancourt, M., Birtles, R., Claverie, J. M. & Raoult, D. (2003). A giant virus in amoebae. *Science* **299**, 2033.

Lawrence, J. G. (2002). Gene transfer in bacteria: speciation without species? *Theor Popul Biol* **61**, 449–460.

Lepp, P. W., Brinig, M. M., Ouverney, C. C., Palm, K., Armitage, G. C. & Relman, D. A. (2004). Methanogenic archaea and human periodontal disease. *Proc Natl Acad Sci U S A* **101**, 6176–6181.

Lindh, J. M., Terenius, O. & Faye, I. (2005). 16S rRNA gene-based identification of midgut bacteria from field-caught *Anopheles gambiae* sensu lato and *A. funestus* mosquitoes reveals new species related to known insect symbionts. *Appl Environ Microbiol* **71**, 7217–7223.

Lobitz, B., Beck, L., Huq, A., Wood, B., Fuchs, G., Faruque, A. S. G. & Colwell, R. (2000). Climate and infectious disease: use of remote sensing for detection of *Vibrio cholerae* by indirect measurement. *Proc Natl Acad Sci U S A* **97**, 1438–1443.

Lomolino, M. V., Riddle, B. R. & Brown, J. H. (2005). *Biogeography*, 3rd edn. Sunderland, MA: Sinauer.

Louis, V. R., Russek-Cohen, E., Choopun, N. & 7 other authors (2003). Predictability of *Vibrio cholerae* in Chesapeake Bay. *Appl Environ Microbiol* **69**, 2773–2785.

MacLeod, R. A. (1965). The question of the existence of specific marine bacteria. *Bacteriol Rev* **29**, 2–23.

McGrady-Steed, J., Harris, P. M. & Morin, P. J. (1997). Biodiversity regulates ecosystem predictability. *Nature* **390**, 162–165.

Michaelis, W., Seifert, R., Nauhaus, K. & 14 other authors (2002). Microbial reefs in the Black Sea fueled by anaerobic oxidation of methane. *Science* **297**, 1013–1015.

Mills, H. J., Martinez, R. J., Story, S. & Sobecky, P. A. (2005). Characterization of microbial community structure in Gulf of Mexico gas hydrates: comparative analysis of DNA- and RNA-derived clone libraries. *Appl Environ Microbiol* **71**, 3235–3247.

Moran, M. A., Buchan, A., Gonzalez, J. M. & 32 other authors (2004). Genome sequence of *Silicibacter pomeroyi* reveals adaptations to the marine environment. *Nature* **432**, 910–913.

Neufeld, J. D. & Mohn, W. W. (2005). Unexpectedly high bacterial diversity in Arctic tundra relative to boreal forest soils, revealed by serial analysis of ribosomal sequence tags. *Appl Environ Microbiol* **71**, 5710–5718.

Ochsenreiter, T., Selezi, D., Quaiser, A., Bonch-Osmolovskaya, L. & Schleper, C. (2003). Diversity and abundance of Crenarchaeota in terrestrial habitats studied by 16S RNA surveys and real time PCR. *Environ Microbiol* **5**, 787–797.

Oremland, R. S., Capone, D. G., Stolz, J. F. & Fuhrman, J. (2005). Whither or wither geomicrobiology in the era of 'community metagenomics'. *Nat Rev Microbiol* **3**, 572–578.

Øvreås, L. (2000). Population and community level approaches for analysing microbial diversity in natural environments. *Ecol Lett* **3**, 236–251.

Parkes, R. J., Cragg, B. A., Bale, S. J., Getliff, J. M., Goodman, K., Rochelle, P. A., Fry, J. C., Weightman, A. J. & Harvey, S. M. (1994). Deep bacterial biosphere in Pacific Ocean sediments. *Nature* **371**, 410–413.

Pfeifer, F., Palm, P. & Schleifer, K. H. (editors) (1994). *Molecular Biology of Archaea*. Stuttgart: Gustav Fischer.

Ridley, C. P., Faulkner, D. J. & Haygood, M. G. (2005). Investigation of *Oscillatoria spongeliae*-dominated bacterial communities in four dictyoceratid sponges. *Appl Environ Microbiol* **71**, 7366–7375.

Robertson, C. E., Harris, J. K., Spear, J. R. & Pace, N. R. (2005). Phylogenetic diversity and ecology of environmental archaea. *Curr Opin Microbiol* **8**, 638–642.

Salzberg, S. L., Hotopp, J. C., Delcher, A. L., Pop, M., Smith, D. R., Eisen, M. B. & Nelson, W. C. (2005). Serendipitous discovery of *Wolbachia* genomes in multiple *Drosophila* species. *Genome Biol* **6**, R23.

Schloss, P. D. & Handelsman, J. (2004). Status of the microbial census. *Microbiol Mol Biol Rev* **68**, 686–691.

Schloss, P. D. & Handelsman, J. (2005). Metagenomics for studying unculturable microorganisms: cutting the Gordian knot. *Genome Biol* **6**, 229.

Schmeisser, C., Stockigt, C., Raasch, C. & 8 other authors (2003). Metagenome survey of biofilms in drinking-water networks. *Appl Environ Microbiol* **69**, 7298–7309.

Schmidt, T. M., DeLong, E. F. & Pace, N. R. (1991). Analysis of a marine picoplankton community by 16S rRNA gene cloning and sequencing. *J Bacteriol* **173**, 4371–4378.

Simon, H. M., Dodsworth, J. A. & Goodman, R. M. (2000). Crenarchaeota colonize terrestrial plant roots. *Environ Microbiol* **2**, 495–505.

Spiegelman, D., Whissell, G. & Greer, C. W. (2005). A survey of the methods for the characterization of microbial consortia and communities. *Can J Microbiol* **51**, 355–386.

Stanier, R. Y. & van Niel, C. B. (1962). The concept of a bacterium. *Arch Mikrobiol* **42**, 17–35.

Stanier, R. Y., Doudoroff, M. & Adelberg, E. A. (1957). *The Microbial World*. Engelwood Cliffs, NJ: Prentice-Hall.

Stanier, R. Y., Doudoroff, M. & Adelberg, E. A. (1963). *The Microbial World*, 2nd edn. Engelwood Cliffs, NJ: Prentice-Hall.

Streit, W. R. & Schmitz, R. A. (2004). Metagenomics – the key to the uncultured microbes. *Curr Opin Microbiol* **7**, 492–498.

Tettelin, H., Masignani, V., Cieslewicz, M. J. & 43 other authors (2005). Genome analysis of multiple pathogenic isolates of *Streptococcus agalactiae*: implications for the microbial "pan-genome". *Proc Natl Acad Sci U S A* **102**, 13950–13955.

Thompson, J. R., Pacocha, S., Pharino, C., Klepac-Ceraj, V., Hunt, D. E., Benoit, J., Sarma-Rupavtarm, R., Distel, D. L. & Polz, M. F. (2005). Genotypic diversity within a natural coastal bacterioplankton population. *Science* **307**, 1311–1313.

Venter, J. C., Remington, K., Heidelberg, J. F. & 20 other authors (2004). Environmental genome shotgun sequencing of the Sargasso Sea. *Science* **304**, 66–74.

Ward, N. & Fraser, C. M. (2005). How genomics has affected the concept of microbiology. *Curr Opin Microbiol* **8**, 564–571.

Ward, D. M., Weller, R. & Bateson, M. M. (1990). 16S rRNA sequences reveal numerous uncultured microorganisms in a natural community. *Nature* **345**, 63–65.

Wayne, L. G., Brenner, D. J., Colwell, R. R. & 9 other authors (1987). Report of the ad hoc committee on reconciliation of approaches to bacterial systematics. *Int J Syst Bacteriol* **37**, 463–464.

Webster, G., Parkes, R. J., Fry, J. C. & Weightman, A. J. (2004). Widespread occurrence of a novel division of bacteria identified by 16S rRNA gene sequences originally found in deep marine sediments. *Appl Environ Microbiol* **70**, 5708–5713.

Whitby, C., Bailey, M., Whiteley, A., Murrell, C., Kilham, K., Prosser, J. & Lappin-Scott, H. (2005). Stable isotope probing links taxonomy with function in microbial communities. *ASM News* **71**, 169–173.

Whitfield, J. (2005). Biogeography: is everything everywhere? *Science* **310**, 960–961.

Whitman, W. B., Coleman, D. C. & Wiebe, W. J. (1998). Prokaryotes: the unseen majority. *Proc Natl Acad Sci U S A* **95**, 6578–6583.

Woese, C. R. (1987). Bacterial evolution. *Microbiol Rev* **51**, 221–271.

Woese, C. R. (2004). A new biology for a new century. *Microbiol Mol Biol Rev* **68**, 173–186.

Woese, C. R. & Fox, G. E. (1977a). The concept of cellular evolution. *J Mol Evol* **10**, 1–6.

Woese, C. R. & Fox, G. E. (1977b). Phylogenetic structure of the prokaryotic domain: the primary kingdoms. *Proc Natl Acad Sci U S A* **74**, 5088–5090.

Woese, C. R., Olsen, G. J., Ibba, M. & Soll, D. (2000). Aminoacyl-tRNA synthetases, the genetic code, and the evolutionary process. *Microbiol Mol Biol Rev* **64**, 202–236.

Xiao, C., Chipman, P. R., Battisti, A. J., Bowman, V. D., Renesto, P., Raoult, D. & Rossmann, M. G. (2005). Cryo-electron microscopy of the giant Mimivirus. *J Mol Biol* **353**, 493–496.

Yannarell, A. C. & Triplett, E. W. (2004). Within- and between-lake variability in the composition of bacterioplankton communities: investigations using multiple spatial scales. *Appl Environ Microbiol* **70**, 214–223.

Yannarell, A. C. & Triplett, E. W. (2005). Geographic and environmental sources of variation in lake bacterial community composition. *Appl Environ Microbiol* **71**, 227–239.

Patterns in prokaryotic biodiversity

M. Claire Horner-Devine,[1] Jessica Green[2] and Brendan J. M. Bohannan[3]

[1]School of Aquatic and Fishery Sciences, University of Washington, Seattle, WA 98195, USA

[2]School of Natural Sciences, University of California, Merced, CA 95344, USA

[3]Department of Biological Sciences, Stanford University, Stanford, CA 94305, USA

INTRODUCTION

The variety of life has long fascinated biologists. One of the most intriguing aspects of this diversity is that it is distributed heterogeneously across the Earth, with some places harbouring a myriad of different forms of life and others supporting depauperate communities. There appear to be regularities in this heterogeneous distribution, *patterns* in the distribution of life's diversity, for many well-studied macro-organisms. Until recently, relatively few patterns in the distribution of microbial life have been documented, in large part because microbiologists and ecologists have just recently begun to look for such patterns.

Diversity patterns have played a major role in the development of the science of general ecology (i.e. theoretical, plant, animal and ecosystem ecology). It is reasonable to assume that the study of diversity patterns could play a similar role in the development of microbial ecology. Where should one look for such patterns in microbial biodiversity? Given our limited knowledge of the distribution of microbial diversity, it is reasonable to start by looking for patterns in microbial diversity that are commonly observed for macro-organisms. In this chapter, we begin by discussing how diversity in general and microbial diversity in particular is estimated. We then describe a number of diversity patterns commonly observed for macro-organisms and review recent attempts to document such patterns in microbial diversity. We focus primarily on prokaryotic micro-organisms; however, patterns in the diversity of eukaryotic microbes have also been documented (e.g. Green *et al.*, 2004; Smith *et al.*, 2005).

SGM symposium 66: Prokaryotic diversity: mechanisms and significance.
Editors N. A. Logan, H. M. Lappin-Scott & P. C. F. Oyston. Cambridge University Press. ISBN 0 521 86935 8

WHAT IS BIODIVERSITY?

A diversity of diversities

Most simply, biological diversity (or 'biodiversity') is the variety of life. This variety is expressed on many levels of biological organization and can be measured in many ways. Biodiversity can be studied at the level of genetic, organismal and ecological (e.g. functional) diversity. These three levels of organization are interrelated, both directly through dependence of one level on another and indirectly through dependence on common processes responsible for the generation and maintenance of diversity. Historically, most studies of plant and animal biodiversity have been at the organismal level, although studies at the genetic and ecological levels are becoming increasingly more common.

There are a number of different approaches to measuring diversity. These different approaches rely on different units of diversity and emphasize different components of diversity. The unit of diversity used is often determined by the level of organization (genetic, organismal, ecological) at which one chooses to examine the diversity of a community. Such units include genotypes, phylotypes, species and functional groups. Biodiversity at any level of organization is composed of two key components: taxonomic richness (the number of different taxa) and the relative abundances of the taxa present. Different approaches to quantifying biodiversity emphasize these different components to differing degrees. The most common unit of biodiversity at the organismal level is the species. Historically, the best-studied patterns in plant and animal biodiversity have been patterns in species richness (the number of different species), although the most robust of these patterns are also evident when one uses other units or components to measure diversity. Finally, composition (the identities of types present, as opposed to how many types are present) is another component of biodiversity and is intimately related to patterns in other components of biodiversity. For example, changes in richness over space or time are an expression of changes in composition over space or time (e.g. an increase in species richness as one samples across space reflects changes in the species composition of communities across space). Of course, composition can also change in space and time without associated changes in richness.

Estimating microbial biodiversity

There are two major differences between describing microbial diversity and macro-organism diversity. Firstly, for many prokaryotes, genetic diversity is the only level of biodiversity that is readily accessible. Organismal and ecological diversity must be inferred from genetic information. Secondly, in most environments, microbial taxonomic richness is believed to be very high, and most studies of microbial biodiversity must contend with substantial undersampling of total richness. These differences are described in more detail below.

In contrast to most macro-organisms, observation in the field or the study of morphology in the laboratory cannot yield the information necessary to define the units of ecological or organismal diversity for most prokaryotes. Ideally, this information can be gathered through studying phenotypic traits other than morphology (such as the utilization of specific substrates). To determine such traits usually requires that the organism be cultured in the laboratory, and the vast majority of prokaryotic micro-organisms cannot yet be successfully cultured (Amann *et al.*, 1995). Thus, alternative approaches have been developed. The most common of these approaches involve using information from genetic markers to make inferences regarding organismal and ecological diversity.

For organismal diversity, the most common of these techniques use rRNA gene sequences as indicators of microbial diversity, although other genes, including protein-coding genes, have also been used. The use of these molecular techniques and their drawbacks and biases has been reviewed in detail elsewhere (e.g. von Wintzingerode *et al.*, 1997). These molecular approaches have enabled the detection of non-culturable species and allowed a more complete and detailed picture of prokaryotic communities (reviewed by Head *et al.*, 1998; Mlot, 2004). Taxa are usually defined as sequence similarity groups using this approach. DNA sequence similarity of 97 % between 16S rRNA genes is the most common definition of taxa and is often considered to approximate the species level of resolution defined using culture-dependent methods (Stackebrandt & Rainey, 1995). Taxa defined using such approaches are probably not comparable to macro-organism species; for example, they may be composed of groups of organisms with a broader range of ecological traits than a macro-organism species. It has been noted that, if the 97 % rRNA gene sequence similarity definition were applied to animals, all the primates (from lemurs to humans) would be considered to belong to the same species (Staley, 1997).

The sequence similarity approach can also been used to define ecological or functional diversity. Where a specific function is phylogenetically constrained (e.g. ammonia-oxidizing bacteria; Kowalchuk & Stephen, 2001), ecological diversity can be targeted using phylogenetically informative genes such as the 16S rRNA gene. However, this is not the case for most prokaryotes and many functions, and there is evidence that substantial ecological diversity is often hidden within taxa defined using 16S rRNA genes (e.g. Moore *et al.*, 1998; Ward *et al.*, 1998). Another commonly used approach to estimate ecological diversity is to use the sequences of 'functional' genes, indicator genes that code for proteins involved directly in the ecological process that defines the group. Appropriate indicator genes have been identified for only a minority of ecologically important microbial processes (e.g. denitrification, nitrification, methane oxidation, sulphate reduction). It is often unclear how the genetic diversity of functional genes corresponds to ecological or functional diversity. Inferences regarding this relationship may come from studies of cultured isolates, from the distribution of genotypes along environmental gradients, from patterns of nucleotide substitutions that suggest recent positive selection or, if a crystal structure is available for the indicator protein, from predicted physical changes in the protein associated with genetic dissimilarity (e.g. Bielawski *et al.*, 2004).

It may soon be possible to use genomic information to estimate organismal and ecological diversity in nature. Recent work has used gene content from 70 closely related and fully sequenced bacterial genomes to help determine species boundaries (Konstantinidis & Tiedje, 2005). This approach holds much promise for helping to distinguish different species and for including both ecological and evolutionary distance in species definitions. However, traditional technology requires that microbes be isolated in pure culture before their genomes can be sequenced. Recent advances in environmental genomics have demonstrated that it is possible to sequence entire genomes from mixed communities. However, this is currently only possible in communities of very low diversity (Tyson *et al.*, 2004) or by sequencing extraordinary amounts of environmental DNA (Venter *et al.*, 2004).

Studies of microbial diversity using genetic techniques have revealed very high levels of diversity (especially taxonomic richness) in most environments. The observed levels are often so high that it is unreasonable to sample the microbial communities exhaustively of all but the most depauperate environments (e.g. Tyson *et al.*, 2004). Microbial ecologists are thus faced with estimating diversity based on small samples of the total diversity. This poses special challenges for detecting biodiversity patterns in microbial communities. However, there are macro-organism communities that are as hyper-diverse as many microbial communities and approaches have been developed by macro-organism ecologists to estimate diversity robustly in such situations (e.g. Hughes *et al.*, 2001). Many of these techniques are applicable to microbial communities (reviewed by Bohannan & Hughes, 2003) and have been used to reveal patterns in microbial biodiversity (e.g. Horner-Devine *et al.*, 2003).

Fundamental patterns in diversity

A number of patterns in the distribution of plant and animal diversity have been documented, some so commonly observed that ecologists often refer to them as fundamental patterns. These include the decrease in taxonomic richness with increasing latitude or altitude, the increase in taxonomic richness observed as the area sampled increases (the 'species–area relationship'), changes in taxonomic richness with changes in primary productivity (the 'species–energy relationship'), the decrease in community similarity observed as the distance between communities increases (the 'distance-decay relationship') and a number of others (reviewed by Rosenzweig, 1995). These patterns have been very influential in the development of the general science of ecology, stimulating the development of theory in community ecology and conservation biology, as well as forming the basis for predicting community responses to global change and habitat loss, among other things.

Patterns in plant and animal biodiversity provide a natural starting point for the study of microbial biodiversity patterns, not necessarily because microbes should exhibit macro-organism-like patterns (although they may), but rather by necessity, because macro-organism ecology is so much better studied. A common objection to this approach is that plants and animals are so different biologically from microbes that such studies are unlikely to be successful. We address this objection below.

ARE MICROBES TOO DIFFERENT?

It is commonly argued that microbes should have qualitatively different patterns in their diversity, relative to those of plants and animals, because of the unique biology of microbes. The unique aspects of their biology most frequently cited to defend this idea are a very high capacity for dispersal, the lack of clear species boundaries and the possibility of horizontal gene exchange ('parasexuality'). Despite claims to the contrary, these aspects do not necessarily eliminate the possibility that microbes have diversity patterns that are at least qualitatively similar to those exhibited by plants and animals.

The existence of very high rates (and long distances) of dispersal among microbes is a hypothesis; it is not clear how widespread such traits are among microbes. Even if we assume that at least some microbes have a very high capacity for dispersal, this does not necessarily prevent them from exhibiting patterns, even biogeographical patterns, in their diversity. The most robust patterns in plant and animal diversity (e.g. the species–area relationship) have been commonly observed at spatial scales where dispersal is not limiting for plants and animals (where 'everything is everywhere'; reviewed by Rosenzweig, 1995). It is also a possibility that, while some microbes have high dispersal capacity, others may not.

Clearly definable species are also not necessarily a requirement for the existence of patterns similar to those of plants and animals. Many plant and animal diversity patterns (e.g. species–area relationships, richness–energy relationships, etc.) have been documented at multiple taxonomic scales (species, genus, family, etc.; e.g. Bennett, 1997; Harcourt, 1999). Such patterns have also been observed at the level of genetic diversity, presumably because the mechanisms that underlie these patterns operate at multiple scales of complexity (Vellend & Geber, 2005). It is also important to note that, while

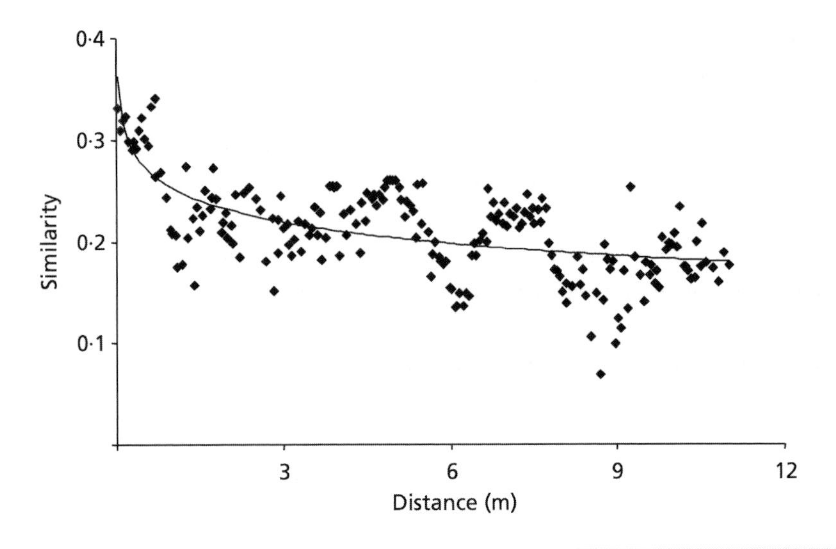

Fig. 1. The distance-decay of similarity. Bacterial community similarity (calculated using the Jaccard index) decreased with increasing distance between soil samples. Bacterial community composition was estimated using amplified fragment length polymorphism (AFLP) analysis. Plotted from data published in Franklin & Mills (2003).

the species concept is better developed for macro-organisms than for micro-organisms, species definitions for macro-organisms are not immutable; for example, they are subject to change as more is learned about the genetics of different populations.

The potentially most important difference between micro-organisms and macro-organisms is the existence of parasexuality. Parasexuality has the potential to conceal or eliminate detectable patterns in microbial biodiversity. The effect that parasexuality might have depends on the rate, extent and heterogeneity of gene exchange, as well as the method used to assess microbial biodiversity. The scenario that is likely to have the greatest impact on biodiversity patterns is one in which the rate of exchange is high (much higher than the mutation rate), wide in extent (not restricted to closely related organisms) and heterogeneous across the genome (i.e. where some genes are much more likely to be exchanged than others). Under such a scenario, it is possible that the phylogenetically informative genes commonly used to define prokaryotic taxa (e.g. the 16S rRNA gene) could be effectively 'decoupled' from genes that code for proteins that interact directly with the environment (such as those involved in nutrient uptake, detoxification, etc.), potentially obscuring biodiversity patterns that are caused by environmental heterogeneity. It is not known how common such a scenario might be in nature. However, the fact that 16S rRNA gene sequences have been observed to vary along environmental gradients (e.g. those reviewed by Horner-Devine *et al.*, 2004b) suggests that such a scenario is not present universally for prokaryotes. It is interesting to note that recent evidence suggests that horizontal gene transfer may also occur among different species of plants (Bergthorsson *et al.*, 2003).

Despite these potential biological differences, patterns have been reported in the distribution of prokaryotic biodiversity, patterns that in some cases are qualitatively similar to those of macro-organisms. We discuss some examples of these patterns below. We have focused on the type of patterns that are the most frequently reported for macro-organisms: spatial patterns in community composition and taxonomic richness.

PATTERNS IN COMMUNITY COMPOSITION

Distance-decay relationships

Plant and animal community composition tends to 'decay' (i.e. becomes increasingly dissimilar) with increasing geographical distance between samples (Condit *et al.*, 2002; Qian *et al.*, 2005; Tuomisto *et al.*, 2003). This 'distance-decay' relationship is a commonly observed pattern in plant and animal communities (Nekola & White, 1999). This decay can be due to dispersal limitation, a decay in environmental similarity with geographical distance or some combination of these two factors. The relative importance of these different underlying mechanisms can be unravelled using a variety of statistical methods (Tuomisto *et al.*, 2003; Green *et al.*, 2004; Horner-Devine *et al.*, 2004a).

To our knowledge, Franklin & Mills (2003) were the first to report a distance-decay relationship for prokaryotes. They used amplified fragment-length polymorphism (AFLP) analysis, a molecular fingerprinting method, to assess bacterial community composition between pairs of soil samples ranging from 2·5 cm to 11 m apart. They observed a significant distance-decay relationship, with a scale-dependent slope that decreased in magnitude at larger scales (Fig. 1). Because the scale of sampling was relatively small, they argued that the observed distance-decay relationship was due to environmental heterogeneity, rather than dispersal limitation. However, they did not test this assertion.

A growing number of studies have examined the relative effects of environmental heterogeneity and dispersal limitation on microbial distance-decay relationships (reviewed by Martiny *et al.*, 2006). Whitaker *et al.* (2003) examined *Sulfolobus* strains across spatial scales of approximately 6 to 6000 km and found that the decline in genetic similarity with distance was explained by geographical distance but not by environmental heterogeneity. This suggested that dispersal limitation was the primary process underlying this relationship. In contrast, sediment bacterial communities studied at smaller spatial scales (3 cm to 300 m) have shown environmental heterogeneity to be the primary factor underlying distance-decay relationships (Horner-Devine *et al.*, 2004a). Together, these studies suggest that the spatial scaling of microbial biodiversity can be influenced by both environmental heterogeneity and dispersal history, with dispersal history likely to become increasingly important at larger spatial scales (Martiny *et al.*, 2006).

Co-occurrence patterns

In communities of macro-organisms, it has been observed that pairs of taxa or species tend to co-occur less often than would be expected by chance (Gotelli & McCabe, 2002). For example, in his studies of bird species of the Bismarck archipelago near New Guinea, Diamond (1975) observed that some pairs of bird species never co-existed: they seemed to be 'forbidden' combinations. In other words, one island might harbour bird species A and another similar island might harbour bird species B, but A and B were never found on the same island. A meta-analysis (Gotelli & McCabe, 2002) revealed a consistent, non-random pattern in macro-organism communities, with significantly less co-occurrence of species than would be expected by chance. A recent meta-analysis of over 100 datasets from microbial communities (including both prokaryotes and microbial eukaryotes) documented a statistically identical pattern to that found for macro-organisms (Horner-Devine *et al.*, 2006; Fig. 2). This pattern of community structure may therefore be a universal feature of biological communities from all three domains of life, suggesting that similar processes may govern the structure of these different communities.

What might drive these patterns? To date, while both macro-organismal and micro-organismal communities have been shown to exhibit significant co-occurrence patterns, we know of no attempts to evaluate the underlying mechanisms. It is possible that interspecific competition could be responsible for the observed non-random patterns of co-occurrence. It is also possible that significant co-occurrence patterns result from habitat affiliations, i.e. from the association of different taxa with different abiotic features of individual sites. A third possible explanation is that historical processes have led to less co-existence than would be expected by chance. In particular, allopatric speciation may lead to a pattern of little or no co-existence among congeners, whether or not there is competition occurring. Finally, it is possible that these hypotheses for non-random co-occurrence (competition, habitat affiliations and historical effects) are not mutually exclusive mechanisms and, in fact, may interact to produce these patterns.

Phylogenetic patterns

Phylogenetic analyses of community composition and structure have proven valuable in studies of plant and animal diversity, because such approaches can lend insight into the relative importance of evolutionary and ecological forces in shaping communities

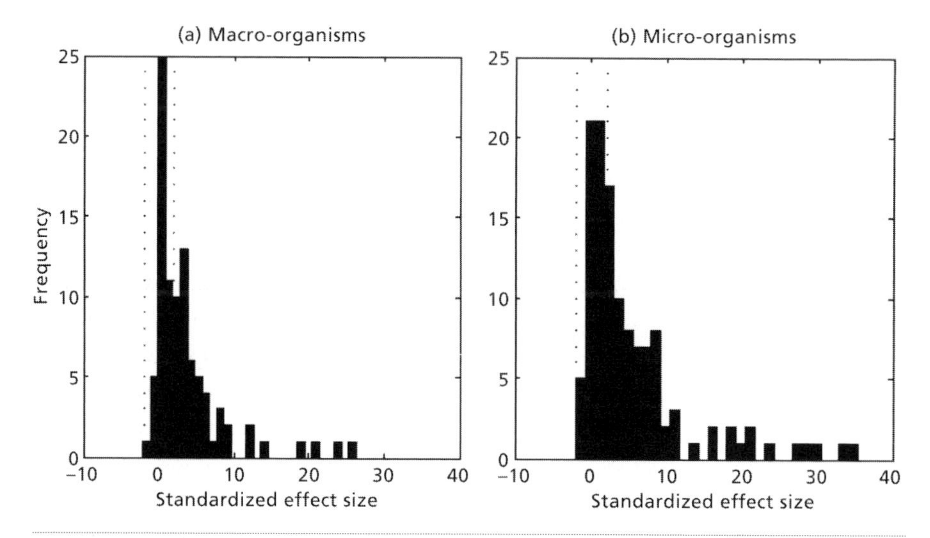

Fig. 2. Frequency histograms of standardized effect sizes for an index of co-occurrence for (a) macro-organism ($n = 93$) datasets and (b) micro-organism ($n = 124$) datasets. The null hypothesis asserts that the mean effect size will equal zero and that 95 % of the observations will be between −2 and 2. Both communities of macro-organisms and micro-organisms exhibited significant patterns of co-occurrence. Reproduced with permission from Horner-Devine *et al.* (2006).

(Elton, 1946; Webb *et al.*, 2002; Cavender-Bares & Wilczek, 2003). For example, co-occurring rain-forest tree species have been observed to be more closely related than would be expected by chance (Webb, 2000); this pattern of phylogenetic clustering may indicate that these closely related tree species share traits that are important for their persistence in a particular environment (Webb *et al.*, 2002). In other words, the habitat may serve as a filter for determining which species may be present in a given location, and such habitat filtering may be more important than competition in maintaining rain-forest tree species diversity (Tofts & Silvertown, 2000; Webb, 2000; Kembel & Hubbell, 2006).

It is also possible that a community may comprise distantly related taxa as a result of current or past competitive exclusion between similar (and thus closely related) taxa and/or as a result of convergent evolution in traits important for persistence in a given environment (Cavender-Bares *et al.*, 2004; Kembel & Hubbell, 2006).

Horner-Devine & Bohannan (2006) identified phylogenetic structure in bacterial communities sampled from four different environments (temperate soil, tropical soil, marine sediment and freshwater), where community structure was estimated using three different indicator genes (16S rRNA, *amoA* and *nirS*). They estimated phylogenetic structure using indices that measure the degree of phylogenetic clustering of taxa across a phylogenetic tree in a given sample relative to the regional pool of taxa

(Webb, 2000). They observed that bacteria tend to co-occur with other closely related bacteria more often than would be expected by chance, as has been observed for some macro-organisms (e.g. Webb, 2000).

These results suggest that habitat filtering may be relatively more important to the assembly of bacterial communities than competition. Why might this be the case? Recent studies of macro-organisms suggest that the greater the degree of environmental heterogeneity over which one samples a community, the more likely it is that phylogenetic clustering rather than overdispersion will be present (Cavender-Bares *et al.*, 2006; Silvertown *et al.*, 2006). The datasets used in the study of Horner-Devine & Bohannan (2006) consisted of samples that were extremely large in volume relative to the volume of an individual bacterium and the spatial scales over which individual bacteria interact (as is the case in most studies of microbial biodiversity) and thus probably included substantial environmental heterogeneity.

Recent studies of macro-organisms have also suggested that phylogenetic resolution should influence the prevalence of clustering; changing phylogenetic resolution (i.e. an increase in 'taxonomic lumping') may result in an increased prevalence of clustering (Silvertown *et al.*, 2006). Substantial ecological diversity has been shown to be present within microbial taxa defined using molecular markers, especially ribosomal markers (e.g. Ward *et al.*, 1998; Rocap *et al.*, 2002). The 'taxonomic lumping' inherent in these molecular approaches may result in the observation of phylogenetic clustering. Finer-scale markers [e.g. internal transcribed spacer (ITS) or multigene approaches] could potentially reveal the presence of increased phylogenetic overdispersion, rather than clustering.

PATTERNS IN TAXONOMIC RICHNESS

Richness–area relationships

One of the most widely studied patterns in ecology is the species–area relationship (SAR), in which species richness (S) increases with the size of the sampled area (A). Although many functional forms of the SAR have been proposed, the power-law form $S \propto A^z$ has withstood the test of time relatively well. Empirical evidence suggests that, for large organisms (i.e. plants and animals) within contiguous areas, z ranges between 0·1 and 0·3. There is evidence for slightly steeper SARs between islands in archipelagos ($0{\cdot}25 < z < 0{\cdot}35$; Rosenzweig, 1995).

Power-law SARs (or more accurately, taxa–area relationships or TARs) have recently been reported for prokaryotes (reviewed by Green & Bohannan, 2006). Some of these studies report z values that are substantially lower than those most frequently reported

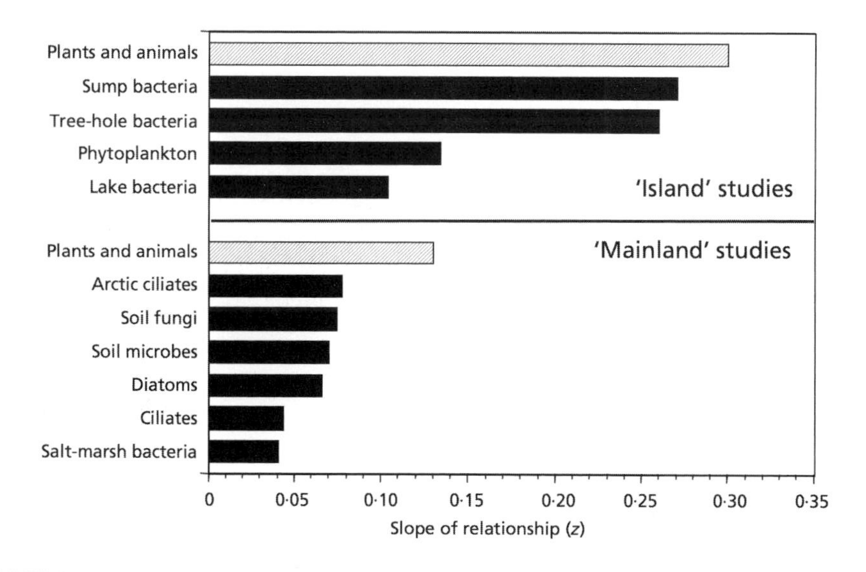

Fig. 3. Slope of the TAR (z) for micro-organisms and macro-organisms. Modified from Bell *et al.* (2005), using data from Franklin & Mills (2003), van der Gast *et al.* (2005), Reche *et al.* (2005), Smith *et al.* (2005), Green *et al.* (2004) and Horner-Devine *et al.* (2004a).

for macro-organisms (i.e. microbial z values $< 0{\cdot}1$), although there are also reports of higher z values consistent with those of macro-organisms (Fig. 3).

Detecting TARs in communities of prokaryotes is especially challenging because a complete census of prokaryotic diversity is not possible in most environments, and observed counts of taxon richness for areas of different sizes may result in biased z values. A recent study of soil prokaryotes highlights the challenge of interpreting z values based solely on observed taxon-richness counts (Noguez *et al.*, 2005). Using a nested sampling design and terminal restriction fragment length polymorphism (T-RFLP) analysis of 16S rRNA genes, Noguez *et al.* (2005) characterized soil prokaryote diversity across the spatial scale of 1 to 64 m^2. They assumed that prokaryote diversity characterized in point samples (one 5 cm^3 soil core) was equal to that in 1 m^2, and they may therefore have underestimated small-scale (1 m^2) richness. In addition, they assumed that the cumulative taxon richness in nested, non-contiguous soil cores adequately represented diversity at larger scales ($1\ m^2 < A \leq 64\ m^2$), resulting in a potential underestimation of taxon richness at all spatial scales studied. Without more information on the spatial distribution of taxa in this system, it is difficult to assess the relative magnitude of biases at different scales.

The relationship between the observed and true TAR slopes, z, in microbial studies will depend on the proportion of undetected taxa at each spatial scale. One approach

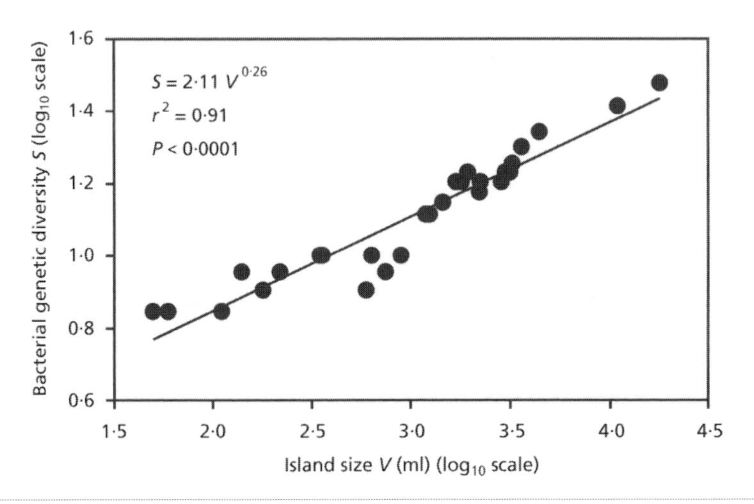

Fig. 4. The TAR for micro-organisms in water-filled tree holes. Reproduced from Bell *et al.* (2005) with permission (© American Association for the Advancement of Science 2005).

to extrapolating microbial taxon richness is to make assumptions about microbial species-abundance distributions and how they vary with spatial scale. This type of approach has been used to extrapolate bacterial diversity (Curtis *et al.*, 2002). Little is currently known about the spatial scaling of microbial species-abundance relationships, and it is therefore difficult to assess the validity of projected extrapolations. Ultimately, our ability to extrapolate richness from microbial TARs will require creative approaches for estimating the total number of taxa at small spatial scales from point samples and methods for extrapolating richness to larger scales using sparse sample data (Bohannan & Hughes, 2003; Cam *et al.*, 2002; Green *et al.*, 2005).

An alternative approach to characterizing microbial TARs is to use the slope of the distance-decay relationship to estimate the slope of the TAR (Krishnamani *et al.*, 2004). The advantage of this approach is that it only requires the sampling of localities spatially in such a way that the decline in similarity with distance can be measured. This method has been applied to estimate the spatial scaling of prokaryotic diversity (Horner-Devine *et al.*, 2004a).

Theoretically, the TAR slope z should vary with taxonomic resolution, and this must be taken into account when comparing micro-organisms and macro-organisms. The analyses of Horner-Devine *et al.* (2004a) found that the slope of the TAR relationship z increased with increasing taxonomic resolution, ranging from $z = 0{\cdot}0019$ at 95 % sequence similarity to $z = 0{\cdot}040$ at 99 % sequence similarity. Their data clearly indicate that spatial biodiversity patterns vary with taxonomic resolution and that the often coarse taxonomic resolutions commonly used for micro-organisms (e.g. morphotypes, molecular 'fingerprints', ribotypes etc.) may result in low z values relative to plant and animal species.

'Island' patterns

The TARs described above were documented for contiguous areas. Studies of plants and animals have shown that the scaling of biodiversity can be different in discontinuous areas (i.e. 'islands' of habitat surrounded by an inhospitable matrix). For example, the TAR slope z tends to be higher when measured for islands of different sizes in an archipelago (reviewed by Rosenzweig, 1995). Three recent studies have attempted to document 'island' patterns in microbial biodiversity. van der Gast *et al.* (2005) studied bacterial communities colonizing metal-cutting fluids from machines of increasing sump tank size, assuming these to be analogous to islands of variable size. Similarly, Bell *et al.* (2005) studied bacterial communities in water-filled tree holes of varying volume (Fig. 4). Both studies measured bacterial taxonomic richness as the observed number of unique bands or ribotypes detected using denaturing gradient gel electrophoresis (DGGE) of 16S rRNA genes. In both systems, the number of taxa detected increased significantly with increasing volume, and the rate of increase was similar to that reported for plants and animals ($z = 0{\cdot}245$ to $0{\cdot}295$ for sump tanks, $z = 0{\cdot}26$ for tree holes). Reche *et al.* (2005) used a similar approach to study bacterial communities in lakes of varying sizes. They observed that taxonomic richness (estimated using techniques similar to those of van der Gast *et al.*, 2005 and Bell *et al.*, 2005) increased significantly with lake area, with an estimated z value of $0{\cdot}104$. All three studies used observed rather than estimated richness to describe the richness from each sample, despite undersampling each individual sample.

Traditional studies of island biogeography for plants and animals are commonly based on richness estimates from surveys or atlases that tally species over multiple years and therefore are more likely to be complete descriptions of the taxa present than are most microbial studies. The studies of prokaryotic diversity described in the previous paragraph quantified bacterial richness in equal volumes sampled from islands assumed to be relatively well mixed, which is analogous to randomly sampling an equal number of individuals from every island. The relationship between z estimated from equal-sized random samples per island versus a cumulative survey of an archipelago will depend on several factors, including population aggregation patterns, the rank-abundance curve on each island and the sampling effort per survey.

Richness–energy relationships

Primary productivity (the rate of energy capture and carbon fixation by primary producers) is thought to be a key determinant of plant and animal biodiversity (Rosenzweig, 1995). Many studies of plants and animals have reported a positive quadratic or hump-shaped relationship between productivity and diversity, where diversity peaks at intermediate productivity (Rosenzweig, 1995), although other patterns have also been observed (Abrams, 1995; Mittelbach *et al*., 2001). There is evidence from both laboratory studies (e.g. Bohannan & Lenski, 2000; Kassen *et al*., 2000) and field studies (e.g. Benlloch *et al*., 1995; Torsvik *et al*., 1998; Horner-Devine *et al*., 2003; De Wever *et al*., 2005) that productivity can influence bacterial diversity. In what follows, we highlight a few of the growing number of field studies that examine this relationship for prokaryotes.

Horner-Devine *et al*. (2003) observed that increasing productivity both increased and decreased taxonomic diversity of bacteria in aquatic mesocosms and that the shape of the relationship between productivity and diversity differed for different taxonomic groups of bacteria (Fig. 5). For example, they observed that the diversity of members of the *Bacteroidetes*, the most common taxon in their study system, exhibited a significant hump-shaped relationship with primary productivity. In contrast, they observed a significant U-shaped relationship between primary productivity and diversity for the *Alphaproteobacteria*, the second most common group, and no discernible relationship between primary productivity and diversity for the *Betaproteobacteria*, the third most common group. Other studies have found that bacterial richness tends to increase with productivity or energy. For example, De Wever *et al*. (2005) observed that bacterial richness (measured as the number of DGGE bands) tended to increase with productivity in the epilimnion of Lake Tanganyika in Africa. In contrast, Torsvik *et al*. (1998) observed that pristine aquatic sediments had much higher bacterial genetic diversity than did sediments below fish farms (which receive a substantial input of nutrients via fish feed), suggesting that increased energy decreased diversity.

Richness–disturbance relationships

The influence of disturbance on communities of macro-organisms has received considerable attention and is thought to play an important role in the maintenance of diversity. Ecologists have shown that the intermediate disturbance hypothesis (IDH), in which diversity may peak at intermediate intensities or frequencies of small-scale disturbance (Connell, 1978; Sousa, 1979), holds for a variety of macro-organisms in a number of different environments (e.g. Brown & Hutchings, 1997; Flöder & Sommer, 1999).

While laboratory studies suggest that the IDH may also be applicable to bacteria (Buckling *et al*., 2000), it has not been tested definitively in the field. A number of field studies have begun to explore the relationship between disturbance and bacterial diversity (e.g. Johnsen *et al*., 2001; Kent & Triplett, 2002; Kozdroj & van Elsas, 2001), most of which addressed the effect of either chemical or physical disturbance of soil on

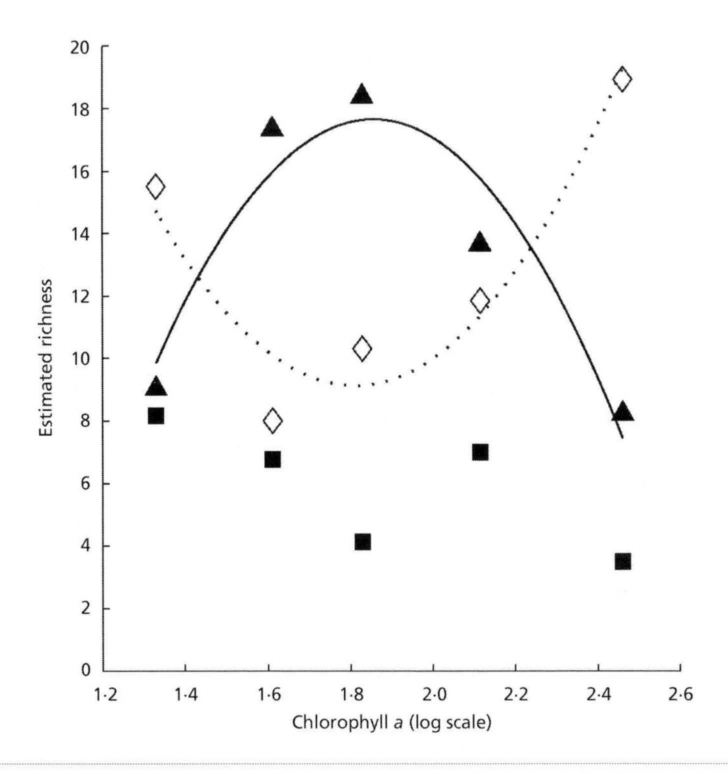

Fig. 5. Bacterial richness–energy relationships. The relationship between bacterial taxonomic richness and primary productivity in aquatic mesocosms varied by taxonomic group. Alphaproteobacterial diversity (◇) exhibited a significant U-shaped relationship with productivity (dotted line), diversity within the *Bacteroidetes* (the *Cytophaga–Flavobacterium–Bacteroides* group) (▲) had a significant unimodal relationship with productivity (solid line) and betaproteobacterial diversity (■) showed no discernible response to increasing productivity. Primary productivity was estimated as chlorophyll *a*; taxonomic diversity was estimated from observed phylotype richness via the Chao1 approach (Hughes *et al.*, 2001). From Horner-Devine *et al.* (2003).

bacteria. Muller *et al.* (2001) examined bacterial diversity along a gradient of mercury contamination in fields and found that sequence diversity decreased with increasing exposure to mercury. In contrast, Bruce *et al.* (1995) examined the diversity of mercury-resistance genes in sedimentary bacteria in sites with varying levels and histories of mercury contamination. They found that the sites exposed to intermediate levels of mercury had the highest genetic diversity, while the most heavily contaminated site and pristine sites had low diversity. Although these studies suggest a relationship between disturbance and bacterial diversity, disturbance was always confounded with difference in other factors (e.g. vegetative cover, land history, soil structure) that may influence bacterial composition and diversity.

Testing the IDH for prokaryotes will require systematically examining how the frequency, intensity and size of disturbance affect prokaryotic diversity, and few studies have accomplished this (but see Bruce *et al.*, 1995; Ibekwe *et al.*, 2002; Muller *et al.*, 2001; Fierer *et al.*, 2003). Fierer *et al.* (2003) exposed two different soil types to water stress at different drying–rewetting cycle frequencies and this provides a strong example of the design needed to assess the IDH. They found that the frequency of the stress event influenced bacterial taxonomic richness (estimated by T-RFLP) in one of the soil types. However, they observed lowest richness at intermediate drying–rewetting frequencies, in conflict with the predictions of the IDH. While bacterial community composition and diversity may respond to gradients in disturbance, there are not currently enough data available to assess the applicability of the IDH to prokaryotes in the field.

CONCLUSIONS

Despite the diversity and ubiquity of prokaryotes, we still know relatively little about patterns in their biodiversity. Given our limited knowledge of microbial diversity, it is reasonable to start by looking for patterns in microbial diversity that appear to be fundamental for macro-organisms. Patterns derived from studies on plants and animals provide a natural starting point for the study of microbial biodiversity patterns.

A number of recent studies have documented patterns in prokaryotic diversity that are qualitatively similar to those commonly observed for plants and animals. The quantitative differences in these patterns can often be attributed to the different approaches used to define the taxa of micro-organisms and macro-organisms as well as to differences in spatial scale. However, despite these differences, common patterns emerge.

The study of microbial biodiversity patterns is still in its infancy. We do not yet know how widespread such patterns are. Ultimately, the study of patterns in microbial biodiversity will shed light on the relative importance of the processes that generate and maintain diversity, such as diversification, extinction, dispersal and species interactions, as well as the potential importance of these processes for maintaining ecosystem functions. In particular, such studies can test the idea that there are universal ecological patterns and processes, based on universal attributes of all life. If microbes are commonly observed to exhibit biodiversity patterns similar to those of macro-organisms, it is likely that such universality exists. Conversely, if microbial diversity can be shown to represent clear exceptions to patterns common in macro-organism diversity, it will highlight the unique features of microbes that have influenced the generation and maintenance of their diversity.

REFERENCES

Abrams, P. A. (1995). Monotonic or unimodal diversity-productivity gradients: what does competition theory predict? *Ecology* **76**, 2019–2027.

Amann, R. I., Ludwig, W. & Schleifer, K. H. (1995). Phylogenetic identification and *in situ* detection of individual microbial cells without cultivation. *Microbiol Rev* **59**, 143–169.

Bell, T., Ager, D., Song, J.-I., Newman, J. A., Thompson, I. P., Lilley, A. K. & van der Gast, C. J. (2005). Larger islands house more bacterial taxa. *Science* **308**, 1884.

Benlloch, S., Rodriguez-Valera, F. & Martinez-Murcia, A. J. (1995). Bacterial diversity in two coastal lagoons deduced from 16S rDNA PCR amplification and partial sequencing. *FEMS Microbiol Ecol* **18**, 267–279.

Bennett, J. P. (1997). Nested taxa-area curves for eastern United States floras. *Rhodora* **99**, 241–251.

Bergthorsson, U., Adams, K. L., Thomason, B. & Palmer, J. D. (2003). Widespread horizontal transfer of mitochondrial genes in flowering plants. *Nature* **424**, 197–201.

Bielawski, J. P., Dunn, K. A., Sabehi, G. & Beja, O. (2004). Darwinian adaptation of proteorhodopsin to different light intensities in the marine environment. *Proc Natl Acad Sci U S A* **101**, 14824–14829.

Bohannan, B. J. M. & Hughes, J. (2003). New approaches to analyzing microbial biodiversity data. *Curr Opin Microbiol* **6**, 282–287.

Bohannan, B. J. M. & Lenski, R. E. (2000). The relative importance of competition and predation varies with productivity in a model community. *Am Nat* **156**, 329–340.

Brown, K. S., Jr & Hutchings, R. W. (1997). Disturbance, fragmentation, and the dynamics of diversity in Amazonian forest butterflies. In *Tropical Forest Remnants: Ecology, Management, and Conservation of Fragmented Communities*, pp. 91–110. Edited by W. F. Laurance & R. O. Bierregaard. Chicago: University of Chicago Press.

Bruce, K. D., Osborn, A. M., Pearson, A. J., Strike, P. & Ritchie, D. A. (1995). Genetic diversity within *mer* genes directly amplified from communities of noncultivated soil and sediment bacteria. *Mol Ecol* **4**, 605–612.

Buckling, A., Kassen, R., Bell, G. & Rainey, P. B. (2000). Disturbance and diversity in experimental microcosms. *Nature* **408**, 961–964.

Cam, E., Nichols, J. D., Hines, J. E., Sauer, J. R., Alpizar-Jara, R. & Flather, C. H. (2002). Disentangling sampling and ecological explanations underlying species–area relationships. *Ecology* **83**, 1118–1130.

Cavender-Bares, J. & Wilczek, A. (2003). Integrating micro- and macroevolutionary processes in community ecology. *Ecology* **84**, 592–597.

Cavender-Bares, J., Ackerly, D. D., Baum, D. A. & Bazzaz, F. A. (2004). Phylogenetic overdispersion in Floridian oak communities. *Am Nat* **163**, 823–843.

Cavender-Bares, J., Keen, A. & Miles, B. (2006). Phylogenetic structure of Floridian plant communities depends on spatial and taxonomic scale. *Ecology* (in press).

Condit, R., Pitman, N., Leigh, E. G., Jr & 10 other authors (2002). Beta-diversity in tropical forest trees. *Science* **295**, 666–669.

Connell, J. H. (1978). Diversity in tropical rain forests and coral reefs. *Science* **199**, 1302–1310.

Curtis, T. P., Sloan, W. T. & Scannell, J. W. (2002). Estimating prokaryotic diversity and its limits. *Proc Natl Acad Sci U S A* **99**, 10494–10499.

De Wever, A., Muylaert, K., Van der Gucht, K., Pirlot, S., Cocquyt, C., Descy, J.-P.,

Plisnier, P.-D. & Vyverman, W. (2005). Bacterial community composition in Lake Tanganyika: vertical and horizontal heterogeneity. *Appl Environ Microbiol* **71**, 5029–5037.

Diamond, J. M. (1975). Assembly of species communities. In *Ecology and Evolution of Communities*, pp. 342–444. Edited by M. L. Cody & J. M. Diamond. Cambridge, MA: Harvard University Press.

Elton, C. S. (1946). Competition and the structure of ecological communities. *J Anim Ecol* **15**, 54–68.

Fierer, N., Schimel, J. P. & Holden, P. A.. (2003). Influence of drying-rewetting frequency on soil bacterial community structure. *Microb Ecol* **45**, 63–71.

Flöder, S. & Sommer, U. (1999). Diversity in planktonic communities: an experimental test of the intermediate disturbance hypothesis. *Limnol Oceanogr* **44**, 1114–1119.

Franklin, R. B. & Mills, A. L. (2003). Multi-scale variation in spatial heterogeneity for microbial community structure in an eastern Virginia agricultural field. *FEMS Microbiol Ecol* **44**, 335–346.

Gotelli, N. J. & McCabe, D. J. (2002). Species co-occurrence: a meta-analysis of J. M. Diamond's assembly rules model. *Ecology* **83**, 2091–2096.

Green, J. L. & Bohannan, B. J. M. (2006). Spatial scaling of microbial biodiversity. In *Scaling Biodiversity*. Edited by D. Storch, P. A. Marquet & J. H. Brown. Cambridge: Cambridge University Press (in press).

Green, J. L., Holmes, A. J., Westoby, M., Oliver, I., Briscoe, D., Dangerfield, M., Gillings, M. & Beattie, A. J. (2004). Spatial scaling of microbial eukaryote diversity. *Nature* **432**, 747–750.

Green, J. L., Hastings, A., Arzberger, P. & 8 other authors (2005). Complexity in ecology and conservation: mathematical, statistical, and computational challenges. *Bioscience* **55**, 501–510.

Harcourt, A. H. (1999). Biogeographic relationships of primates on South-East Asian islands. *Biogeography* **8**, 55–61.

Head, I. M., Saunders, J. R. & Pickup, R. W. (1998). Microbial evolution, diversity, and ecology: a decade of ribosomal RNA analysis of uncultivated microorganisms. *Microb Ecol* **35**, 1–21.

Horner-Devine, M. C. & Bohannan, B. J. M. (2006). Phylogenetic clustering and overdispersion in bacterial communities. *Ecology* (in press).

Horner-Devine, M. C., Leibold, M. A., Smith, V. H. & Bohannan, B. J. M. (2003). Bacterial diversity patterns along a gradient of primary productivity. *Ecol Lett* **6**, 613–622.

Horner-Devine, M., Lage, M., Hughes, J. & Bohannan, B. J. M. (2004a). A taxa-area relationship for bacteria. *Nature* **432**, 750–753.

Horner-Devine, M. C., Carney, K. M. & Bohannan, B. J. M. (2004b). An ecological perspective on bacterial biodiversity. *Proc Biol Sci* **271**, 113–122.

Horner-Devine, M. C., Silver, J., Leibold, M. A. & 11 other authors (2006). Patterns of taxon co-occurrence: a comparison of assembly rules for macro and microorganisms. *Ecology* (in press).

Hughes, J. B., Hellman, J. J., Ricketts, T. H. & Bohannan, B. J. M. (2001). Counting the uncountable: statistical approaches to estimating microbial diversity. *Appl Environ Microbiol* **67**, 4399–4406.

Ibekwe, A. M., Kennedy, A. C., Frohne, P. S., Papiernik, S. K., Yang, C.-H. & Crowley, D. E. (2002). Microbial diversity along a transect of agronomic zones. *FEMS Microbiol Ecol* **39**, 183–191.

Johnsen, K., Jacobsen, C. S., Torsvik, V. & Sørensen, J. (2001). Pesticide effects on bacterial diversity in agricultural soils: a review. *Biol Fertil Soils* **33**, 443–453.

Kassen, R., Buckling, A., Bell, G. & Rainey, P. B. (2000). Diversity peaks at intermediate productivity in a laboratory microcosm. *Nature* **406**, 508–512.

Kembel, S. W. & Hubbell, S. P. (2006). The phylogenetic structure of a neotropical forest tree community. *Ecology* (in press).

Kent, A. D. & Triplett, E. W. (2002). Microbial communities and their interactions in soil and rhizosphere ecosystems. *Annu Rev Microbiol* **56**, 211–236.

Konstantinidis, K. T. & Tiedje, J. M. (2005). Towards a genome-based taxonomy for prokaryotes. *J Bacteriol* **187**, 6258–6264.

Kowalchuk, G. A. & Stephen, J. R. (2001). Ammonia-oxidizing bacteria: a model for molecular microbial ecology. *Annu Rev Microbiol* **55**, 485–529.

Kozdroj, J. & van Elsas, J. D. (2001). Structural diversity of microorganisms in chemically perturbed soil assessed by molecular and cytochemical approaches. *J Microbiol Methods* **43**, 197–212.

Krishnamani, Kumar, R. A. & Harte, J. (2004). Estimating species richness at large spatial scales using data from small discrete plots. *Ecography* **27**, 637–642.

Martiny, J. B. H., Bohannan, B. J. M., Brown, J. H. & 13 other authors (2006). Microbial biogeography: putting microbes on the map. *Nat Rev Microbiol* **4**, 103–111.

Mittelbach, G. G., Steiner, C. F., Scheiner, S. M., Gross, K. L., Reynolds, H. L., Waide, R. B., Willig, M. R., Dodson, S. I. & Gough, L. (2001). What is the observed relationship between species richness and productivity? *Ecology* **82**, 2381–2396.

Mlot, C. (2004). Microbial diversity unbound: what DNA-based techniques are revealing about the planet's hidden biodiversity. *Bioscience* **54**, 1064–1068.

Moore, L. R., Rocap, G. & Chisholm, S. W. (1998). Physiology and molecular phylogeny of coexisting *Prochlorococcus* ecotypes. *Nature* **393**, 464–467.

Muller, A. K., Westergaard, K., Christensen, S. & Sørensen, S. J. (2001). The effect of long-term mercury pollution on the soil microbial community. *FEMS Microbiol Ecol* **36**, 11–19.

Nekola, J. C. & White, P. S. (1999). The distance decay of similarity in biogeography and ecology. *J Biogeogr* **26**, 867–878.

Noguez, A. M., Arita, H. T., Escalante, A. E., Forney, L. J., Garcia-Oliva, F. & Souza, V. (2005). Microbial macroecology: highly structured prokaryotic soil assemblages in a tropical deciduous forest. *Global Ecol Biogeogr* **14**, 241–248.

Qian, H., Ricklefs, R. E. & White, P. S. (2005). Beta diversity of angiosperms in temperate floras of eastern Asia and eastern North America. *Ecol Lett* **8**, 1–15.

Reche, I., Pulido-Villena, E., Morales-Baquero, R. & Casamayor, E. O. (2005). Does ecosystem size determine aquatic bacterial richness? *Ecology* **86**, 1715–1722.

Rocap, G., Distel, D. L., Waterbury, J. B. & Chisholm, S. W. (2002). Resolution of *Prochlorococcus* and *Synechococcus* ecotypes by using 16S-23S ribosomal DNA internal transcribed spacer sequences. *Appl Environ Microbiol* **68**, 1180–1191.

Rosenzweig, M. L. (1995). *Species Diversity in Space and Time*. Cambridge: Cambridge University Press.

Silvertown, J., Dodd, M. & Lawson, C. (2006). Phylogeny and the hierarchical organization of plant diversity. *Ecology* (in press).

Smith, V. H., Foster, B. L., Grover, J. P., Holt, R. D., Leibold, M. A. & deNoyelles, F., Jr (2005). Phytoplankton species richness scales consistently from laboratory microcosms to the world's oceans. *Proc Natl Acad Sci U S A* **102**, 4393–4396.

Sousa, W. P. (1979). Disturbance in marine intertidal boulder fields: the nonequilibrium maintenance of species diversity. *Ecology* **60**, 1225–1239.

Stackebrandt, E. & Rainey, F. A. (1995). Partial and complete 16S rDNA sequences, their use in generation of 16S rDNA phylogenetic trees and their implications in molecular ecological studies. In *Molecular Microbial Ecology Manual*, pp. 3.1.1.1–3.1.1.17. Edited by A. D. L. Akkermans, J. D. van Elsas & F. J. de Bruijn. Dordrecht: Kluwer Academic.

Staley, J. T. (1997). Biodiversity: are microbial species threatened? *Curr Opin Biotechnol* **8**, 340–345.

Tofts, R. & Silvertown, J. (2000). A phylogenetic approach to community assembly from a local species pool. *Proc Biol Sci* **267**, 363–369.

Torsvik, V., Daae, F. L., Sandaa, R. A. & Ovreas, L. (1998). Novel techniques for analysing microbial diversity in natural and perturbed environments. *J Biotechnol* **64**, 53–62.

Tuomisto, H., Ruokolainen, K. & Yli-Halla, M. (2003). Dispersal, environment, and floristic variation of western Amazonian forests. *Science* **299**, 241–244.

Tyson, G. W., Chapman, J., Hugenholtz, P. & 7 other authors (2004). Community structure and metabolism through reconstruction of microbial genomes from the environment. *Nature* **428**, 37–43.

van der Gast, C. J., Lilley, A. K., Ager, D. & Thompson, I. P. (2005). Island size and bacterial diversity in an archipelago of engineering machines. *Environ Microbiol* **7**, 1220–1226.

Vellend, M. & Geber, M. A. (2005). Connections between species diversity and genetic diversity. *Ecol Lett* **8**, 767–781.

Venter, J. C., Remington, K., Heidelberg, J. F. & 20 other authors (2004). Environmental shotgun sequencing of the Sargasso Sea. *Science* **304**, 66–74.

von Wintzingerode, F., Göbel, U. B. & Stackebrandt, E. (1997). Determination of microbial diversity in environmental samples: pitfalls of PCR-based rRNA analysis. *FEMS Microbiol Rev* **21**, 213–229.

Ward, D. M., Ferris, M. J., Nold, S. C. & Bateson, M. M. (1998). A natural view of microbial biodiversity within hot spring cyanobacterial mat communities. *Microbiol Mol Biol Rev* **62**, 1353–1370.

Webb, C. O. (2000). Exploring the phylogenetic structure of ecological communities: an example for rain forest trees. *Am Nat* **156**, 145–155.

Webb, C. O., Ackerly, D. D., McPeek, M. A. & Donoghue, M. J. (2002). Phylogenies and community ecology. *Annu Rev Ecol Syst* **33**, 475–505.

Whitaker, R. J., Grogan, D. W. & Taylor, J. W. (2003). Geographic barriers isolate endemic populations of hyperthermophilic archaea. *Science* **301**, 976–978.

A putative RNA-interference-based immune system in prokaryotes: the epitome of prokaryotic genomic diversity

Eugene V. Koonin,[1] Kira S. Makarova,[1] Nick V. Grishin[2] and Yuri I. Wolf[1]

[1]National Center for Biotechnology Information, National Library of Medicine, National Institutes of Health, Bethesda, MD 20894, USA

[2]Department of Biochemistry, University of Texas Southwestern Medical Center, 5323 Harry Hines Blvd, Dallas, TX 75390-9050, USA

INTRODUCTION

The extreme diversity and plasticity of prokaryotic genomes, manifest both at the level of gene loss and acquisition via horizontal gene transfer and at the level of gene order rearrangement, are arguably among the major generalizations of comparative genomics (Coenye *et al.*, 2005; Koonin & Galperin, 1997, 2002; Snel *et al.*, 2002). Bacteria and archaea have numerous partially conserved operons, probably thanks, in part, to the 'selfish' behaviour of operons, but little conservation of genome organization is seen at large evolutionary distances beyond the operon level (Dandekar *et al.*, 1998; Lawrence, 1999; Mushegian & Koonin, 1996; Watanabe *et al.*, 1997; Wolf *et al.*, 2001). To detect traces of such long-range conservation, specially designed computational methods for detecting 'überoperons', or partially conserved gene neighbourhoods, have been developed (Lathe *et al.*, 2000; Rogozin *et al.*, 2002).

As a case study for testing the methods for conserved neighbourhood analysis that we have developed, we characterized an extensive gene set that included several proteins related to DNA or RNA metabolism and was, mostly, specific to thermophiles (Makarova *et al.*, 2002). These genes comprise a complex array of overlapping neighbourhoods that are partially conserved but highly diversified, in terms of both gene composition and gene order, and are represented in all archaeal and many bacterial genomes. At the time, we hypothesized that these genes encoded an uncharacterized, versatile repair system, largely associated with the thermophilic lifestyle.

SGM symposium 66: Prokaryotic diversity: mechanisms and significance.
Editors N. A. Logan, H. M. Lappin-Scott & P. C. F. Oyston. Cambridge University Press. ISBN 0 521 86935 8

Independently and almost simultaneously, Jansen *et al.* (2002) found that at least several genes from this gene neighbourhood are tightly associated with the so-called 'clustered regularly interspaced short palindrome repeats' (CRISPR); the acronym *cas* (for CRISPR-associated) genes was thus coined. The CRISPR are a distinct class of repetitive elements that are present in numerous prokaryotic genomes. A CRISPR element consists of a direct repeat of about 28–40 bp, with the copies separated by a unique sequence of about 25–40 bp. Typically, CRISPR form tandem arrays containing from four to more than 100 elements. Most of the genomes contain a single array of CRISPR in which the sequences of the repeats are (nearly) identical; some, however, possess multiple CRISPR cassettes that may have substantially different sequences (Haft *et al.*, 2005; Jansen *et al.*, 2002). The repeats in CRISPR from different genomes show only limited similarity, but often retain distinct, conserved motifs (Mojica *et al.*, 2000; Tang *et al.*, 2002). There seems to be a strict link between CRISPR and *cas* genes, suggestive of a (nearly) mutualistic relationship: the great majority of genomes that contain CRISPR also have at least a minimal set of *cas* genes and vice versa.

Recently, Mojica *et al.* (2005) reported that the unique inserts in some of the CRISPRs are homologous to fragments of bacteriophage and plasmid genes. This led to the hypothesis that the CRISPR might have a function in the defence of prokaryotes against invading foreign replicons and that there could be functional analogies between this putative defence system and eukaryotic RNA interference (RNAi). Similar findings have been reported independently by two other groups (Bolotin *et al.*, 2005; Pourcel *et al.*, 2005).

With the recent dramatic increase in the number and diversity of sequenced prokaryotic genomes, the complexity of the *cas* gene systems has become mind-boggling (Haft *et al.*, 2005). Here, we describe the results of an exhaustive sequence analysis of the Cas protein sequences which resulted in a classification of these proteins, a number of functional predictions and a reconstruction of evolutionary relationship between these genes. We propose that the *cas* genes encode the protein machinery of prokaryotic RNAi, which primarily, but perhaps not exclusively, performs defence functions loosely analogous to those of the vertebrate immune system. This system seems to be functionally analogous, but not homologous, to the protein apparatus involved in eukaryotic RNA-mediated gene silencing. Finally, we outline possible molecular mechanisms of the predicted prokaryotic RNAi.

IDENTIFICATION, CLASSIFICATION AND EVOLUTIONARY ANALYSIS OF *cas* GENES

In the original study of the *cas* gene neighbourhoods, which was performed with about 40 genomes, we identified about 20 genes tightly or more loosely associated with the

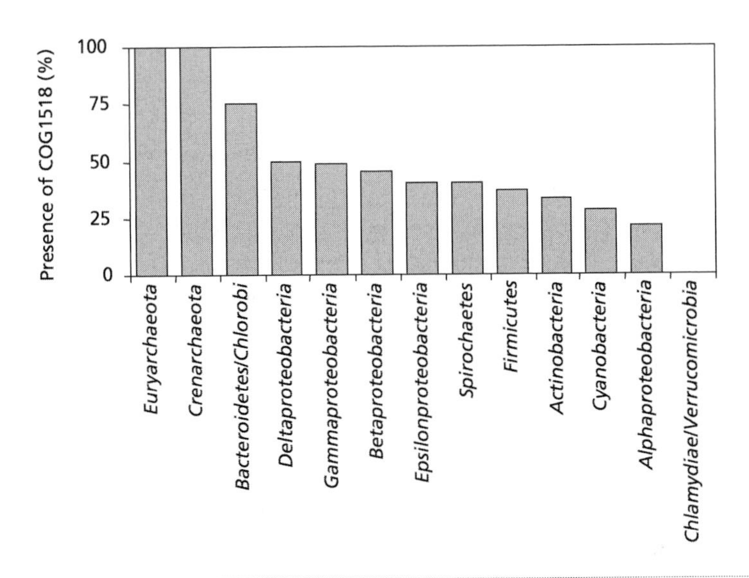

Fig. 1. Distribution of COG1518 genes and, by implication, CASS among prokaryotic lineages. Percentages of genomes containing COG1518 (*cas1*) genes in each taxon are shown.

system we now call CASS (CRISPR-associated system). A recent update by Haft *et al.* (2005) on more than 200 genomes yielded a diverse 'guild' of about 45 genes. Many of the *cas* genes show very low sequence conservation, which makes identification of homologous relationships between them a non-trivial task. We employed an iterative approach to the exhaustive analysis of the Cas protein sequences, whereby the sequences of each *cas* family were compared to the protein sequences from all available prokaryotic genomes using PSI-BLAST (Altschul & Koonin, 1998; Altschul *et al.*, 1997), the proteins encoded in the neighbourhoods of all identified candidates were used as queries for further searches and the process was iterated until convergence. The amino acid sequences of Cas proteins were, in addition, carefully compared to each other in an attempt to identify possible traces of common origins of some of these genes that have so far eluded detection. In agreement with previous observations (Makarova *et al.*, 2002), we found that Cas1 (COG1518 in the Clusters of Orthologous Group of proteins classification system; Tatusov *et al.*, 2003) is the best marker of the CASS. This gene encodes a highly conserved protein and is the only gene represented in all *cas* neighbourhoods. A PSI-BLAST search for COG1518 members in completely sequenced prokaryotic genomes revealed the presence of at least one representative of this COG in 77 of the 177 analysed genomes. The distribution of COG1518 and, by implication, CASS among prokaryotic lineages is distinctly non-uniform (Fig. 1). All sequenced archaeal genomes, including, notably, the tiny genome of '*Nanoarchaeum equitans*', encode this protein. In contrast, in each of the bacterial lineages with a representative set of sequenced genomes, half or more of the species do not have COG1518. We came across several cases of differences in the presence of CASS among closely related bacterial species. Thus, *Corynebacterium diphtheriae* has a COG1518 gene but *Corynebacterium efficiens* does not; similarly, CASS was detected in *Mycobacterium tuberculosis* but not in *Mycobacterium bovis* or *Mycobacterium leprae*. Significant differences in the composition of CASS were noticed even between strains of several bacterial species, e.g. *Thermus thermophilus* (Omelchenko *et al*., 2005). These observations demonstrate the extraordinary evolutionary mobility of CASS.

Phylogenetic analysis of COG1518 combined with gene neighbourhood comparisons yields a clear classification of CASS based on gene composition and (predicted) operon organization (Fig. 2) which is, generally, in agreement with the recent work of Haft *et al.* (2005). Altogether, we delineated seven distinct versions of CASS gene systems which show substantial differences in gene composition and predicted operon organization. Only four CASS variants (CASS1–4) have a stable operon organization. While a distinct operon organization motif is also recognizable in the other CASS variants (Fig. 2), these versions have undergone numerous, complex rearrangements in different prokaryotic genomes including gene duplications, fusion with other CRISPR-related operons, insertion of additional genes and others. Thus, in line with the observation that CASS is often eliminated from genomes during evolution, this plasticity of CASS operon organization suggests that the CRISPR-associated genomic regions are 'hot spots' of recombination and ensuing genome rearrangement.

The recent work of Haft *et al.* (2005) has identified a 'guild' of over 45 protein families that comprise the CASS. Here we refine this classification by unifying many of the families into superfamilies and also expand the list of CASS-linked genes to include those that are found in *cas* operons less commonly and comprise the 'cloud' surrounding the core CASS components. Only COG1518 and COG1343 (see below) are invariably present in all versions of CASS. A few genomes that have a cassette of CRISPR-associated genes without COG1518 also possess another CRISPR-associated operon in a different genomic location that does include this gene. Altogether, we end up with about 25 gene families that are tightly associated with CASS.

COG1343 (*cas2*) is another gene that is common in CASS. Typically, this gene is located immediately downstream of COG1518 (Fig. 2). Exhaustive PSI-BLAST searches starting from COG1343 proteins identified proteins of COG3512 as homologues of COG1343, so that these COGs could be unified in a single superfamily. The members of this superfamily are small (80–120 amino acids) proteins with distinct structural motifs, in particular, an N-terminal β-strand followed by a polar amino acid, most often aspartate or asparagine. In CASS2, which is typified by the *cas* operon of *Escherichia coli*, there is an uncharacterized gene encoding a small protein immediately downstream of the COG1518 gene. Analysis of the multiple alignment of the homologues of this protein revealed motifs highly similar to those in COG1343, suggesting that these proteins are actually diverged COG1343 homologues. Only CASS3 does not seem to contain a gene for a small protein that potentially could be a COG1343 homologue. However, we found that the next gene after COG1518, which encodes the CASS helicase (*cas3* or COG1203), is unusually long and contains a small domain preceding the HD hydrolase N-terminal domain characteristic of many COG1203 proteins. We analysed this domain separately and found that its size and motifs are consistent with a homologous relationship with COG1343. Thus, it appears that the COG1343 domain is present in all CASS and, accordingly, could be essential for the CASS function. Furthermore, searches starting with the sequences of many proteins of COG3512 revealed similarity to *vapD* (COG3309 family), a family of uncharacterized proteins that are functionally linked to the *vapBC* operon encoding a variant of the bacterial toxin–antitoxin (TA) module which includes a helix–turn–helix (HTH)-containing protein and PIN-like nuclease protein. Interestingly, the PIN domain has been shown to possess RNase activity that, in eukaryotes, is involved in pre-rRNA processing, nonsense-mediated decay (NMD) of aberrant mRNAs and RNAi (Arcus *et al.*, 2005; Clissold & Ponting, 2000; Fatica *et al.*, 2004). It has been suggested that there is an evolutionary connection between eukaryotic NMD and bacterial TA, and that the functioning of the TA module might involve mRNA degradation as well (Anantharaman & Aravind, 2003).Together, these observations seem to establish links between CASS and the TA system and, through the latter, between CASS and eukaryotic NMD and RNAi, but do not shed light directly on the function of the COG1343 domain. However, the COG1343 proteins and VapD show some generic similarity to the PIN domain in terms of size and the nature of the signature motifs (COG1343 proteins contain a constellation of partially conserved aspartates), which makes it tempting to speculate that these proteins represent yet another family of nucleases.

Another conserved CASS family that resists straightforward functional prediction is COG1857. This gene is present in all versions of CASS, with the exception of the 'minimal' variants, CASS4, 6 and 7. The COG1857 proteins consist of approximately 350 amino acids and typically have an N-terminal β-strand followed by a loop containing a conserved glutamine. Although no other motifs containing potential catalytic residues seem to be conserved throughout COG1857, perhaps because of the extensive divergence between the subfamilies of this family, the observed conservation pattern suggests that this protein is an enzyme, perhaps yet another nuclease. COG1857 proteins are typically encoded immediately upstream of another widespread CASS gene, COG1688 (named *cas5* by Haft *et al.*, 2005), which we previously identified as a member of the repair-associated mysterious protein (RAMP) superfamily (Makarova

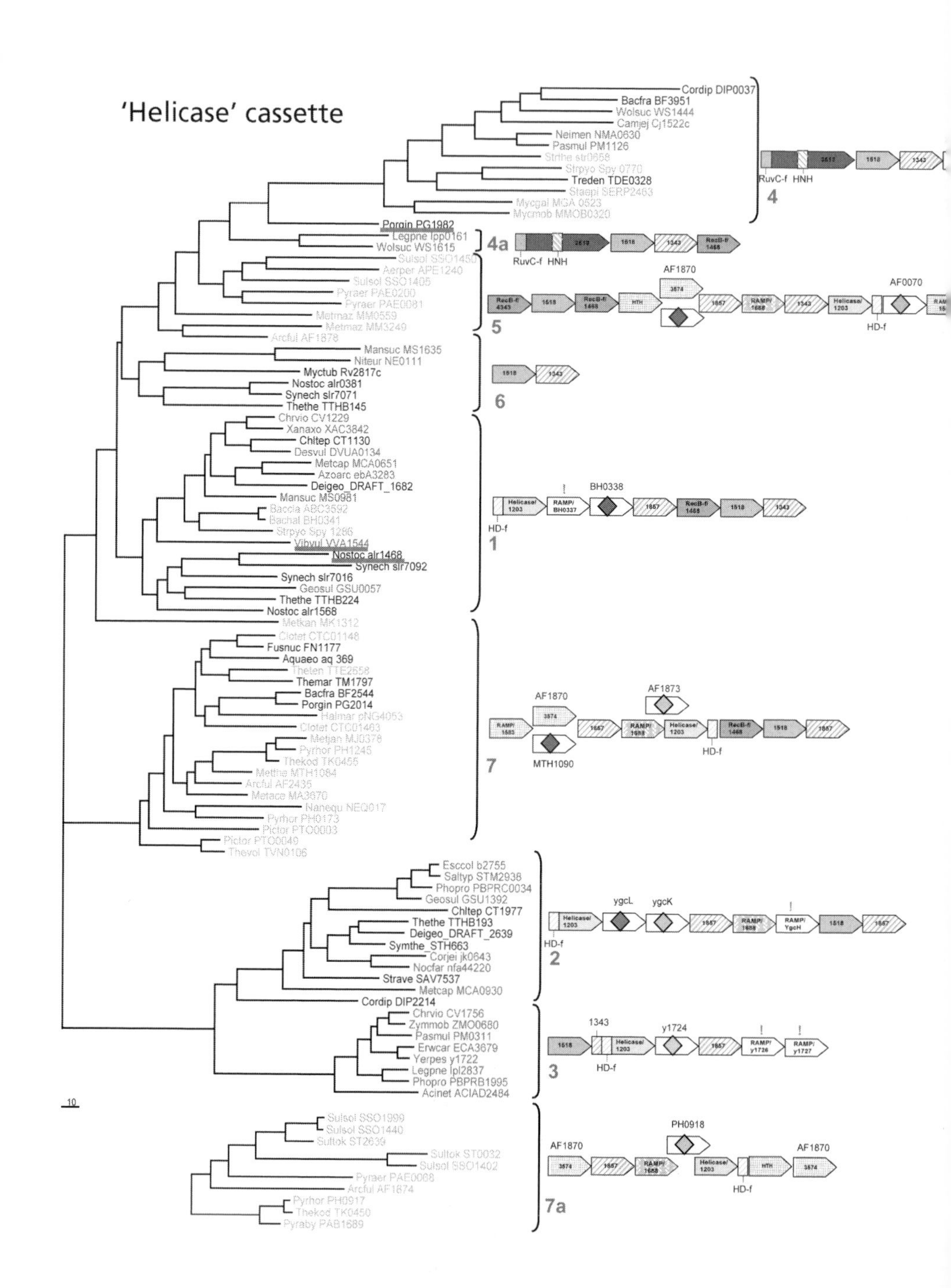
'Helicase' cassette
Cordip DIP0037
Bacfra BF3951
Wolsuc WS1444
Camjej Cj1522c
Neimen NMA0630
Pasmul PM1126
Treden TDE0328
Porgin PG1982
Legpne lpp0161
Wolsuc WS1615
4
4a
RuvC-f
HNH
5
AF1870
AF0070
HD-f
Mansuc MS1635
Niteur NE0111
Myctub Rv2817c
Nostoc alr0381
Synech slr7071
Thethe TTHB145
6
Chrvio CV1229
Xanaxo XAC3842
Chltep CT1130
Desvul DVUA0134
Metcap MCA0651
Azoarc ebA3283
Deigeo_DRAFT_1682
Mansuc MS0981
Vibvul VVA1544
Nostoc alr1468
Synech slr7092
Synech slr7016
Geosul GSU0057
Thethe TTHB224
Nostoc alr1568
BH0338
1
Fusnuc FN1177
Aquaeo aq 369
Themar TM1797
Bacfra BF2544
Porgin PG2014
AF1870
AF1873
MTH1090
7
Esccol b2755
Saltyp STM2938
Phopro PBPRC0034
Geosul GSU1392
Chltep CT1977
Thethe TTHB193
Deigeo_DRAFT_2639
Symthe_STH663
Corjei jk0643
Nocfar nfa44220
Strave SAV7537
Metcap MCA0930
Cordip DIP2214
ygcL
ygcK
2
Chrvio CV1756
Zymmob ZMO0680
Pasmul PM0311
Erwcar ECA3679
Yerpes y1722
Legpne lpl2837
Phopro PBPRB1995
Acinet ACIAD2484
1343
y1724
3
10
PH0918
AF1870
7a

'Polymerase' cassette

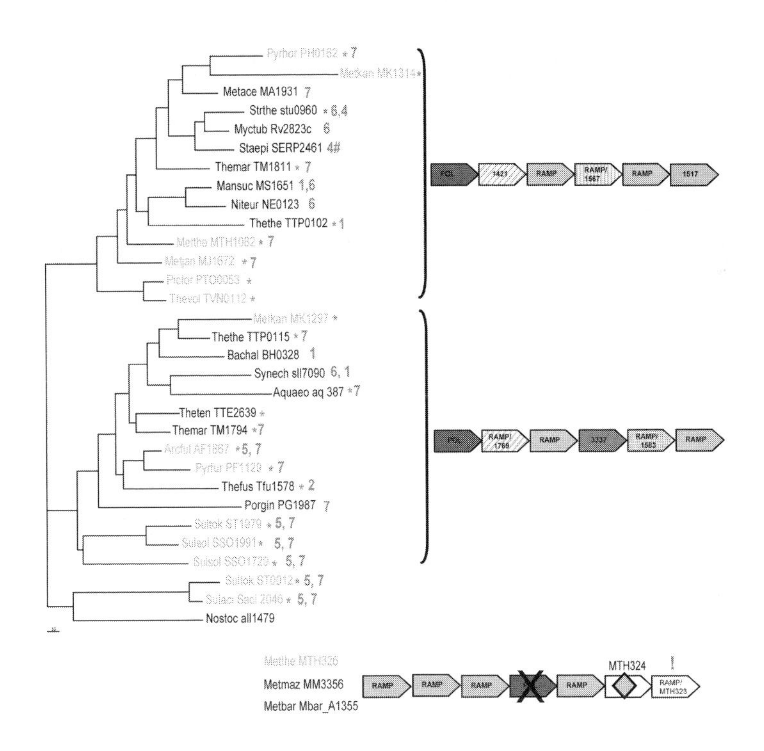

Fig. 2. Phylogenies of the key *cas* genes and organization of *cas* operons. Opposite page, the phylogenetic tree for COG1518 proteins; this page, the phylogenetic tree for the predicted CASS polymerase (COG1353). Prokaryotic lineages are coded by shading, from lightest to darkest: archaea, low-G+C-content Gram-positive bacteria, proteobacteria, other bacteria. In the operon organization schematics, orthologous genes are denoted by either the predicted function or the COG number. Exclamation marks denote previously undetected RAMPs. The names of species that have a reverse transcriptase (RT) gene within one of the *cas* operons are underlined. In the left panel, the distinct versions of CASS are numbered and, in the right panel, these numbers are given at tree leaves to indicate the helicase cassette(s) that co-occurs with the given polymerase cassette.

et al., 2002). In CASS3, COG1688 is replaced by the y1726 family, which we found to include another set of RAMPs distantly related to COG1688.

The RAMPs are the most diverse class of CASS genes. In addition to the five previously identified distinct families of RAMPs, we detected several more, namely BH0337-like, y1726-like, YgcH-like, y1727-like and MJ0978-like families, as well as numerous diverged members of the previously described families (Fig. 3a). Despite the dramatic sequence divergence, all these proteins contain the RAMP signature, the glycine-rich loop at the C terminus (Fig. 3a). With the identification of these new families as RAMPs, it now can be seen that all CASS versions, with the apparent exception of the minimal CASS4, include at least one RAMP. The crystal structure of one of the RAMPs from the newly detected YgcH-like family has been solved within one of the structural-genomics projects (PDB: 1wj9). A comparison of this structure to the structures in the PDB database using the VAST and DALI programs revealed a duplication of a distinct version of the ferredoxin fold, with the glycine-rich loop positioned between the two domains (Fig. 3b).

Thus, at least six gene (super)families seem to comprise the stable core of the CASS: COG1518 (*cas1*); COG1343 (*cas2*); COG1203, a helicase often fused to an HD-family hydrolase (*cas3*) plus free-standing versions of the HD hydrolase; COG1468 (*cas4*), *recB*-family nucleases usually containing a C-terminal Zn cluster; COG1857; and RAMP. This exact set of genes is seen in only a few genomes; most versions of CASS have substantial variations around this core – loss of some core genes in the minimal versions and addition of other genes and whole gene cassettes in others (Fig. 2).

The most notable non-core CASS module, which in a sense constitutes a second, even if non-ubiquitous, central part of CASS, may be called the '*pol* cassette' after the predicted palm-domain RNA or DNA polymerase of COG1353. The *pol* cassette also includes several distinct RAMPs and a few uncharacterized genes. The *pol* cassette is strictly linked only to CASS5 and CASS7 and is also, in some instances, found in CASS1, 2, 4 and 6 (Fig. 2), although, in some genomes, the *pol* cassette is not adjacent to the CASS core gene array. The phylogenetic tree for the predicted polymerase (COG1353), which consisted of two major branches corresponding to two distinct operon organizations (Fig. 2), showed essentially no topological congruence with the COG1518 tree (for those CASS that have both components; compare the two trees in Fig. 2). Thus, it appears that the *pol* cassette comprises a distinct evolutionary unit that is often transferred horizontally independently of the CASS core. Notably, the *pol* cassette is strongly, although not strictly, associated with thermophily; the great majority of the species containing this module, typically associated with CASS5 and CASS7, are thermophiles. Additionally, several species possess a third module containing a diverged

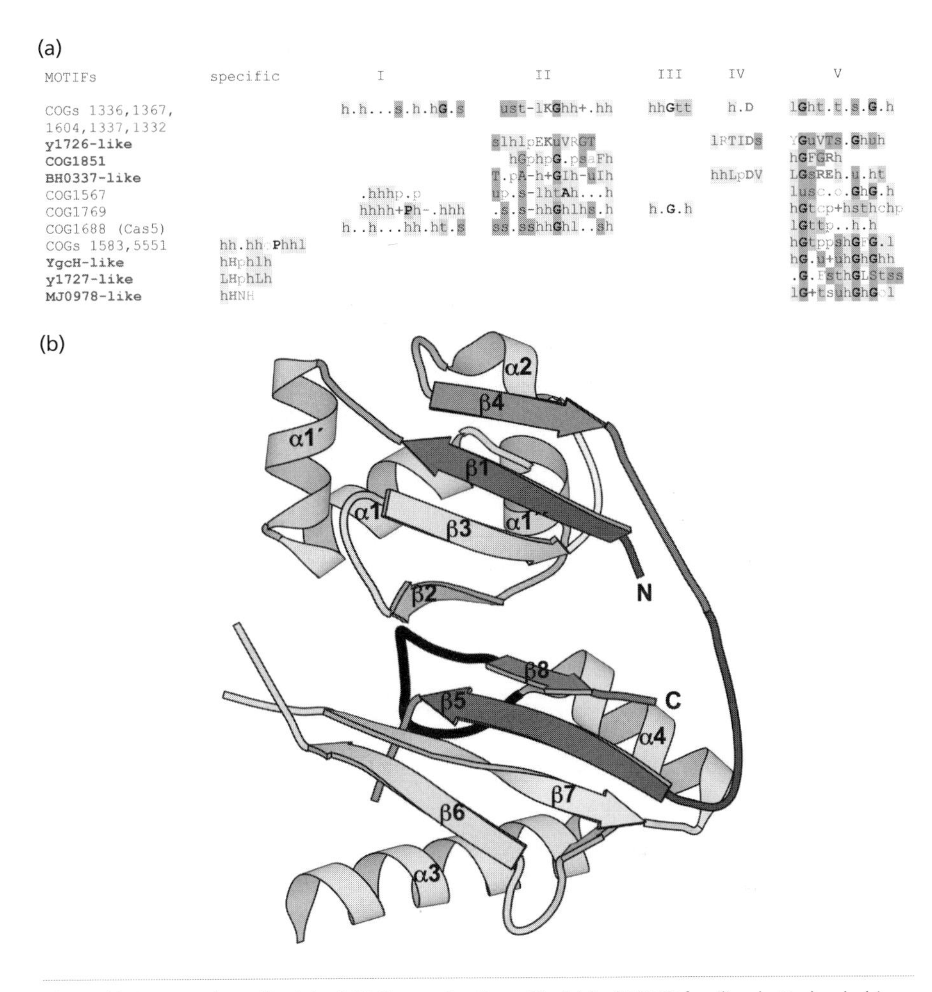

Fig. 3. (a) Conserved motifs of the RAMP superfamily and individual RAMP families. h, Hydrophobic residue; p, polar residue; t, residue with high turn-forming propensity; +, positively charged residue. (b) Ribbon model of the RAMP structure. The conserved glycine-rich loop (motif V) is in black.

form of COG1353 with a seemingly intact HD hydrolase domain but inactivated catalytic polymerase (PALM) domain.

The CASS core and the *pol* cassette together comprise the extended central components of CASS, which consist of six COGs plus the RAMP superfamily (Table 1). In addition, 19 other families were consistently found in CASS, even if in a minority of the CASS-containing species. It has to be taken into account that most of the CASS proteins have highly diverged sequences, and further unification as well as expansion and the ensuing rise in status of some of the families remains a possibility. For example, COG1517 is such a growing family. Typically, the COG1517 genes are located on the periphery of

Table 1. Protein components of CASS

Subfamilies are named by the corresponding COG number or by a protein ID.

Family	Subfamilies	Phyletic distribution*	Comments
COG1518	COG1518 (*cas1*)	All	Putative novel nuclease/integrase. Mostly α-helical protein
COG1343	COG1343 (*cas2*), COG3512, YgbF-like, MTH324-like, y1723_N-like		Small protein related to VapD, fused to helicase (COG1203) in y1723-like proteins
COG1203	COG1203 (*cas3*)	All	DNA helicase. Most proteins have fusion to HD nuclease
RecB-like nuclease	COG1468 (*cas4*), COG4343	All	RecB-like nuclease. Contains three-cysteine C-terminal cluster
RAMP	COG1688 (*cas5*), COG1769, COG1583, COG1567, COG1336, COG1367, COG1604, COG1337, COG1332, COG5551, BH0337-like, MJ0978-like, YgcH-like, y1726-like, y1727-like	All	Belong to RAMP family. Possibly RNA-binding protein, structurally related to duplicated ferredoxin fold (PDB: 1wj9)
COG1857	COG1857, COG3649, YgcJ-like, y1725-like	All	α/β protein. Predicted nuclease or integrase
HD-like nuclease	COG1203 (N terminus), COG2254	All	HD-like nuclease
BH0338	BH0338-like, MTH1090-like	All, mostly archaea and F	Large Zn-finger-containing proteins, possibly nucleases (nuclease activity reported for MTH1090; Guy *et al.*, 2004)
YgcL	YgcL	Bacteria, mostly P	Large Zn-finger-containing proteins
COG1353	COG1353, MTH326-like, alr1562, slr7011	All, mostly archaea	Putative novel polymerase. Multidomain protein with permuted HD nuclease domain, palm domain, polymerase-thumb-like domain and Zn ribbon. MTH326-like has inactivated polymerase catalytic domain. alr1562 and slr7011 predicted only on the basis of size, presence of HD domain and location with RAMPs in one operon
COG2462	COG2462	All, mostly archaea	HTH-type transcriptional regulator. Possible regulator of expression of CASS system in archaea
COG1421	COG1421	All, mostly archaea	~150 aa protein. Has a few motifs similar to YgcK-like. Mostly α-helical protein

YgcK	YgcK-like	Bacteria, mostly P	~180 aa protein. Has a few motifs similar to COG1421; mostly α-helical protein
COG3337	COG3337	All, mostly archaea	~110 aa. No prediction. Mostly α-helical protein
COG1517	COG1517, COG4006	All, mostly archaea	Often fused to HTH domain, some proteins have the domain duplication. Structure is available (1XMX); domain appears to have a Rossmann-like fold
COG3513	COG3513	Bacteria, mostly P	Huge protein. Contains McrA/HNH nuclease-related domain and RuvC-like nuclease domain
PH0918	PH0918-like	All, mostly archaea	Specific to *Pyrococcus* and *Thermococcus*. No prediction. Pair ST0031/SSO1401 and AF1873 probably belong to the same family because they have similar lengths and are located in identical positions in an operon but, due to low conservation, are not alignable
AF1870	AF1870-like	Archaea	Former COG3574; ~150 aa protein
AF0070	AF0070-like	Archaea	~420 aa protein. No prediction
y1724	y1724-like	Bacteria, mostly P	~450 aa protein. No prediction
Spy1049	Spy1049-like	Bacteria, mostly F	~220 aa protein. No prediction
TTE2665	TTE2665-like	Bacteria, mostly C	~130 aa protein. No prediction
LA3191	LA3191-like	Few bacteria	~650 aa protein. No prediction

*All, Family is widespread in all major prokaryotic lineages; P, *Proteobacteria*; F, *Firmicutes*; C, *Chlorobi*.

pol cassettes and, in many cases, contain fused HTH domains. The core domain of COG1517 is about 150 amino acids; several proteins, e.g. TM1812, comprise a duplication of this domain and several fusions, in addition to the one with HTH, were detected. The sequences of COG1517 proteins tend to be highly diverged and some of these were detected in genomes with no CASS, indicating that the association of this gene with CASS is not as tight as it is for the extended core of CASS. For one such protein from *Vibrio cholerae*, the three-dimensional structure was solved, again, in one of the structural genomics projects (PDB: 1xmx). It turned out that the N-terminal domain of this protein, which corresponds to the COG1517 core domain, has a Rossmann-like fold with an apparent permutation. The C-terminal domain of this protein belongs to the restriction endonuclease-like fold (the same fold that is seen in RecB-like endonucleases), with an inserted DNA-binding winged-helix domain. The endonuclease domain in this protein is predicted to be active as it retains all catalytic residues.

Several other CASS gene families remain mysterious. For example, CASS1, 5 and 7 contain genes upstream of COG1857 that encode large (500–600 amino acids), homologous proteins; the best conserved family in this set of proteins is represented by BH0338 and its orthologues. Some members of the BH0338 family contain a Zn ribbon in the middle of the sequence but otherwise have no recognizable domains or motifs. Among several conserved motifs of these proteins are two conserved aspartates and a distal conserved glycine, a combination that resembles the motifs seen in the PALM polymerase domain. Although we were unable to obtain additional evidence for the potential connection of this family with polymerases, it is tempting to speculate that these proteins might contain an extremely diverged version of the PALM domain. Similar, albeit less pronounced, motifs are detectable in the MTH1090 family proteins, which are present in CASS5 and 7. This similarity and the fact that the respective genes occupy the same position in the corresponding operons suggest that the BH0338-family and MTH1090-family proteins are highly diverged homologues. CASS2 also includes a large protein (YgcL family), some of which contain Zn clusters; however, the conserved motifs of these proteins and their position in the respective operons are different from those of the MTH1090 and BH0338 families. CASS4 contains another huge protein (COG3513, approx. 1150–1400 amino acids) with two recognizable domains, an McrA/HNH nuclease and an RuvC-like nuclease. These observations emphasize the striking diversity of still poorly characterized CASS components, particularly, the plethora of predicted nucleases of various classes and potential novel ones.

HYPOTHESIS: CRISPR-CASS IS A PROKARYOTIC DEFENCE SYSTEM THAT FUNCTIONS ON THE RNAi PRINCIPLE

Based on the properties of CRISPR and CASS, we speculate that this system is a functional analogue of the eukaryotic systems of RNAi and propose possible mechanisms of the putative prokaryotic small RNAi. The crucial observation reported independently by Mojica *et al.* (2005), Pourcel *et al.* (2005) and Bolotin *et al.* (2005) is that a certain fraction (about 10 %) of the unique inserts in CRISPR units are homologous to fragments of bacteriophage or plasmid genes. Since only a miniscule fraction of the existing phage and plasmid sequences is currently available, it is not far-fetched to propose that all CRISPR inserts have this property (Pourcel *et al.* 2005).

Should that be the case, the general principle of action of the CRISPR seems more or less obvious: the inserts are transcribed and silence the cognate phage or plasmid genes via an antisense mechanism, i.e. the formation of a duplex between the prokaryotic small interfering (psi) RNA and the target mRNA followed by cleavage of the duplex or translation repression. Indeed, Mojica *et al.* (2005) mention the analogy between the CRISPR and eukaryotic RNAi systems but propose no specific mechanisms for the action of the putative defence systems and, crucially, do not explore the connection between the putative psiRNA and the predicted activities of *cas* gene products. Important supporting evidence has been obtained independently through analysis of the small non-messenger RNA expression in the archaeon *Archaeoglobus fulgidus*, which showed that CRISPRs are transcribed, apparently, in the form of a multiunit precursor that is subsequently cleaved into CRISPR monomers and oligomers (Tang *et al.*, 2002). Furthermore, as noticed by Pourcel *et al.* (2005), one of the unique CRISPR inserts in the MIGAS string of the bacterium *Streptococcus pyogenes* is homologous to a prophage present in other strains of the same bacterium that, conversely do not carry the CRISPR. This is compatible with the possibility that the insert makes the bacterium immune to the given phage.

Here, we pursue these lines of thought in an attempt to present a coherent, even if speculative, description of prokaryotic RNAi. Circumstantial but crucial evidence in support of the psiRNA hypothesis comes from analogies between the predicted functions of CASS proteins and the protein components of eukaryotic RNAi systems, in particular, the RNA-induced silencing complexes (RISCs) (Filipowicz, 2005; Sontheimer, 2005; Tang, 2005) (Table 2). The core parts of RISCs are a helicase combined with RNase III domains, dicer, and the exonuclease of the argonaute family, slicer (Hammond, 2005; Miyoshi *et al.*, 2005; Sontheimer, 2005). Both dicer and slicer are represented by variable numbers of paralogues in eukaryotes, and different paralogues are included into RISCs with distinct functions (Hammond, 2005; Sontheimer, 2005). Various RISCs also contain additional RNA-binding proteins and nucleases (Scadden, 2005; Sontheimer, 2005). Putative functional analogues of all these proteins can be gleaned among the CASS proteins. The dicer analogue is immediately apparent: the COG1203 helicase that is either fused or encoded next to a predicted nuclease of the HD family (Fig. 2); we tentatively designate this protein p-dicer (prokaryotic dicer).

Table 2. Functional and structural parallels between CASS and eukaryotic RNAi machinery

Eukaryotic RNAi	Domains/function	CASS	Domains/function
Dicers	Helicase/RNase III; processing of long dsRNA into siRNA and pre-miRNA into miRNA; involves unwinding	Helicase (COG1203) + HD nuclease (COG2254), fused or adjacent genes, possibly COG1343 (a novel nuclease?)	SFII helicase + HD nuclease (+COG1343). Pre-psiRNA processing?
Argonautes/slicers	Ferredoxin-fold-PAZ-PIWI; endonuclease, target degradation	RecB-family nuclease (COG1468, 4343); COG1857, a novel nuclease?	Target degradation
R2D2/RDE-4	dsRNA-binding domain; interacts with Dicer	RAMPs	Ferredoxin-fold duplication. Size-specific psiRNA-binding, pre-psiRNA-binding, other RNA-binding functions?
Fmr1/Fxr	RGG, KH; ssRNA binding	RAMPs	Ferredoxin-fold duplication. Size-specific psiRNA-binding, pre-psiRNA-binding, other RNA-binding functions?
Tsn	Tudor-SN; RNA binding	RAMPs	Ferredoxin-fold duplication. Size-specific psiRNA-binding, pre-psiRNA-binding, other RNA-binding functions?
Vig	RGG; RNA binding	RAMPs	Ferredoxin-fold duplication. Size-specific psiRNA-binding, pre-psiRNA-binding, other RNA-binding functions?
RNA-dependent RNA polymerase (RdRp)	RdRp domain related to DdRp; second-strand synthesis for siRNA production	Predicted RdRp/RT (COG1353)	Palm polymerase domain. Second-strand synthesis for psiRNA production, reverse transcription for CRISPR formation

Identification of the slicer counterpart (p-slicer) is less straightforward because of the diversity of predicted nucleases within the CASS. One candidate is the predicted RecB-family nuclease. Alternatively, the slicer function could be performed by a novel, still unidentified nuclease, such as COG1857. We further propose that RAMP proteins, by far the most diverse components of the CASS, play a major role in prokaryotic RNAi as RNA-binding proteins that display a degree of specificity to the psiRNAs, most likely by specifically binding psiRNAs of different sizes. The RAMPs, diverse representatives of the same protein superfamily, can be functional analogues of the more structurally diverse RNA-binding proteins of eukaryotic RISCs (Table 2). The presence of two ferredoxin-fold domains in RAMPs (Fig. 3b) is compatible with this proposal, given that this fold is present in a broad variety of RNA-binding proteins, such as the ribosomal proteins S6 and S10, several spliceosomal subunits and others.

Fig. 4(a) presents our current hypothesis on the basic mechanism of the functioning of the prokaryotic RNAi system. We speculate that CRISPR regions are transcribed from a promoter located in the AT-rich CRISPR leader. Transcription might be regulated with the participation of one of the CASS proteins and, conceivably, would be stimulated by stress, such as phage infection, although the results of expression analysis in *A. fulgidus* suggest some level of constitutive transcription (Tang *et al.*, 2002). The work of Tang *et al.* (2002) further indicates that the primary transcript is likely to encompass the entire CRISPR repeat region. This transcript would be cleaved into 70–100 nt pre-psiRNA, conceivably by the hypothetical p-dicer, the COG1203 protein (Fig. 5a). We further postulate that p-dicer catalyses the second, perhaps slower (judging by the results of Tang *et al.*, 2002) processing step that releases mature psiRNA species (Fig. 5a). The psiRNA molecules would bind to RAMPs in a size-specific manner and anneal to the target mRNA. The resulting complex would recruit p-slicer, forming the minimal form of the prokaryotic analogue of RISC (pRISC) that would cleave the mRNA and could be recycled to attack the next target molecule, thus silencing the respective gene (Fig. 5a).

Fig. 4(b) shows a version of this pathway that involves the activity of the CASS polymerase, by analogy with the eukaryotic RNA-dependent RNA polymerase (RdRp), which participates in some RNAi pathways in most eukaryotes but has apparently been lost in arthropods and chordates (Lipardi *et al.*, 2001; Sijen *et al.*, 2001; Smardon *et al.*, 2000). The initial steps in this scheme are the same as in the basic one (Fig. 4a), transcription and processing of the CRISPR RNA, but at the next step, psiRNA is postulated to serve as the primer for elongation by the CASS polymerase, yielding an extended double-stranded form of the target (Fig. 4b). This form would be cleaved by p-dicer, which might interact with the respective RAMP to form a distinct version of pRISC. This could be the end point of the pathway or, alternatively, the dsRNA degradation

(a)

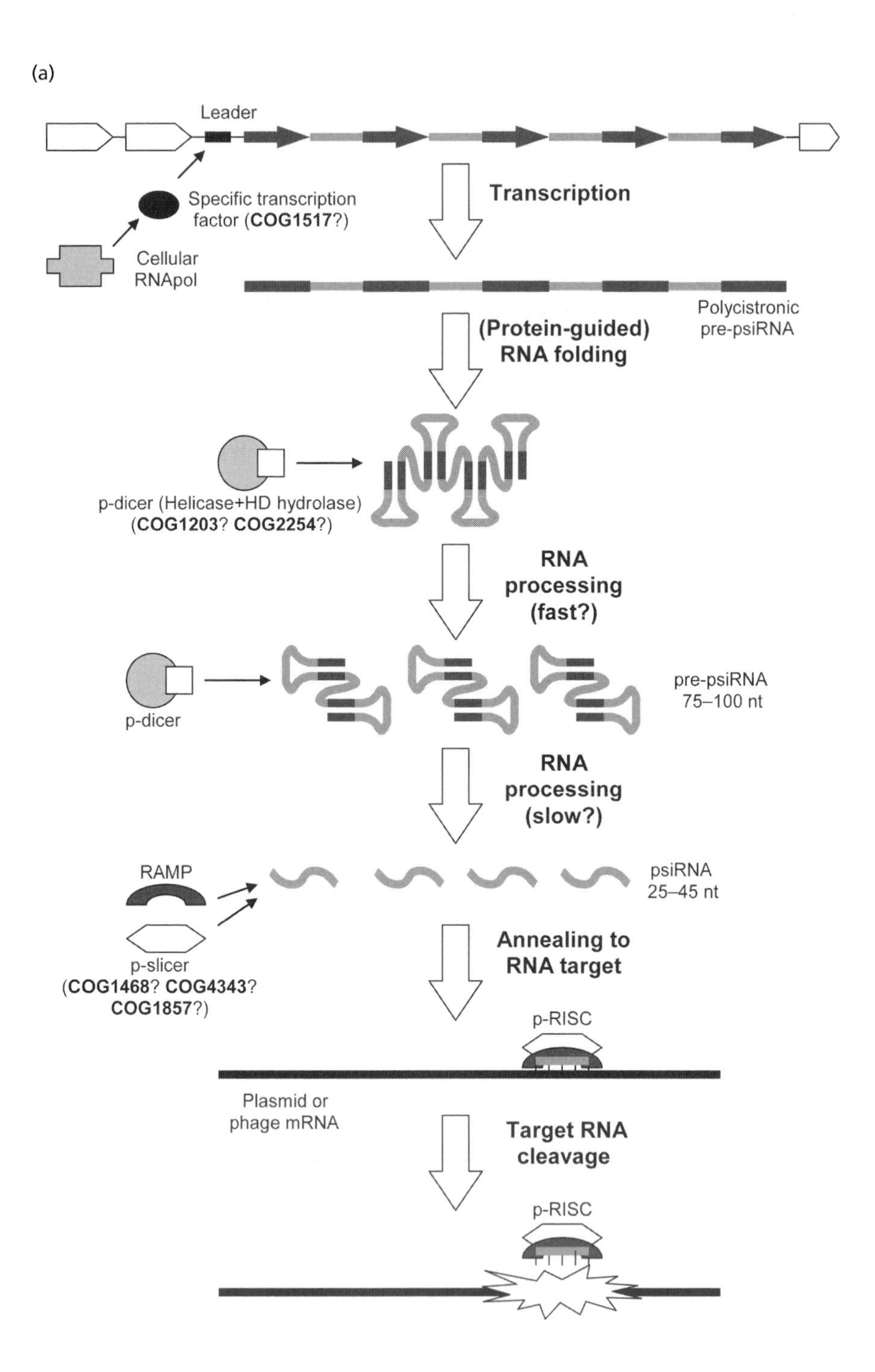
Leader
Transcription
Specific transcription factor (**COG1517**?)
Cellular RNApol
Polycistronic pre-psiRNA
(Protein-guided) RNA folding
p-dicer (Helicase+HD hydrolase) (**COG1203**? **COG2254**?)
RNA processing (fast?)
p-dicer
pre-psiRNA 75–100 nt
RNA processing (slow?)
RAMP
psiRNA 25–45 nt
p-slicer (**COG1468**? **COG4343**? **COG1857**?)
Annealing to RNA target
p-RISC
Plasmid or phage mRNA
Target RNA cleavage
p-RISC

(b)

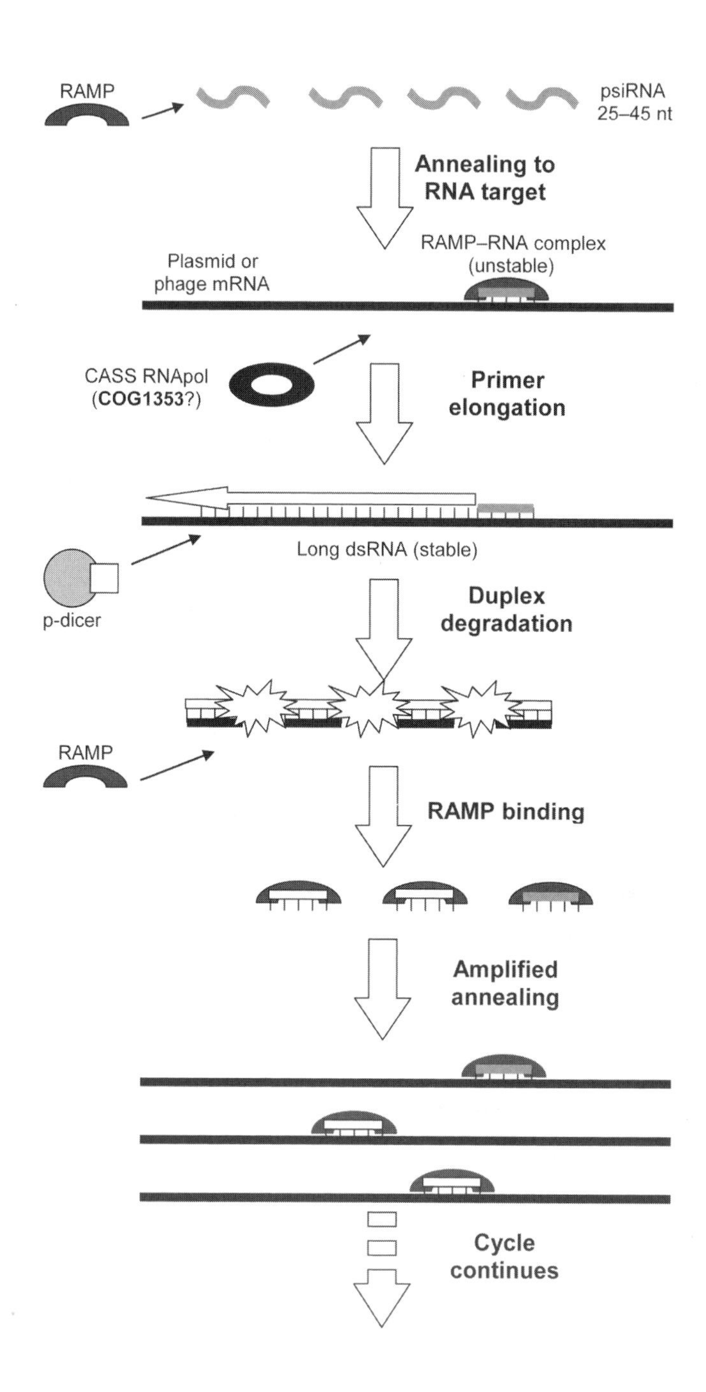
RAMP
psiRNA
25–45 nt
Annealing to
RNA target
RAMP–RNA complex
(unstable)
Plasmid or
phage mRNA
CASS RNApol
(COG1353?)
Primer
elongation
Long dsRNA (stable)
p-dicer
Duplex
degradation
RAMP
RAMP binding
Amplified
annealing
Cycle
continues

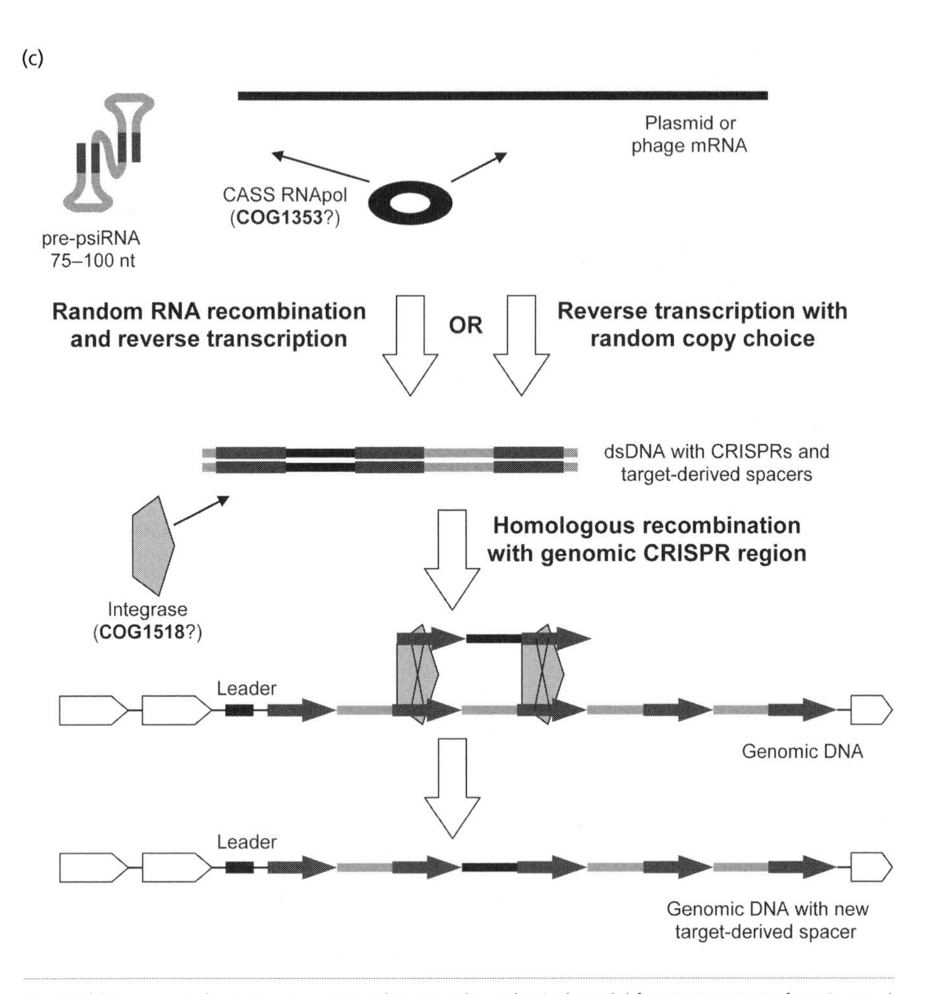

Fig. 4. *(This page and previous two pages.)* Current hypothetical model for CRISPR-CASS function and CRISPR formation. (a) Basic model of CRISPR-CASS function. (b) Variant of CRISPR-CASS function involving the CASS polymerase. (c) Formation of new CRISPR with unique inserts.

products could be utilized as new psiRNAs, resulting in amplification of the silencing effect (Fig. 4b). The CASS polymerase is most common in thermophiles, and it is tempting to speculate that the prevalence of this form of the psiRNA pathway has to do with the instability of the psiRNA–target duplex under the high ambient temperatures of these organisms.

The most complex and uncertain aspect of the putative prokaryotic RNAi system discussed here is the formation of new CRISPR units containing unique psiRNA genes specific for new targets encountered by the organism (Fig. 4c). The path to the creation of new psiRNAs starts just like the response pathway, i.e. with transcription of the

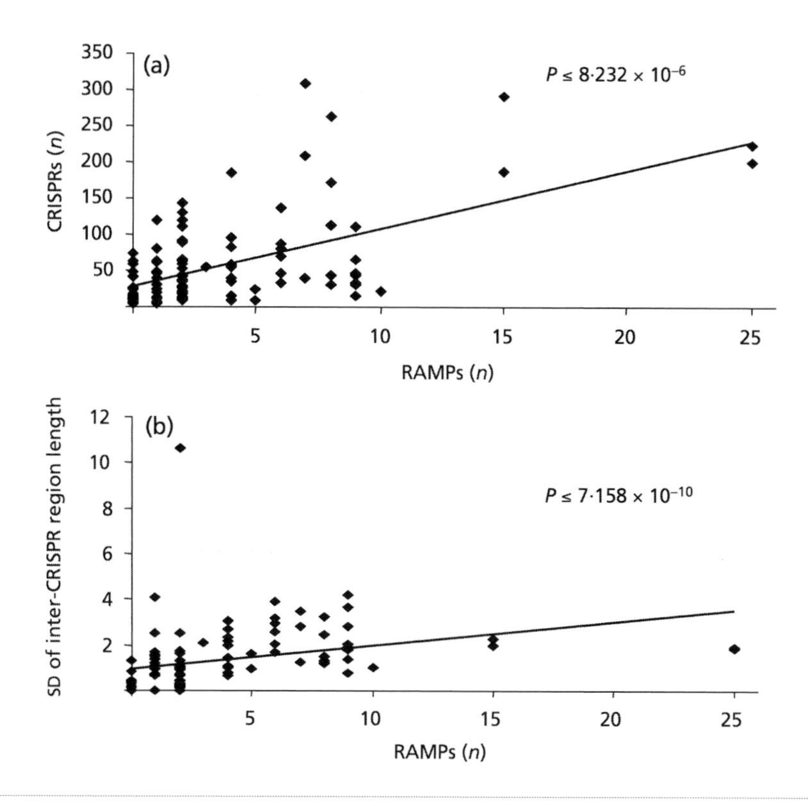

Fig. 5. CRISPR and RAMPs. (a) Correlation between the number of encoded RAMPs and the number of CRISPR units in prokaryotic genomes. (b) Correlation between the number of encoded RAMPs and the variance of unique insert lengths in CRISPR in prokaryotic genomes.

CRISPR locus and the first processing step yielding the 70–100 nt psiRNA precursors (compare Figs 4c and 4a). At the next step, however, there must be a mechanism to replace the unique insert within the pre-psiRNA with a new fragment of foreign (e.g. phage) RNA. The nature of this mechanism remains unclear. In principle, two possibilities can be envisaged: (i) reverse transcription with copy choice whereby a reverse transcriptase (RT), most likely the CASS polymerase (COG1353), switches from using the pre-psiRNA as a template to using a phage mRNA and then switches back and (ii) direct RNA recombination between a pre-psiRNA and a foreign mRNA followed by reverse transcription of the resulting recombinant RNA (Fig. 4c). Both mechanisms are non-trivial in their molecular choreography and are unlikely to occur with high efficiency. Nevertheless, there are precedents for both in the molecular biology of retroviruses and other RNA viruses. In particular, RT switches templates in each cycle of retrovirus first-strand cDNA synthesis although, in this case, copy choice is facilitated by the spatial juxtaposition of the two templates within the virus particle; a similar mechanism is responsible for recombination in retroviruses (Negroni & Buc, 2001). In addition, and perhaps of more relevance to the psiRNA case, copy choice is thought to be involved in the incorporation of copies of cellular genes, such as oncogenes, in retroviral genomes (Huang *et al.*, 1986; Swain & Coffin, 1992). The alternative, namely direct recombination between RNA molecules, might seem far fetched, but such a process has been demonstrated by several groups independently to occur in RNA viruses, apparently via a protein-independent mechanism (Gmyl *et al.*, 2003; Raju *et al.*, 1995). During the formation of new psiRNA species, these low-frequency processes might be facilitated by the involvement of abundant phage mRNAs. Indeed, it has been shown that the unique inserts in CRISPR most often correspond to fragments of essential, highly conserved phage genes that are typically expressed at a high level in infected bacteria (Mojica *et al.*, 2005). Once the dsDNA molecule consisting of a CRISPR unit with the new, unique insert is produced, by one of the mechanisms outlined here or, perhaps, via a different pathway, it must insert into the CRISPR array via homologous recombination (Fig. 5c). We strongly suspect that this process is mediated by the COG1518 protein, the universal marker of CASS containing conserved motifs resembling those of different nucleases (Makarova *et al.*, 2002). It seems likely that this protein functions as the CRISPR integrase/recombinase, perhaps in cooperation with the COG1343 protein, another universal component of CASS.

ADDITIONAL LINES OF EVIDENCE COMPATIBLE WITH AN RNAi FUNCTION OF CASS

Genes loosely associated with CASS

In addition to the bona fide (even if not ubiquitous) CASS components, a variety of genes were found in association with CRISPR and CASS in only one or a few closely related genomes. Thus, functional association of these genes with CASS appears uncertain. However, examination of the list of such genes clearly shows that they are not a random set (Table 3). Similarly to the common CASS components, there are several nucleases on this list. More notable is the presence of RT which, on at least two independent occasions, has been fused to the COG1518 gene (Fig. 2 and Table 3). This fusion suggests the intriguing possibility that, in the respective variants of CASS, the RT (probably derived from retron-type elements) takes over the proposed function of the CASS polymerase (Fig. 4c); this appears to provide indirect but unequivocal support for the reverse-transcription-mediated mechanism of CASS function. Perhaps the most remarkable observation is the presence of the archaeal homologue of the eukaryotic argonaute protein (the slicer) in one of the CASS operons of the archaeon *Methanopyrus kandleri* (Table 3). This suggests the possibility of a functional association of this protein with CASS, at least in some archaea, and provides the only putative (even if weak, until other archaeal genomes with such a gene arrangement are found) link between CASS and the eukaryotic RISCs at the level of homologous proteins.

Table 3. Genes loosely associated with CASS

Family	Example(s) of genes associated with CASS	Comments
Reverse transcriptase (RT)	VVA1544, PG1982, alr1468	Fused to COG1518 on three occasions and a remnant of RT (Mbar_A1351 and MM3360) in *Methanosarcina barkeri* and *Methanosarcina mazei* genomes is located close to CAS_G
PIN domain	alr1560, ST0017, Ava_4168	RNase
COG1432	MS0983	Large family of proteins, predicted to be a phosphatase or a nuclease on the basis of sequence motifs which are shared by all three domain of life. In multidomain proteins in plants, associated with C2H2 Zn-finger domain
PA2117-like	MS0982, MS0989	Enzymic domain located in an operon with restriction–modification systems or in association with a diverged helicase
COG3645-like	ACIAD2479	Homologues of phage anti-repressor Ant, which is known to be inhibited by an antisense RNA
Argonaute	MK1311	Homologue of eukaryotic argonaute protein, which is a key player in RNA-guided post-transcriptional regulation by siRNA and miRNA
COG1598/COG4226/HicB	MCA0653, MTH321	Probably has an RNase H-like fold, often fused to CopG family of transcriptional regulators; forms a conserved operon with COG1724/*hicA*
PUA domain	LIC10933	RNA-binding domain
3′–5′ Exonuclease		Fused to COG1343 in *Lactobacillus bulgaricus* and *Lactobacillus casei*

A connection between CRISPR and RAMPs

We searched for CRISPR in all available complete genome sequences of prokaryotes and found remarkable diversity in both the number of repeat units and the spacer size, even among closely related species and strains, which is in agreement with previous observations (Bolotin *et al.*, 2005; Jansen *et al.*, 2002; Mojica *et al.*, 2005; Pourcel *et al.*, 2005). Given our hypothesis that RAMPs are size-specific psiRNA-binding proteins, we examined possible connections between the number of RAMP genes, the number of CRISPR units and the length heterogeneity of the unique inserts in prokaryotic genomes. The number of RAMPs showed strong positive correlations both with the number of CRISPRs (Fig. 5a) and with the variance of the insert lengths (Fig. 5b), which seems to be compatible with our hypothesis of the psiRNA-binding function of RAMPs.

IMPLICATIONS AND CONCLUSIONS

Obviously, at this stage, the whole concept of the prokaryotic CRISPR-CASS defence system functioning on the RNAi principle remains a hypothesis. However, we believe that three lines of evidence make such a mechanism, in its broad outline, more or less a logical inevitability: (i) the indisputable origin of at least some of the unique CRISPR inserts from phage and plasmid genes, (ii) the demonstration of transcription and processing of the CRISPR locus in *A. fulgidus* and (iii) the abundance of CASS components that are clearly implicated in nucleic acid degradation, processing and, possibly, recombination. The only reasonable alternative to an RNAi-type mechanism seems to be an antisense mechanism acting on DNA. However, to our knowledge, this would be completely unprecedented so, given the deep analogies between CRISPR-CASS and the eukaryotic RNAi systems as well as the demonstrated instances of antisense RNA regulation in bacteria, it seems that RNAi-mediated defence against invading alien replicons is, indeed, the most likely function of CRISPR-CASS. This being said, we must emphasize that the mechanisms shown in Fig. 4 are just rough outlines of some of the ways in which this system could function. There is no doubt that experimental studies will reveal mechanisms that differ from these, at least in detail.

The predicted psiRNA system resembles its eukaryotic counterparts not only in its functional principle but also in the general characteristics of the repertoires of implicated proteins. What is most striking is the comparable complexity, diversity and plasticity of the protein machineries involved. Both systems consist of one or more helicases, a broad spectrum of nucleases, a specific polymerase and a variety of RNA-binding proteins. In both CASS and RISCs, only two or three protein subunits appear to be truly indispensable; the rest come and go, resulting in a variety of RISCs with their distinct functions, many of them still poorly understood (Filipowicz, 2005; Sontheimer,

2005; Tang, 2005), and, presumably, in a comparable diversity of CASS. It is remarkable, however, that not a single protein belonging to the bona fide CASS has an orthologue in eukaryotes, involved in RNAi or otherwise. The single direct link could be the argonaute protein, which is the central active moiety of eukaryotic RISCs (slicer) and might have some functional connection to CASS in archaea, as tentatively suggested by the *Methanopyrus kandleri* CASS operon structure.

The eukaryotic RNAi systems come in two distinct varieties: (i) small interfering (si) RNAs that are produced from dsRNAs of viruses and transposons and protect the host from the respective agents via perfect base-pairing with the respective target mRNAs and (ii) microRNAs (miRNAs) that regulate translation of endogenous genes via either exact (plants) or imperfect (animals) base-pairing. CRISPR-CASS appears to be the functional counterpart of the siRNA mechanism inasmuch as it seems to be involved in defence against infecting agents and the psiRNAs seem to be derived from the invading genome and are predicted to function via perfect base-pairing with the target. Prokaryotes also have apparent functional counterparts of the miRNA pathways, i.e. regulation of bacterial gene expression by small antisense RNAs. The best characterized of these pathways employ the RNA-binding protein Hfq for small RNA presentation and RNase E for target degradation (Gottesman, 2004, 2005; Majdalani *et al.*, 2005; Storz *et al.*, 2004). Thus, prokaryotes seem to have at least two distinct RNAi systems, neither of which is operated by homologues of eukaryotic RNAi protein machinery components. Furthermore, unlike the case of eukaryotes, where the siRNA and miRNA systems are controlled by substantially overlapping sets of proteins (Tang, 2005), the prokaryotic systems seem to be completely independent.

All the analogies notwithstanding, the predicted psiRNA system is substantially different from the eukaryotic analogue in at least one crucial principle of its action: the coding segments for the putative psiRNAs are derived from genes of invading agents and incorporated into the host genome to confer inheritable immunity to the respective agent. As an acquired immunity mechanism, CRISPR-CASS resembles more the vertebrate immune system than the eukaryotic RNAi pathways but, again, with the crucial difference in that animal immunity is not inheritable. Interestingly, as a mechanism of inheritance of acquired traits, CRISPR-CASS seems to come closest to a true Lamarckian mode of evolution among all known systems of heredity.

Finally, a practical note. It seems that, once the psiRNA mechanism described here is investigated experimentally, it could be exploited to silence any gene in organisms that encode CRISPR-CASS. The simple design of such experimental gene silencing in prokaryotes would involve transfection with a plasmid containing the desired psiRNA inserted between CRISPR to facilitate homologous recombination.

REFERENCES

Altschul, S. F. & Koonin, E. V. (1998). Iterated profile searches with PSI-BLAST – a tool for discovery in sequence databases. *Trends Biochem Sci* **23**, 444–447.

Altschul, S. F., Madden, T. L., Schaffer, A. A., Zhang, J., Zhang, Z., Miller, W. & Lipman, D. J. (1997). Gapped BLAST and PSI-BLAST: a new generation of protein database search programs. *Nucleic Acids Res* **25**, 3389–3402.

Anantharaman, V. & Aravind, L. (2003). New connections in the prokaryotic toxin-antitoxin network: relationship with the eukaryotic nonsense-mediated RNA decay system. *Genome Biol* **4**, R81.

Arcus, V. L., Rainey, P. B. & Turner, S. J. (2005). The PIN-domain toxin-antitoxin array in mycobacteria. *Trends Microbiol* **13**, 360–365.

Bolotin, A., Quinquis, B., Sorokin, A. & Ehrlich, S. D. (2005). Clustered regularly interspaced short palindrome repeats (CRISPRs) have spacers of extrachromosomal origin. *Microbiology* **151**, 2551–2561.

Clissold, P. M. & Ponting, C. P. (2000). PIN domains in nonsense-mediated mRNA decay and RNAi. *Curr Biol* **10**, R888–R890.

Coenye, T., Gevers, D., Van de Peer, Y., Vandamme, P. & Swings, J. (2005). Towards a prokaryotic genomic taxonomy. *FEMS Microbiol Rev* **29**, 147–167.

Dandekar, T., Snel, B., Huynen, M. & Bork, P. (1998). Conservation of gene order: a fingerprint of proteins that physically interact. *Trends Biochem Sci* **23**, 324–328.

Fatica, A., Tollervey, D. & Dlakic, M. (2004). PIN domain of Nob1p is required for D-site cleavage in 20S pre-rRNA. *RNA* **10**, 1698–1701.

Filipowicz, W. (2005). RNAi: the nuts and bolts of the RISC machine. *Cell* **122**, 17–20.

Gmyl, A. P., Korshenko, S. A., Belousov, E. V., Khitrina, E. V. & Agol, V. I. (2003). Nonreplicative homologous RNA recombination: promiscuous joining of RNA pieces? *RNA* **9**, 1221–1231.

Gottesman, S. (2004). The small RNA regulators of *Escherichia coli*: roles and mechanisms. *Annu Rev Microbiol* **58**, 303–328.

Gottesman, S. (2005). Micros for microbes: non-coding regulatory RNAs in bacteria. *Trends Genet* **21**, 399–404.

Guy, C. P., Majernik, A. I., Chong, J. P. & Bolt, E. L.(2004). A novel nuclease-ATPase (Nar71) from archaea is part of a proposed thermophilic DNA repair system. *Nucleic Acids Res* **32**, 6176–6186.

Haft, D. H., Selengut, J., Mongodin, E. F. & Nelson, K. E. (2005). A guild of 45 CRISPR-associated (Cas) protein families and multiple CRISPR/Cas subtypes exist in prokaryotic genomes. *PLoS Comput Biol* **1**, e60.

Hammond, S. M. (2005). Dicing and slicing: the core machinery of the RNA interference pathway. *FEBS Lett* **579**, 5822–5829.

Huang, C. C., Hay, N. & Bishop, J. M. (1986). The role of RNA molecules in transduction of the proto-oncogene c-fps. *Cell* **44**, 935–940.

Jansen, R., Embden, J. D., Gaastra, W. & Schouls, L. M. (2002). Identification of genes that are associated with DNA repeats in prokaryotes. *Mol Microbiol* **43**, 1565–1575.

Koonin, E. V. & Galperin, M. Y. (1997). Prokaryotic genomes: the emerging paradigm of genome-based microbiology. *Curr Opin Genet Dev* **7**, 757–763.

Koonin, E. V. & Galperin, M. Y. (2002). *Sequence – Evolution – Function. Computational Approaches in Comparative Genomics*. New York: Kluwer Academic.

Lathe, W. C., III, Snel, B. & Bork, P. (2000). Gene context conservation of a higher order than operons. *Trends Biochem Sci* **25**, 474–479.

Lawrence, J. (1999). Selfish operons: the evolutionary impact of gene clustering in prokaryotes and eukaryotes. *Curr Opin Genet Dev* **9**, 642–648.

Lipardi, C., Wei, Q. & Paterson, B. M. (2001). RNAi as random degradative PCR: siRNA primers convert mRNA into dsRNAs that are degraded to generate new siRNAs. *Cell* **107**, 297–307.

Majdalani, N., Vanderpool, C. K. & Gottesman, S. (2005). Bacterial small RNA regulators. *Crit Rev Biochem Mol Biol* **40**, 93–113.

Makarova, K. S., Aravind, L., Grishin, N. V., Rogozin, I. B. & Koonin, E. V. (2002). A DNA repair system specific for thermophilic archaea and bacteria predicted by genomic context analysis. *Nucleic Acids Res* **30**, 482–496.

Miyoshi, K., Tsukumo, H., Nagami, T., Siomi, H. & Siomi, M. C. (2005). Slicer function of *Drosophila* Argonautes and its involvement in RISC formation. *Genes Dev* **19**, 2837–2848.

Mojica, F. J., Diez-Villasenor, C., Soria, E. & Juez, G. (2000). Biological significance of a family of regularly spaced repeats in the genomes of archaea, bacteria and mitochondria. *Mol Microbiol* **36**, 244–246.

Mojica, F. J., Diez-Villasenor, C., Garcia-Martinez, J. & Soria, E. (2005). Intervening sequences of regularly spaced prokaryotic repeats derive from foreign genetic elements. *J Mol Evol* **60**, 174–182.

Mushegian, A. R. & Koonin, E. V. (1996). Gene order is not conserved in bacterial evolution. *Trends Genet* **12**, 289–290.

Negroni, M. & Buc, H. (2001). Retroviral recombination: what drives the switch? *Nat Rev Mol Cell Biol* **2**, 151–155.

Omelchenko, M. V., Wolf, Y. I., Gaidamakova, E. K., Matrosova, V. Y., Vasilenko, A., Zhai, M., Daly, M. J., Koonin, E. V. & Makarova, K. S. (2005). Comparative genomics of *Thermus thermophilus* and *Deinococcus radiodurans*: divergent routes of adaptation to thermophily and radiation resistance. *BMC Evol Biol* **5**, 57.

Pourcel, C., Salvignol, G. & Vergnaud, G. (2005). CRISPR elements in *Yersinia pestis* acquire new repeats by preferential uptake of bacteriophage DNA, and provide additional tools for evolutionary studies. *Microbiology* **151**, 653–663.

Raju, R., Subramaniam, S. V. & Hajjou, M. (1995). Genesis of Sindbis virus by in vivo recombination of nonreplicative RNA precursors. *J Virol* **69**, 7391–7401.

Rogozin, I. B., Makarova, K. S., Murvai, J., Czabarka, E., Wolf, Y. I., Tatusov, R. L., Szekely, L. A. & Koonin, E. V. (2002). Connected gene neighborhoods in prokaryotic genomes. *Nucleic Acids Res* **30**, 2212–2223.

Scadden, A. D. (2005). The RISC subunit Tudor-SN binds to hyper-edited double-stranded RNA and promotes its cleavage. *Nat Struct Mol Biol* **12**, 489–496.

Sijen, T., Fleenor, J., Simmer, F., Thijssen, K. L., Parrish, S., Timmons, L., Plasterk, R. H. & Fire, A. (2001). On the role of RNA amplification in dsRNA-triggered gene silencing. *Cell* **107**, 465–476.

Smardon, A., Spoerke, J. M., Stacey, S. C., Klein, M. E., Mackin, N. & Maine, E. M. (2000). EGO-1 is related to RNA-directed RNA polymerase and functions in germ-line development and RNA interference in *C. elegans*. *Curr Biol* **10**, 169–178.

Snel, B., Bork, P. & Huynen, M. A. (2002). Genomes in flux: the evolution of archaeal and proteobacterial gene content. *Genome Res* **12**, 17–25.

Sontheimer, E. J. (2005). Assembly and function of RNA silencing complexes. *Nat Rev Mol Cell Biol* **6**, 127–138.

Storz, G., Opdyke, J. A. & Zhang, A. (2004). Controlling mRNA stability and translation with small, noncoding RNAs. *Curr Opin Microbiol* **7**, 140–144.

Swain, A. & Coffin, J. M. (1992). Mechanism of transduction by retroviruses. *Science* **255**, 841–845.

Tang, G. (2005). siRNA and miRNA: an insight into RISCs. *Trends Biochem Sci* **30**, 106–114.

Tang, T. H., Bachellerie, J. P., Rozhdestvensky, T., Bortolin, M. L., Huber, H., Drungowski, M., Elge, T., Brosius, J. & Huttenhofer, A. (2002). Identification of 86 candidates for small non-messenger RNAs from the archaeon *Archaeoglobus fulgidus*. *Proc Natl Acad Sci U S A* **99**, 7536–7541.

Tatusov, R. L., Fedorova, N. D., Jackson, J. D. & 14 other authors (2003). The COG database: an updated version includes eukaryotes. *BMC Bioinformatics* **4**, 41.

Watanabe, H., Mori, H., Itoh, T. & Gojobori, T. (1997). Genome plasticity as a paradigm of eubacteria evolution. *J Mol Evol* **44** (Suppl. 1), S57–S64.

Wolf, Y. I., Rogozin, I. B., Kondrashov, A. S. & Koonin, E. V. (2001). Genome alignment, evolution of prokaryotic genome organization, and prediction of gene function using genomic context. *Genome Res* **11**, 356–372.

The significance of prokaryote diversity in the human gastrointestinal tract

Harry J. Flint

Microbial Ecology Group, Rowett Research Institute, Greenburn Road, Bucksburn, Aberdeen AB21 9SB, UK

INTRODUCTION

Gut microbial communities have existed since the earliest multicellular life forms developed a digestive tract. In invertebrates and vertebrates that rely largely on plant material as their main source of energy, the gut microbial community plays a crucial nutritional role in supplying energy to the host through anaerobic fermentation of plant structural polysaccharides. This type of symbiotic association has been studied particularly in ruminants, where the host derives around 75 % of its energy from the diet via microbial fermentation (Hungate, 1966), and in the termite gut (Brune, 1998). In all hosts, however, including carnivores and omnivores, where their nutritional contribution is less important, the gut microbiota exert a major influence on health as a potential source of infectious agents, as a barrier against infectious agents and as determinants of the gut environment, gut metabolism and immune development. This chapter will consider the extent and significance of microbial diversity in gut communities, with particular reference to the microbiota of the human large intestine. The best-studied gut inhabitants are of course pathogens such as *Escherichia coli*, but there is increasing awareness of the importance of the numerically predominant commensal colonizers of the gastrointestinal tract.

MICROBIAL DIVERSITY IN THE HUMAN LARGE INTESTINE

Diversity revealed by 16S rRNA gene sequences

The communities of the human large intestine and rumen show the highest prokaryote cell density of any microbial ecosystem, approaching or exceeding 10^{11} cells g^{-1}

SGM symposium 66: Prokaryotic diversity: mechanisms and significance.
Editors N. A. Logan, H. M. Lappin-Scott & P. C. F. Oyston. Cambridge University Press. ISBN 0 521 86935 8 ©SGM 2006

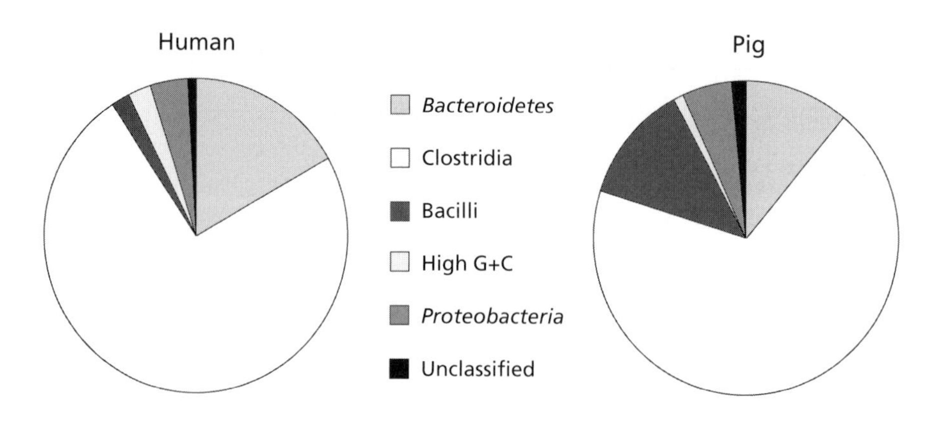

Fig. 1. Distribution of phylotypes identified by sequencing of amplified 16S rRNA genes from samples of human faeces and colon (Eckburg *et al.*, 2005) and from the porcine intestine (Leser *et al.*, 2002). 'Unclassified' refers to unclassified organisms and members of other, unlisted phyla.

(Whitman *et al.*, 1998). Despite past cultural work, studies on 16S rRNA diversity in human stool (Wilson & Blitchington, 1996; Suau *et al.*, 1999; Hayashi *et al.*, 2002a) and colon (Hold *et al.*, 2002; Eckburg *et al.*, 2005) samples have shown that 75 % of directly amplified sequences do not correspond closely to known cultured bacteria. The predominant phyla detected are the Gram-negative *Bacteroidetes* (CFB) and low-G+C-content Gram-positive *Firmicutes*. In the largest study of this type, Eckburg *et al.* (2005) analysed 13 355 16S rRNA gene sequences amplified from colonic and stool samples of three individuals. *Clostridium*-related firmicutes, most belonging to clostridial clusters XIVa and IV, accounted for 69 % of the total of 395 phylotypes detected (Fig. 1) and for 73 % of the 244 novel phylotypes that did not correspond to sequences from named cultured species. Archaeal sequences recovered corresponded to a single phylotype, the methanogen *Methanobrevibacter smithii*. No study has yet come close to revealing all of the eubacterial phylotypes present in gut communities; however, this has been estimated by Eckburg *et al.* (2005) to exceed 500 for the three subjects that they studied.

While PCR-amplified libraries reveal diversity, the potential for bias, in particular against high-G+C-content groups, means that they do not reliably estimate relative abundance. Surveys of the human faecal microbiota using 16S rRNA-targeted fluorescence *in situ* hybridization (FISH) probes have, however, provided a fairly consistent picture of the major groups that confirms the abundance of the CFB phylum and of firmicutes belonging to clostridial clusters IV and XIVa. High-G+C-content Gram-positive *Bifidobacterium* and *Atopobium* groups were estimated at around 4 and 3 % of total eubacteria, respectively (Table 1). FISH may still underestimate particle-attached bacteria, however, which may be difficult to detach from their substrate as intact cells.

Comparison with other mammalian gut communities

The same bacterial phyla and broad clusters have been shown to dominate other anaerobic gut communities including the rumen (Tajima *et al.*, 1999; Whitford *et al.*, 1998) and the large intestine of horses (Daly *et al.*, 2001) and pigs (Leser *et al.*, 2002; Pryde *et al.*, 1999). In a survey of 4270 cloned bacterial 16S rRNA gene sequences from the pig large intestine, for example, 304 out of 375 phylotypes (81 %) belonged to the *Firmicutes* and 42 (11·2 %) to the *Bacteroidetes* (Leser *et al.*, 2002) (Fig. 1); 83 % of phylotypes were reported as novel.

Although the major groups of anaerobic gut bacteria thus appear well conserved between different mammalian host species, this uniformity quickly disappears at a more detailed level. A major cluster of *Prevotella* sequences, for example, appears to be largely confined to the rumen (Ramšak *et al.*, 2000), while several subclusters of *Clostridium*-related bacteria identified from the equine colon appear to be unique (Daly *et al.*, 2001).

Thus, while gut communities exhibit a high degree of diversity at the species level, this diversity mainly reflects radiation within a few dominant phyla of prokaryotes (Hugenholtz *et al.*, 1998). In the gut of herbivores, especially in the rumen, microbial diversity is greatly increased by significant populations of eukaryotic micro-organisms, the anaerobic fungi and protozoa (Hespell *et al.*, 1996).

Culturability, oxygen sensitivity and syntrophy

Despite the large proportions of novel 16S rRNA gene sequences detected, it does not follow that the majority of gut bacteria are unculturable. The phyla that are most dominant in gut communities contain many cultivated representatives (Hugenholtz *et al.*, 1998) and, in contrast with other environments, e.g. soils, the constant turnover of gut contents imposes a certain minimum growth rate for survival. Early investigations successfully cultured a wide range of anaerobic bacteria for most of which 16S rRNA gene sequences are unfortunately unavailable (Finegold *et al.*, 1983). Oxygen sensitivity is the most likely reason why some groups, notably clostridial cluster XIVa bacteria, are less easy to recover (Hayashi *et al.*, 2002b). Bacteria can be recovered anaerobically from human faeces, however, whose viable count is reduced by 8 logs following exposure to air for 1 min on an agar surface (S. H. Duncan and H. J. Flint, unpublished). Since the colon is not completely anoxic (Cummings & Macfarlane, 1991), it must be presumed that micro-niches and the proximity of more oxygen-tolerant bacteria provide protection to such ultra-strict anaerobes *in vivo*. The existence of close nutritional interdependence (syntrophy) (Schink, 1992) also means that, for some organisms, growth requirements cannot be met by standard culture media.

Table 1. Abundance of different bacterial groups in adult human faecal samples estimated with targeted FISH probes

Values are percentages expressed relative to counts for a broad eubacterial probe (Eub 338) in each case. Empty cells mean that the relevant probe was not tested in that study.

Bacterial group	Probe	Lay *et al.* (2005)	Harmsen *et al.* (2002)	Franks *et al.* (1998)	Walker *et al.* (2005)	Other studies*
Number of subjects		91	11	10	2	
Low-G+C-content Gram-positives						
Clostridial cluster XIVa, b	Erec 482	28·0	22·7	29	28·4	16[a], 10·8[b]
Ruminococcus obeum	Urobe 63					2·5[a]
Roseburia/Eubacterium rectale	Rrec 584				8·8	
	Rint 623					2·3[b]
Eubacterium hallii	Ehal1469		3·8			0·6[b]
Lachnospira	Lach 571		3·6			
Clostridial cluster IV	Clep 866	25·2				
Faecalibacterium prausnitzii	Fprau 645 (a)	15·4			5	3·8[b]
	Elgc 01 (b)		10·8	12		
Ruminococcus bromii	Rbro 730	6·3	10·3 (*R. bromii* + *R. flavefaciens* combined)		1·7 (*R. bromii* + *R. flavefaciens* combined)	
Ruminococcus flavefaciens	Rfla 729	0·8				
Clostridial cluster XVI						
Eubacterium cylindroides	Ecyl 387	1·1	1·4			0·4[b]
Clostridial cluster IX						
Selenomonas/Megasphaera	Prop 853				7·1	
Veillonella	Veil 223	1·3	0·1			

Bacillus relatives						
Lactobacillus/Enterococcus	Lab 158	1·8	0·01		<0·3	
Streptococcus/Lactococcus	Strc 493	0·6		<0·1		
High-G+C-content Gram-positives						
Bifidobacterium	Bif 164	4·4	4·8	3·5	4·9	2·5[c]
Atopobium	Ato 291	3·1	11·9		2·1	2·6[c]
Gram-negatives						
Bacteroidetes (CFB)	Bac 303	8·5	27·7	21·9	9·3	
Proteobacteria	Enter 1432 (a)	0·1				
	E. coli 1531 (b)		0·2			
	Enterobact D (c)				<0·3	

*Results from: *a*, Zoetendal *et al.* (2002b) (3 subjects); *b*, Hold *et al.* (2003) (10 subjects); *c*, Harmsen *et al.* (2000b) (10 subjects in 25–55 year age group).

Zonation, microenvironments and biofilms within the gut

Different regions of the gut differ markedly in turnover, pH, secretions and mucosal anatomy. Within the large intestine, the proximal colon is the site of most active microbial fermentation of residual dietary material, resulting in a slight lowering of the pH (Bown *et al.*, 1974). Microbial metabolic activity decreases and pH rises in the distal colon (Macfarlane *et al.*, 1992). The upper gut shows less heavy microbial colonization than the large intestine and the composition of these microbial communities is distinct from that of the colon, being dominated by facultative anaerobes (Wang *et al.*, 2005).

At a microscopic level, epithelial surfaces, mucin layers and substrate particles all provide potential substrates for biofilm formation, although they clearly differ from inert sites such as tooth surfaces and catheters that are known to develop biofilms (O'May *et al.*, 2005) in being subject to rapid turnover. There is evidence for distinct particle-associated communities in the rumen (Tajima *et al.*, 1999) and for mucosal biofilms in the human colon (Swidsinski *et al.*, 2005). The mucosa of the distal ileum and the proximal and distal colon in healthy persons carry similar communities (Zoetendal *et al.*, 2002a; Lepage *et al.*, 2005; Wang *et al.*, 2005). The profiles of these communities, and of the caecum (Marteau *et al.*, 2001), differ from those for faeces from the same individuals.

Flow through the gut means that representatives of different micro-communities are constantly being intermixed. Faecal bacteria, most commonly studied for obvious reasons of accessibility, cannot be regarded as a 'community'. Rather they are an amalgam of micro-organisms of diverse origin and varying physiological state, as revealed by studies on cells recovered by flow cytometry (Ben-Amor *et al.*, 2005). Nevertheless, faecal samples generally provide the only way of tracking events in the gut lumen.

COMPETITION FOR SUBSTRATES

Utilization of dietary carbohydrates

Gut micro-organisms in the large intestine have access to those dietary compounds that have not been hydrolysed by host enzymes and/or transported by the intestinal mucosa in the upper gut. The concentrations of most free monosaccharides in the colon are therefore assumed to be very low, and the main diet-derived energy sources are plant glycosides, oligosaccharides, storage polysaccharides and cell-wall polysaccharides that are resistant to host enzymes (Cummings & Macfarlane, 1991; Macfarlane & Gibson, 1995a).

Glycoside hydrolases, esterases and other activities required to degrade dietary plant polysaccharides are found in a wide range of gut bacteria (Table 2) and are encoded by

Table 2. Polysaccharide-utilizing abilities of isolated anaerobic bacteria from the human colon

Based largely on Salyers *et al.* (1977a, b), Hoskins (1993), Derrien *et al.* (2004), Robert & Bernalier-Donadille (2003) and Duncan *et al.* (2002b, 2003).

Substrate	Bifidobacteria	*Bacteroidetes*	*Firmicutes*		Other
			Cluster XIVa	Cluster IV	
Starch	*Bifidobacterium adolescentis*, *Bifidobacterium breve*	*Bacteroides fragilis*, *Bacteroides thetaiotaomicron*, *Bacteroides vulgatus*, *Bacteroides eggerthii*, *Bacteroides ovatus*	*Roseburia* spp., *Eubacterium rectale*	*Ruminococcus bromii*	*Clostridium butyricum*, *Collinsella aerofaciens*
Xylan	*Bifidobacterium adolescentis*, *Bifidobacterium infantis*	*Bacteroides ovatus*	*Roseburia intestinalis*		
Pectin		*Bacteroides fragilis*, *Bacteroides thetaiotaomicron*, *Bacteroides vulgatus*, *Bacteroides ovatus*	*Eubacterium eligens*		
Cellulose				*Ruminococcus* sp.	
Inulin	*Bifidobacterium* spp.		*Roseburia* spp.		
Mucin (porcine)	*Bifidobacterium bifidum*		*Ruminococcus torques*		*Akkermansia muciniphila*

multiple gene copies in the few genomes from gut anaerobes that have been sequenced (Xu *et al.*, 2003; Schell *et al.*, 2002). The genomes of *Bacteroides thetaiotaomicron* and *Bifidobacterium longum* also show large numbers of ABC-type transporters that are likely to be involved in oligosaccharide transport. For gut bacteria, oligosaccharide uptake, especially if linked to phosphorylation (Lou *et al.*, 1996), provides considerable advantages both energetically and in the competition for substrate, compared with the uptake of monosaccharides.

The organization and regulation of these gene products is crucial in determining the niches occupied by different carbohydrate-degrading species (Flint, 2004). The small number of genuine cellulolytic rumen bacteria that have been studied are characterized by highly specialized enzyme systems held on the cell surface. In the cellulosome of *Ruminococcus flavefaciens*, a clostridial cluster IV bacterium, a wide range of enzyme specificities is organized via multiple cohesin–dockerin interactions into a large cellulosome complex (Ding *et al.*, 2001; Rincon *et al.*, 2003). This complex has recently been shown to anchor to the cell surface via a sortase-mediated attachment mechanism (Rincon *et al.*, 2005) (Fig. 2). Related cellulolytic bacteria are reported from the human large intestine (Robert & Bernalier-Donadille, 2003), but the organization of their enzyme systems has yet to be established.

In addition to the primary polysaccharide degraders, many other gut bacteria have highly efficient mechanisms for processing polysaccharides and oligosaccharides that are released from plant material. Cross-feeding of oligosaccharides has been demonstrated between members of the rumen community, with *Prevotella/Bacteroides* strains able to compete for breakdown products of hemicellulose and pectin released by cellulolytic species (Dehority, 1991). The Gram-negative human colonic species *Bacteroides thetaiotaomicron* has been shown to sequester starch breakdown products via outer-membrane substrate-attachment proteins (Reeves *et al.*, 1997; Cho & Salyers, 2001). An outer-membrane glycoside hydrolase (neopullulanase) is thought to perform limited hydrolysis of starch molecules, allowing them to enter the periplasm, where most of the hydrolytic activity is located (Anderson & Salyers, 1989; Shipman *et al.*, 1999) (Fig. 2). In the related rumen species *Prevotella bryantii*, most of the assayable xylanase activity was also found to be cell associated and not expressed on the cell surface, suggesting a similar mechanism (Miyazaki *et al.*, 1997). The generality of such sequestration mechanisms is suggested by the detection of large numbers (163) of *susC*- or *D*- (two genes that encode starch-binding proteins) related genes in the genome of *Bacteroides thetaiotaomicron* (Xu *et al.*, 2003). These sequestration mechanisms are likely to be crucial in allowing *Bacteroides* to scavenge carbohydrates solubilized by the activities of other community members, while also retaining the products of its own degradative enzymes.

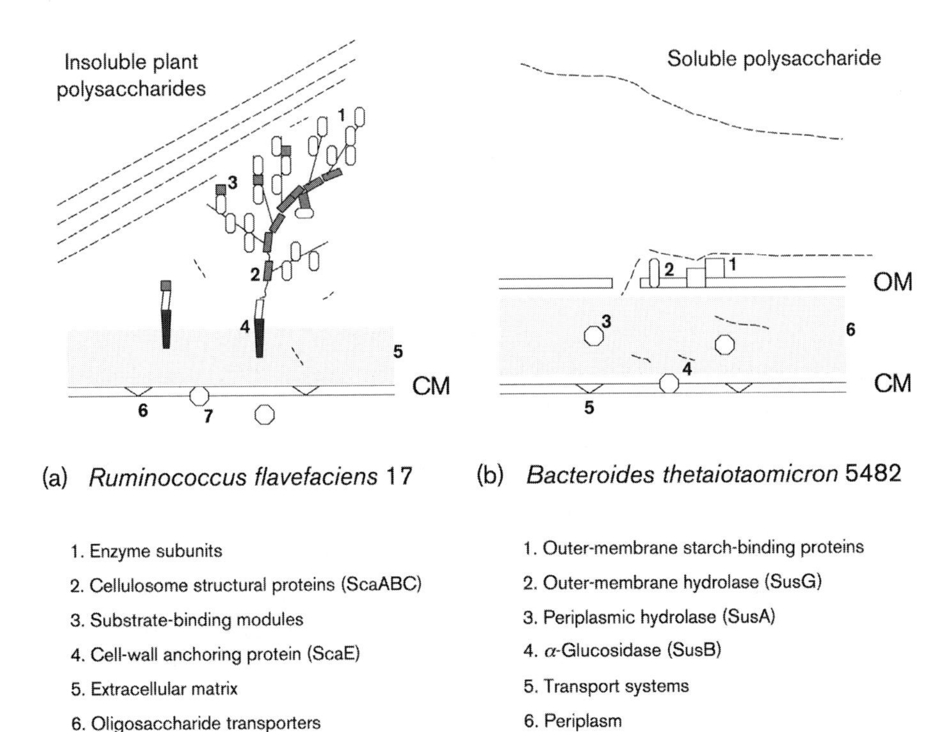

Fig. 2. Schematic representation of two contrasting systems for polysaccharide utilization in anaerobic gut bacteria. (a) Cellulosome complex of the Gram-positive *Ruminococcus flavefaciens* 17, which degrades plant cell-wall polysaccharides (based on Ding *et al.*, 2001; Rincon *et al.*, 2003, 2005). (b) Starch utilization system of the Gram-negative *Bacteroides thetaiotaomicron* 5482 (based on Anderson & Salyers, 1989; Cho & Salyers, 2001; Reeves *et al.*, 1997; Shipman *et al.*, 1999). CM, Cell membrane; OM, outer membrane.

The substrate-utilizing capabilities are known for many representative colonic anaerobes when in pure culture (Table 2). It remains unclear, however, which groups of bacteria act as the primary colonizers and degraders of insoluble substrates such as plant cell walls and resistant starch, and which derive energy mainly by cross-feeding, in the human gut. Surface-bound enzyme systems as found in some Gram-positive bacteria (type a; Fig. 2) might be expected to typify primary colonizers that rely on close proximity to insoluble substrates to gain first access to polysaccharides. On the other hand, the systems so far demonstrated in Gram-negative *Bacteroides* (type b) appear better suited to sequestration of soluble polysaccharides. Future research, supported by further genomic analyses, will undoubtedly reveal many variations on these models. Studies based on 16S rRNA gene sequence analysis, however, support the view that certain key groups of Gram-positive anaerobes are found closely attached to fibrous faecal material, whereas *Bacteroides* are abundant in the planktonic phase

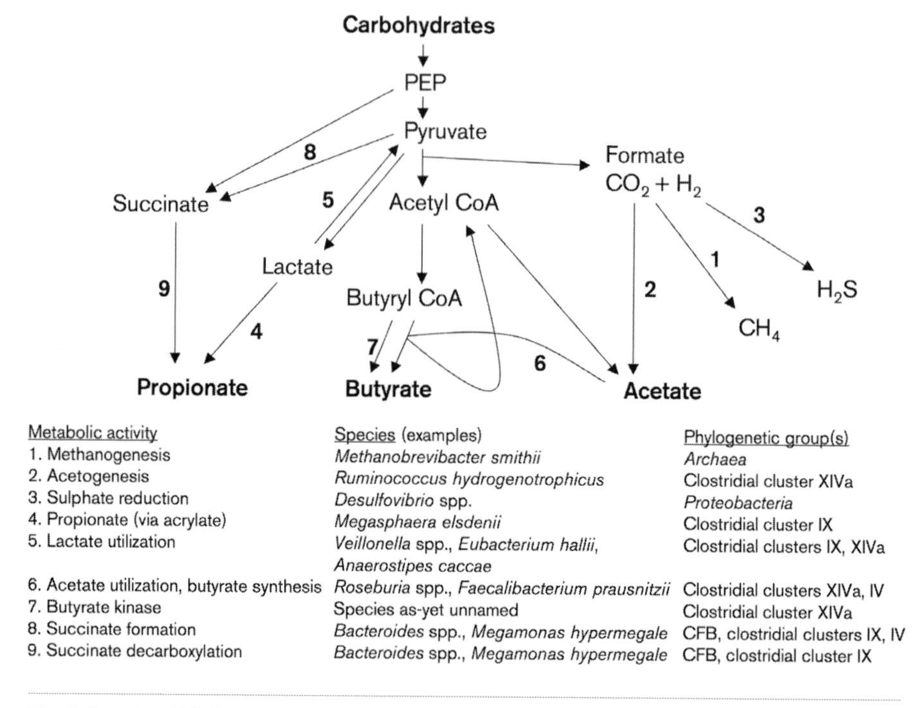

Metabolic activity	Species (examples)	Phylogenetic group(s)
1. Methanogenesis	*Methanobrevibacter smithii*	*Archaea*
2. Acetogenesis	*Ruminococcus hydrogenotrophicus*	Clostridial cluster XIVa
3. Sulphate reduction	*Desulfovibrio* spp.	*Proteobacteria*
4. Propionate (via acrylate)	*Megasphaera elsdenii*	Clostridial cluster IX
5. Lactate utilization	*Veillonella* spp., *Eubacterium hallii*, *Anaerostipes caccae*	Clostridial clusters IX, XIVa
6. Acetate utilization, butyrate synthesis	*Roseburia* spp., *Faecalibacterium prausnitzii*	Clostridial clusters XIVa, IV
7. Butyrate kinase	Species as-yet unnamed	Clostridial cluster XIVa
8. Succinate formation	*Bacteroides* spp., *Megamonas hypermegale*	CFB, clostridial clusters IX, IV
9. Succinate decarboxylation	*Bacteroides* spp., *Megamonas hypermegale*	CFB, clostridial cluster IX

Fig. 3. Functional/phylogenetic groups involved with the formation and conversion of SCFA in the human large intestine. Almost all saccharolytic anaerobes produce acetate, but in *Bifidobacterium* species this involves the fructose 6-phosphate shunt (Macfarlane & Gibson, 1995b). PEP, Phosphoenolpyruvate.

(E. C. McWilliam Leitch, A. W. Walker, S. H. Duncan and H. J. Flint, unpublished results). A similar partitioning was reported previously for rumen contents (Tajima *et al.*, 1999).

Non-carbohydrate, and host-derived, substrates

In addition to carbohydrates, gut communities are involved in metabolism of a huge array of diet- and host-derived compounds, and many of these transformations have consequences for the host (Conway, 1995; Gill & Rowland, 2002). In most cases, the phylogenetic groups responsible have not yet been identified, and only selected examples are given here.

Release and conversion of plant metabolites. Many metabolites of plant origin, particularly aromatic compounds, exert biological effects upon the host that range from toxicity to anti-oxidant, anti-cancer and hormonal activities. In many cases, the bioactive compounds are released (e.g. phenolics such as ferulate) or formed (e.g. phyto-oestrogens) by the action of gut micro-organisms, often acting in concert. The cluster XIVa bacterium *Eubacterium ramulus*, for example, is a degrader of the phenolic compound quercitin, but quercitin is first released by the action of glycosidases from other species (Schneider *et al.*, 1999). Certain plant metabolites can also show differential toxicity against pathogens (Duncan *et al.*, 1998). Xenobiotic compounds, including many plant metabolites, are subject to conjugation in the liver and release into the gut; deconjugation by gut micro-organisms can result in an enterohepatic circulation.

Utilization of host products. An important consequence of carbohydrate fermentation in the colon is that significant amounts of nitrogen are incorporated into microbial protein and eliminated. In the absence of carbohydrates, the microbiota come to depend to a far greater extent on endogenously derived substrates and on fermentation of protein and amino acids. A wide range of secreted products, sloughed epithelial cells, bile components and mucins provide substrates of endogenous origin for microbial growth. Certain specialist bacteria have been identified as mucin degraders, including *Akkermansia muciniphila*, a member of the *Verrucomicrobia* (Derrien *et al.*, 2004), *Bifidobacterium bifidum* and *Ruminococcus torques* (Hoskins, 1993; Salyers *et al.*, 1977b) (Table 2). Mucin and epithelial cells carry carbohydrate chains that act as receptors for binding and as potential energy sources. Evidence from gnotobiotic animal studies suggests that *Bacteroides thetaiotaomicron*, which is able to utilize the host-derived sugar fucose, can induce increased expression of fucose on gut receptors (Bry *et al.*, 1996; Hooper *et al.*, 1999).

Degradation of proteins. Protein from endogenous sources certainly contributes to microbial fermentation, but it is also suggested that some dietary protein can survive to the large intestine (Magee *et al.*, 2000). Fermentation of amino acids yields a range of products including indole, cresols, phenols and branched-chain fatty acids (Gill & Rowland, 2002; Cummings & Macfarlane, 1991).

CONTRIBUTIONS OF DIFFERENT BACTERIAL GROUPS TO GUT METABOLISM

The main products of anaerobic microbial metabolism in the large intestine are the short-chain fatty acids (SCFA) acetate, propionate and butyrate (which achieve combined concentrations of 90–130 mM in gut contents) and the gases H_2, CO_2 and CH_4 (Cummings & Macfarlane, 1991; Macfarlane & Gibson, 1995b) (Fig. 3). These products have a major influence on the gut environment and on the colonic mucosa and play a crucial role in linking diet with the prevention and causation of gut disease (Topping & Clifton, 2001; Gill & Rowland, 2002). Anaerobic systems are characterized by a high degree of metabolic cross-feeding, with many of the organic acids and much of the hydrogen formed by fermentation being reutilized. Recent progress in phylogenetic analysis discussed earlier makes it possible to start to relate metabolic activities to phylogenetic groups.

Hydrogen production and consumption

Most of the dominant *Clostridium*-related gut anaerobes are active hydrogen-producers, and gas metabolism has important consequences for colonic health (Levitt *et al.*, 1995). Hydrogen is either released as gas or consumed by hydrogen-utilizing species, including methanogenic archaea, sulphate reducers and acetogens (Macfarlane & Gibson, 1995a) (Fig. 3). Acetogenesis from CO_2 and H_2 has been demonstrated for a number of species, such as *Ruminococcus hydrogenotrophicus*, belonging to clostridial cluster XIVa (Bernalier *et al.*, 1996a). Although acetogenesis is less favoured thermodynamically than methanogenesis (Macfarlane & Gibson, 1995a; Bernalier *et al.*, 1996b), acetogens are thought to compete for hydrogen better at reduced pH, e.g. during active fermentation in the proximal colon. Hydrogen removal can increase the energy yield for hydrogen-producing, carbohydrate-utilizing bacteria by allowing increased acetate formation (Latham & Wolin, 1977).

SCFA metabolism

Butyrate. Microbial supply of butyrate, which is stimulated by certain dietary carbohydrates, is considered to help in preventing colorectal cancer and colitis (Wachtershauser & Stein, 2000; Perrin *et al.*, 2001; McIntyre *et al.*, 1993; Pryde *et al.*, 2002). This is related to the role of butyrate as the preferred energy source for the colonic mucosa and its influence on the development and gene expression in colonic epithelial cells (Mariadason *et al.*, 2000). The majority (>90 %) of butyrate-producing anaerobes isolated from human faeces on carbohydrate-containing media belong to clostridial clusters XIVa or IV (Barcenilla *et al.*, 2000; Pryde *et al.*, 2002) including novel species and genera such as *Roseburia intestinalis* (Duncan *et al.*, 2002b) and *Anaerostipes caccae* (Schwiertz *et al.*, 2002a). The two most abundant groups appear to be the cluster IV species *Faecalibacterium prausnitzii* (formerly *Fusobacterium prausnitzii*) (Duncan *et al.*, 2002c) and a cluster XIVa group related to *Roseburia intestinalis* and *Eubacterium rectale* (Table 1). Additional butyrate-producers belonging to clostridial clusters I, III, XI, XV and XVI are found in smaller numbers (Schwiertz *et al.*, 2002b; Pryde *et al.*, 2002).

Butyrate kinase has been assumed to catalyse the final steps in butyrate formation in most bacteria. A survey of 38 butyrate-producing strains from human faeces using degenerate PCR primers designed to recognize butyrate kinase/phosphotrans-butyrylase revealed only four strains, however, which grouped into two pairs based on their 16S rRNA gene sequences, that possessed these genes. These were also the only strains to exhibit assayable butyrate kinase activity (Louis *et al.*, 2004). It appears therefore that the majority of human faecal bacteria involved in butyrate production, including *Roseburia* species and *Faecalibacterium prausnitzii*, rely instead on butyryl CoA : acetate CoA transferase activity (Duncan *et al.*, 2002a; Diez-Gonzalez *et al.*, 1999) for the last step in butyrate synthesis. The gene for this enzyme has now been identified in *Roseburia* sp. A2-183; the purified recombinant enzyme has affinity for butyryl CoA and propionyl CoA and defines a new class of CoA transferase that appears to play a central role in gut metabolism (Charrier *et al.*, 2006).

Acetate. Acetate is the SCFA that reaches the highest concentrations in colonic contents and is formed by nearly all heterotrophic gut anaerobes, but with a significant contribution (up to one-third) also coming from reductive acetogenesis (Miller & Wolin, 1996). Approximately 50 % of the butyrate-producing bacteria isolated by Barcenilla *et al.* (2000) proved to be net consumers of acetate when grown in SCFA-supplemented media. In contrast, net acetate utilization was evident in only 1 % of butyrate-non-producing human faeces isolates. Butyrate-producing *Roseburia* and *Faecalibacterium prausnitzii* strains displayed extremely rapid interchange between external ^{13}C-labelled acetate and internal acetyl CoA, and up to 90 % of butyrate carbon was derived from labelled exogenous acetate in glucose-grown cultures. Similarly high rates of acetate incorporation into butyrate were found for mixed human faecal bacteria (Duncan *et al.*, 2004a). Acetate is therefore subject to rapid turnover in the mixed gut system not only because of rapid uptake by the gut mucosa, but also because of net conversion to butyrate by acetate-utilizing bacteria. Acetogens are able both to supply acetate and to consume hydrogen in co-culture with *Roseburia intestinalis in vitro* (Chassard & Bernalier-Donadille, 2006).

Propionate. *Bacteroides* species produce succinate that is partially converted to propionate in pure culture depending on growth conditions (Macfarlane & Gibson, 1995a). In addition, most cultured representatives of clostridial cluster IX, recently shown to be an abundant group of faecal bacteria (Table 1), are propionate producers able to decarboxylate succinate. Succinate concentrations in faeces are normally low in comparison with propionate.

Lactate. Many gut bacteria have the potential to produce lactate, but faecal lactate concentrations are usually much lower (<2 mM) than those of the three major SCFA products. Lactate is known to accumulate, however, in disease conditions such as severe colitis (Vernia *et al.*, 1988). Low lactate concentrations could simply reflect low rates of production arising from low growth rates *in vivo* (Macfarlane & Gibson, 1995a). By analogy with the rumen (Counotte *et al.*, 1981; Hashizume *et al.*, 2003), however, it has been assumed that lactate utilization, in particular by propionate-forming bacteria, is important in limiting its accumulation. Incubations with isotopically labelled lactate recently revealed that butyrate rather than propionate was the main product of lactate utilization for two out of three human faecal samples examined (Bourriaud *et al.*, 2005). Three clostridial cluster XIVa species from human faeces were recently shown to produce butyrate as an end product from lactate and acetate (Duncan *et al.*, 2004b). At least one of these, *Eubacterium hallii*, is abundant in the colonic microbiota (Table 1) and may have an important role in the conversion of both D- and L-lactate to butyrate. Although *Eubacterium hallii* uses sugars in preference to lactate, co-cultures of *Eubacterium hallii* and *Bifidobacterium adolescentis* with starch (which *Eubacterium hallii* alone cannot utilize) as an energy source produced butyrate via lactate cross-feeding (Duncan *et al.*, 2004b). The extent and importance of lactate cross-feeding *in vivo*, however, remains to be established. D-Lactate is more toxic to humans than the L-form, and can cause neurological symptoms at high concentrations (Bustos *et al.*, 1994). Conversely, the L-form is apparently more toxic to *Escherichia coli* at acidic pH (McWilliam Leitch & Stewart, 2002).

Branched-chain SCFA. 2-Methyl butyrate, isobutyrate and isovalerate are formed as products of isoleucine, valine and leucine fermentation, respectively (Macfarlane & Gibson, 1995b). They are estimated to represent a higher proportion of SCFA formed in the distal colon, where proportionately more amino acids are being fermented compared with carbohydrates. It is not yet known which are the main species responsible.

Potential impact of pH on bacterial populations and SCFA production

As noted earlier, the presence of fermentable carbohydrates in the proximal colon tends to decrease the luminal pH. The potential impact of a one-unit pH shift upon human faecal communities has recently been studied in anaerobic continuous flow fermenters supplied with a mixed carbohydrate substrate (Walker *et al.*, 2005). At pH 5·5, the butyrate-producing *Roseburia*/*Eubacterium rectale* group accounted for 10–20 % of total eubacteria, and butyrate concentrations exceeded 20 mM (Walker *et al.*, 2005) (Fig. 4). On switching to pH 6·5, however, the community came to be dominated by *Bacteroides* (CFB) and the *Roseburia* group became undetectable, with an accompanying increase in propionate and decrease in butyrate concentration (Walker *et al.*, 2005). This suggests that *Bacteroides* can readily outcompete other bacterial groups for supplied soluble carbohydrate (in this case mainly amylopectin) at pH 6·5. *In vivo*, however, a plentiful supply of dietary carbohydrate is most likely to be found in the proximal colon, where active fermentation tends to drive the luminal pH down, often to below 6·0 (Bown *et al.*, 1974). It appears that *Bacteroides* compete less well at these slightly acidic pHs, allowing other groups including the butyrate-producing *Roseburia* to proliferate. These observations may help to explain why many carbo-

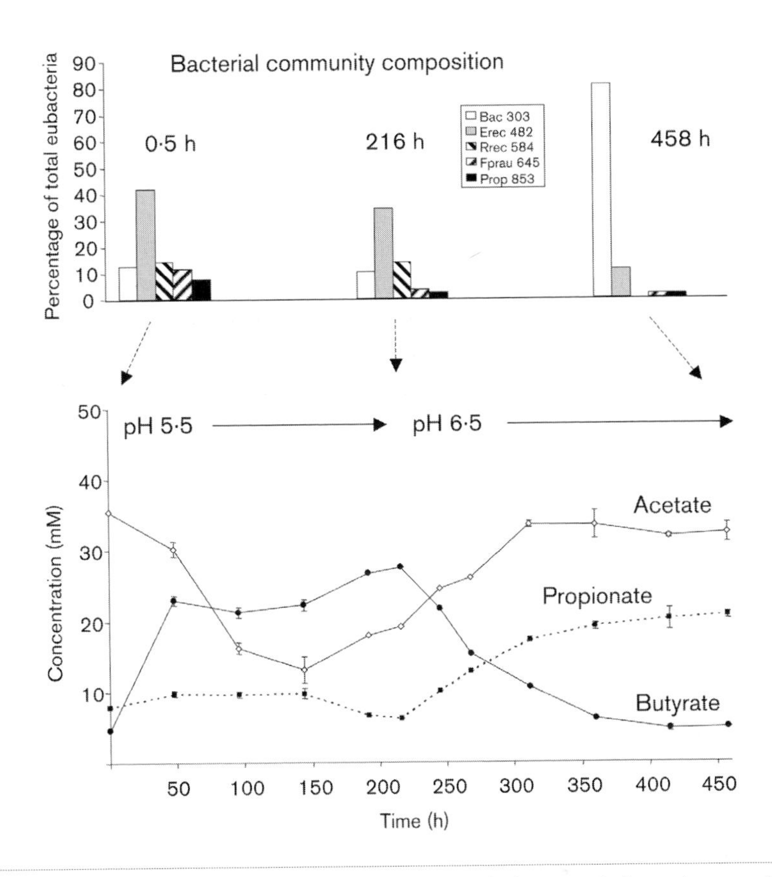

Fig. 4. Impact of a one-unit pH shift upon a human gut microbial community in continuous culture *in vitro* under anaerobic conditions (data are from Walker *et al.*, 2005). The energy source comprised a mixture of plant polysaccharides, with starch as the main ingredient. Populations of the dominant bacterial groups were determined by FISH at the beginning (0·5 h) and after periods of equilibration at pH 5·5 (216 h) and pH 6·5 (458 h). Probes shown are: Bac 303, *Bacteroides*; Erec 482, clostridial cluster XIVa; Rrec 584, the *Roseburia/Eubacterium rectale* group within cluster XIVa; Fprau 645, *Faecalibacterium prausnitzii*; Prop 853, clostridial cluster IX. Concentration changes in the fermenter are shown for the three main SCFA products.

hydrates that are fermentable, but non-digestible by mammalian enzymes, are reported to be butyrogenic (Cummings, 1995).

It should be noted that SCFA ratios resembling those seen in Fig. 4 are seldom reported for human faecal samples, where typical acetate : propionate : butyrate ratios are around 3 : 1 : 1. Since most fermentation occurs in the proximal colon, and 95 % of SCFA produced are normally taken up by the gut mucosa, however, faecal SCFA do not reflect production rates. It is entirely possible that the conditions created by the experiment shown in Fig. 4 occur at particular locations and times within the gut.

CHANGES IN GUT MICROBIOTA WITH HOST, DIET AND DISEASE STATES

Development and individual variation

Gut micro-organisms are assumed to be acquired from the mother during and after normal birth. Early development is strongly influenced by breast milk or bottle feeding, with the former favouring bifidobacteria (Harmsen *et al.*, 2000a). Gram-positive strict anaerobes only appear in detectable numbers at weaning, whereas proteobacteria (coliforms) and *Bacteroides* can colonize within the first few weeks of life (Wang *et al.*, 2004). By 12 months, the microbiota composition begins to resemble that in adults, and remains relatively stable until the onset of old age (Hopkins *et al.*, 2001).

Denaturing gradient gel electrophoresis profiling of 16S rRNA genes suggests that, in each adult, the intestinal community is unique in its overall composition (Zoetendal *et al.*, 2001). Nevertheless, the same species and very similar strains are recovered regularly from different people. Thus, it remains unclear to what extent individual differences reflect differing proportions of the same common strains, as opposed to absolute differences in strain composition. An interesting example is provided by the division of human subjects into methane producers and methane non-producers, where the former typically harbour $> 10^8$ methanogenic bacteria (ml faeces)$^{-1}$ and the latter only 10^2–10^3 ml^{-1} (Doré *et al.*, 1995). Acetogen numbers appear to correlate negatively with methanogen numbers, suggesting that they are outcompeted in methanogenic subjects (Doré *et al.*, 1995; Bernalier *et al.*, 1996b). Potential competitive interactions with sulphate-reducers are also suggested, but are less clear, with some studies showing no difference in numbers of sulphate-reducing bacteria between methanogenic and non-methanogenic subjects (Doré *et al.*, 1995). Methanogenic/non-methanogenic status may also correlate with the types of dominant fibre-degrading populations present (Robert & Bernalier-Donadille, 2003). It remains unclear whether this variation mainly reflects diet, inocula or host factors.

Impact of diet

Studies on prebiotics such as inulin and fructo-oligosaccharides (e.g. Kleessen *et al.*, 1997) have demonstrated that the composition of the gut microbiota responds to dietary manipulation. Not surprisingly, such changes may be complex and are not necessarily limited to the 'target' species (Duncan *et al.*, 2003). Effects can be expected to include influence via the gut environment (e.g. pH) and indirect stimulation of certain groups via cross-feeding, as discussed earlier. Human subjects switching from high- to low-carbohydrate diets showed a significant decrease of around fourfold in both faecal butyrate concentrations and *Roseburia*/*Eubacterium rectale* populations

(S. H. Duncan. A. M. Johnstone, G. E. Lobley, G. Holtrop and H. J. Flint, unpublished results). This may be associated with the positive effect of low pH on *Roseburia* populations noted *in vitro* (Fig. 4) and also suggests that polysaccharide utilization constitutes an important nutritional niche for the *Roseburia*/*Eubacterium rectale* group in the large intestine. Genetically obese (*ob*/*ob*) mice were recently shown to have increased proportions of clostridial cluster XIVa and reduced CFB bacteria in faeces (Ley *et al.*, 2005). This could reflect increased feed intake, again possibly influencing caecal pH and thus community balance.

It has also been demonstrated that the faecal population of the quercitin-degrader *Eubacterium ramulus* decreased dramatically in subjects who eliminated plant material from their diet, but was restored when quercitin was consumed (Simmering *et al.*, 2002).

Changes with disease states

Dramatic changes resulting from gut infections, such as diarrhoea, are of course well documented. Perhaps less well known are the longer-term consequences of such episodes, and of treatment regimes, for the normal gut microbial community. Also unclear is the role that the gut microbial community may have in causation and prevention of long-term gut disorders such as inflammatory bowel diseases (IBD), colorectal cancer and irritable bowel syndrome (IBS). Extensive surveys have suggested quantitative compositional differences in the faecal microbiota of persons at risk of colon cancer and of IBS patients with respect to healthy subjects (Moore & Moore, 1995; Malinen *et al.*, 2005). Evidence for changes in the gut microbiota in IBD is, however, more compelling. Increases in faecal *Bacteroides* and coliforms (Mangin *et al.*, 2004) and decreases in clostridial cluster IV bacteria (Manichanh *et al.*, 2006) were detected in Crohn's patients. IBD patients show much higher colonization of the mucosal surface than controls, and clostridial cluster XIVa bacteria were recently reported to comprise only 6–7 % of colonic mucosal biofilm bacteria in IBD [ulcerative colitis (UC) or Crohn's] patients compared with 32 % in control individuals (Swidsinski *et al.*, 2005). *Bacteroides*, on the other hand, made up 71 and 62 % of the mucosal biofilm in Crohn's and UC, respectively, and 20 % in controls.

Not surprisingly, oral antibiotic administration can have dramatic effects on the large intestinal microbiota (Jernberg *et al.*, 2005). A patient taking amoxicillin/clavulanic acid who developed antibiotic-associated diarrhoea showed a complete loss of clostridial cluster XIVa bacteria and a marked decrease in cluster IV bacteria and bifidobacteria during antibiotic therapy (Young & Schmidt, 2004). The clostridial populations were largely restored after withdrawal of antibiotics.

Host–microbe interactions

Many microbial metabolic products are known to influence mammalian gene expression. Recent evidence also indicates that certain host hormones can influence gene expression in pathogenic *Escherichia coli* and that certain bacterial quorum-sensing signals can be recognized by host cells (Sperandio *et al.*, 2003). Interactions with the host's immune system are well studied for pathogens, and are thought to lead to phase variation (Saunders, 1995). Host antibacterial activities, including innate products such as defensins, have the potential to influence commensal bacteria (Hooper & Gordon, 2001), but it remains unclear whether the immune system normally exerts any major control over the composition of the commensal microbiota (Conway, 1995). On the other hand, interactions of receptors on epithelial cells with commensal bacteria appear to be necessary for normal development and maintenance of epithelial cells (Rakoff-Nahoum *et al.*, 2004). In addition, antibody responses to commensal gut bacteria become evident in IBD (Duchmann *et al.*, 1999). There are therefore numerous potential mechanisms for interplay between commensal microbes and host cells (Backhed *et al.*, 2005), the significance of which needs to be more fully assessed.

CONCLUSIONS AND PROSPECTS

The human gut microbiota has been described as the second human genome. Given the extreme complexity of the colonic community, metagenomics may be most valuable when applied to the analysis of functionally identified inserts. Arguably the most useful development would be to sequence the genomes of a far wider range of cultured gut bacteria, chosen with respect to phylogeny and function. This would simultaneously provide information on horizontal gene transfer, substrate utilization mechanisms, fermentation pathways, cell-surface proteins, regulatory circuits, attachment and survival mechanisms. Surprisingly, the major clostridial clusters that account for some 50 % of all human gut anaerobes (Table 1, Fig. 1) have no sequenced representatives, although some genomes are in progress.

The greatest challenge lies in relating microbial diversity to function, in particular interactions with the host, within the complex gut community. Molecular enumeration techniques allied to cultivation are beginning to make a real contribution (Zoetendal *et al.*, 2004), and stable-isotope probing techniques (Radajewski *et al.*, 2000) have yet to be applied extensively to gut ecosystems. The formulation of new functional foods to benefit health and the treatment and prevention of gut disorders and infections all depend on improving our understanding of the gut microbiota and its interactions with diet and with the human host.

ACKNOWLEDGEMENTS

The Rowett Research Institute receives funding from the Scottish Executive Environment and Rural Affairs Department. Thanks are due to Sylvia Duncan and Petra Louis for discussions and for critical reading of the manuscript.

REFERENCES

Anderson, K. L. & Salyers, A. A. (1989). Biochemical evidence that starch breakdown by *Bacteroides thetaiotaomicron* involves outer membrane starch-binding sites and periplasmic starch-degrading enzymes. *J Bacteriol* **171**, 3192–3198.

Backhed, F., Ley, R. E., Sonnenburg, J. L., Peterson, D. A. & Gordon, J. I. (2005). Host-bacterial mutualism in the human intestine. *Science* **307**, 1915–1920.

Barcenilla, A., Pryde, S. E., Martin, J. C., Duncan, S. H., Stewart, C. S., Henderson, C. & Flint, H. J. (2000). Phylogenetic relationships of dominant butyrate-producing bacteria from the human gut. *Appl Environ Microbiol* **66**, 1654–1661.

Ben-Amor, K., Heilig, H., Smidt, H., Vaughan, E. E., Abee, T. & de Vos, W. M. (2005). Genetic diversity of viable, injured, and dead fecal bacteria assessed by fluorescence-activated cell sorting and 16S rRNA gene analysis. *Appl Environ Microbiol* **71**, 4679–4689.

Bernalier, A., Willems, A., Leclerc, M., Rochet, V. & Collins, M. D. (1996a). *Ruminococcus hydrogenotrophicus* sp. nov., a new H_2/CO_2-utilizing acetogenic bacterium isolated from human feces. *Arch Microbiol* **166**, 176–183.

Bernalier, A., Lelait, M., Rochet, V., Grivet, J.-P., Gibson, G. R. & Durand, M. (1996b). Acetogenesis from H_2 and CO_2 by methane- and non-methane-producing human colonic bacterial communities. *FEMS Microbiol Ecol* **19**, 193–202.

Bourriaud, C., Robins, R. J., Martin, L., Kozlowski, F., Tenailleau, E., Cherbut, C. & Michel, C. (2005). Lactate is mainly fermented to butyrate by human intestinal microfloras but inter-individual variation is evident. *J Appl Bacteriol* **99**, 201–212.

Bown, R. L., Gibson, J. A., Sladen, G. E., Hicks, B. & Dawson, A. M. (1974). Effects of lactulose and other laxatives on ileal and colonic pH as measured by a radiotelemetry device. *Gut* **15**, 999–1004.

Brune, A. (1998). Termite guts: the world's smallest bioreactors. *Trends Biotechnol* **16**, 16–21.

Bry, L., Falk, P. G., Midtvedt, T. & Gordon, J. I. (1996). A model of host-microbial interactions in an open mammalian ecosystem. *Science* **273**, 1380–1383.

Bustos, D., Pons, S., Pernas, J. C., Gonzalez, H., Caldarini, M. L., Ogawa, K. & De Paula, J. A. (1994). Fecal lactate and short bowel syndrome. *Dig Dis Sci* **39**, 2315–2319.

Charrier, C., Duncan, G. J.., Reid, M. D. & 7 other authors (2006). A novel class of CoA-transferase involved in short-chain fatty acid metabolism in butyrate-producing human colonic bacteria. *Microbiology* **152**, 179–185.

Chassard, C. & Bernalier-Donadille, A. (2006). H_2 and acetate transfers during xylan fermentation between a butyrate-producing xylanolytic species and hydrogenotrophic micro-organisms from the human gut. *FEMS Microbiol Lett* (in press).

Cho, K. H. & Salyers, A. A. (2001). Biochemical analysis of interactions between outer membrane proteins that contribute to starch utilization by *Bacteroides thetaiotaomicron*. *J Bacteriol* **183**, 7224–7230.

Conway, P. L. (1995). Microbial ecology of the human large intestine. In *Human Colonic Bacteria: Role in Nutrition, Physiology and Pathology*, pp. 1–24. Edited by G. R. Gibson & G. T. Macfarlane. Boca Raton, FL: CRC Press.

Counotte, G. H. M., Prins, R. A., Jansen, R. H. A. M. & De Bie, M. J. A. (1981). Role of *Megasphaera elsdenii* in the fermentation of DL-lactate in the rumen of dairy cattle. *Appl Environ Microbiol* **42**, 649–655.

Cummings, J. H. (1995). Short chain fatty acids. In *Human Colonic Bacteria: Role in Nutrition, Physiology and Pathology*. pp. 101–130. Edited by G. R. Gibson & G. T. Macfarlane. Boca Raton, FL: CRC Press.

Cummings, J. H. & Macfarlane, G. T. (1991). The control and consequences of bacterial fermentation in the human colon. *J Appl Bacteriol* **70**, 443–459.

Daly, K., Stewart, C. S., Flint, H. J. & Shirazi-Beechey, S. P. (2001). Bacterial diversity within the equine large intestine as revealed by molecular analysis of cloned 16S rRNA genes. *FEMS Microbiol Ecol* **38**, 141–151.

Dehority, B. A. (1991). Effects of microbial synergism on fibre digestion in the rumen. *Proc Nutr Soc* **50**, 149–159.

Derrien, M., Vaughan, E. E., Plugge, C. M. & de Vos, W. M. (2004). *Akkermansia muciniphila* gen. nov., sp. nov., a human intestinal mucin-degrading bacterium. *Int J Syst Evol Microbiol* **54**, 1469–1476.

Diez-Gonzalez, F., Bond, D. R., Jennings, E. & Russell, J. B. (1999). Alternative schemes of butyrate production in *Butyrivibrio fibrisolvens* and their relationship to acetate utilization, lactate production, and phylogeny. *Arch Microbiol* **171**, 324–330.

Ding, S. Y., Rincon, M. T., Lamed, R., Martin, J. C., McCrae, S. I., Aurilia, V., Shoham, Y., Bayer, E. A. & Flint, H. J. (2001). Cellulosomal scaffoldin-like proteins from *Ruminococcus flavefaciens*. *J Bacteriol* **183**, 1945–1953.

Doré, J., Pochart, P., Bernalier, A., Goderel, I., Morvan, B. & Rambaud, J. C. (1995). Enumeration of H_2-utilizing archaea, acetogenic and sulfate-reducing bacteria from human feces. *FEMS Microbiol Ecol* **17**, 279–284.

Duchmann, R., May, E., Heike, M., Knolle, P., Neurath, M. & Meyer zum Buschenfelde, K.-H. (1999). T cell specificity and cross reactivity towards enterobacteria, *Bacteroides*, *Bifidobacterium*, and antigens from resident intestinal flora in humans. *Gut* **44**, 812–818.

Duncan, S. H., Flint, H. J. & Stewart, C. S. (1998). Inhibitory activity of gut bacteria against *Escherichia coli* O157 mediated by dietary plant metabolites. *FEMS Microbiol Lett* **164**, 283–288.

Duncan, S. H., Barcenilla, A., Stewart, C. S., Pryde, S. E. & Flint, H. J. (2002a). Acetate utilization and butyryl coenzyme A (CoA) : acetate-CoA transferase in butyrate-producing bacteria from the human large intestine. *Appl Environ Microbiol* **68**, 5186–5190.

Duncan, S. H., Hold, G. L., Barcenilla, A., Stewart, C. S. & Flint, H. J. (2002b). *Roseburia intestinalis* sp. nov., a novel saccharolytic, butyrate-producing bacterium from human faeces. *Int J Syst Evol Microbiol* **52**, 1615–1620.

Duncan, S. H., Hold, G. L., Harmsen, H. J. M., Stewart, C. S. & Flint, H. J. (2002c). Growth requirements and fermentation products of *Fusobacterium prausnitzii*, and a proposal to reclassify it as *Faecalibacterium prausnitzii* gen. nov., comb. nov. *Int J Syst Evol Microbiol* **52**, 2141–2146.

Duncan, S. H., Scott, K. P., Ramsay, A. G., Harmsen, H. J. M., Welling, G. W., Stewart, C. S. & Flint, H. J. (2003). Effects of alternative dietary substrates on competition between human colonic bacteria in an anaerobic fermentor system.

Appl Environ Microbiol **69**, 1136–1142.

Duncan, S. H., Holtrop, G., Lobley, G. E., Calder, A. G., Stewart, C. S. & Flint, H. J. (2004a). Contribution of acetate to butyrate formation by human faecal bacteria. *Br J Nutr* **91**, 915–923.

Duncan, S. H., Louis, P. & Flint, H. J. (2004b). Lactate-utilizing bacteria, isolated from human feces, that produce butyrate as a major fermentation product. *Appl Environ Microbiol* **70**, 5810–5817.

Eckburg, P. B., Bik, E. M., Bernstein, C. N., Purdom, E., Dethlefsen, L., Sargent, M., Gill, S. R., Nelson, K. E. & Relman, D. A. (2005). Diversity of the human intestinal microbial flora. *Science* **308**, 1635–1638.

Finegold, S. M., Sutter, V. L. & Mathison, G. E. (1983). Normal indigenous flora. In *Human Intestinal Microbiology in Health and Disease*, pp. 3–31. Edited by D. J. Hentges. New York: Academic Press.

Flint, H. J. (2004). Polysaccharide breakdown by anaerobic microorganisms inhabiting the mammalian gut. *Adv Appl Microbiol* **56**, 89–120.

Franks, A. H., Harmsen, H. J., Raangs, G. C., Jansen, G. J., Schut, F. & Welling, G. W. (1998). Variations of bacterial populations in human feces measured by fluorescent in situ hybridization with group-specific 16S rRNA-targeted oligonucleotide probes. *Appl Environ Microbiol* **64**, 3336–3345.

Gill, C. I. R. & Rowland, I. R. (2002). Diet and cancer: assessing the risk. *Br J Nutr* **88** (Suppl. 1), S73–S87.

Harmsen, H. J. M., Wildeboer-Veloo, A. C. M., Raangs, G. C., Wagendorp, A. A., Klijn, N., Bindels, J. G. & Welling, G. W. (2000a). Analysis of intestinal flora development in breast-fed and formula-fed infants by using molecular identification and detection methods. *J Pediatr Gastroenterol Nutr* **30**, 61–67.

Harmsen, H. J. M., Wildeboer-Veloo, A. C. M., Grijpstra, J., Knol, J., Degener, J. E. & Welling, G. W. (2000b). Development of 16S rRNA-based probes for the *Coriobacterium* group and the *Atopobium* cluster and their application for enumeration of *Coriobacteriaceae* in human feces from volunteers of different age groups. *Appl Environ Microbiol* **66**, 4523–4527.

Harmsen, H. J. M., Raangs, G. C., He, T., Degener, J. E. & Welling, G. W. (2002). Extensive set of 16S rRNA-based probes for detection of bacteria in human feces. *Appl Environ Microbiol* **68**, 2982–2990.

Hashizume, K., Tsukahara, T., Yamada, K., Koyama, H. & Ushida, K. (2003). *Megasphaera elsdenii* JCM1772^{T} normalizes hyperlactate production in the large intestine of fructooligosaccharide-fed rats by stimulating butyrate production. *J Nutr* **133**, 3187–3190.

Hayashi, H., Sakamoto, M. & Benno, Y. (2002a). Phylogenetic analysis of the human gut microbiota using 16S rDNA clone libraries and strictly anaerobic culture-based methods. *Microbiol Immunol* **46**, 535–548.

Hayashi, H., Sakamoto, M. & Benno, Y. (2002b). Fecal microbial diversity in a strict vegetarian as determined by molecular analysis and cultivation. *Microbiol Immunol* **46**, 819–831.

Hespell, R. B., Akin, D. E. & Dehority, B. A. (1996). Bacteria, fungi and protozoa of the rumen. In *Gastrointestinal Microbiology*, vol. 2, pp. 59–141. Edited by R. I. Mackie, B. R. White & R. E. Isaacson. New York: Chapman & Hall.

Hold, G. L., Pryde, S. E., Russell, V. J., Furrie, E. & Flint, H. J. (2002). Assessment of microbial diversity in human colonic samples by 16S rDNA sequence analysis. *FEMS Microbiol Ecol* **39**, 33–39.

Hold, G. L., Schwiertz, A., Aminov, R. I., Blaut, M. & Flint, H. J. (2003). Oligonucleotide probes that detect quantitatively significant groups of butyrate-producing bacteria in human feces. *Appl Environ Microbiol* **69**, 4320–4324.

Hooper, L. V. & Gordon, J. I. (2001). Commensal host-bacterial relationships in the gut. *Science* **292**, 1115–1118.

Hooper, L. V., Xu, J., Falk, P. G., Midtvedt, T. & Gordon, J. I. (1999). A molecular sensor that allows a gut commensal to control its nutrient foundation in a competitive ecosystem. *Proc Natl Acad Sci U S A* **96**, 9833–9838.

Hopkins, M. J., Sharp, R. & Macfarlane, G. T. (2001). Age and disease related changes in intestinal bacterial populations assessed by cell culture, 16S rRNA abundance, and community cellular fatty acid profiles. *Gut* **48**, 198–205.

Hoskins, L. C. (1993). Mucin degradation in the human gastrointestinal tract and its significance to enteric microbial ecology. *Eur J Gastroenterol Hepatol* **5**, 205–213.

Hugenholtz, P., Goebel, B. M. & Pace, N. R. (1998). Impact of culture-independent studies on the emerging phylogenetic view of bacterial diversity. *J Bacteriol* **180**, 4765–4774.

Hungate, R. E. (1966). *The Rumen and its Microbes*. New York: Academic Press.

Jernberg, C., Sullivan, A., Edlund, C. & Jansson, J. K. (2005). Monitoring of antibiotic-induced alterations in the human intestinal microflora and detection of probiotic strains by use of terminal restriction fragment length polymorphism. *Appl Environ Microbiol* **71**, 501–506.

Kleessen, B., Sykura, B., Zunft, H. J. & Blaut, M. (1997). Effects of inulin and lactose on fecal microflora, microbial activity, and bowel habit in elderly constipated persons. *Am J Clin Nutr* **65**, 1397–1402.

Latham, M. J. & Wolin, M. J. (1977). Fermentation of cellulose by *Ruminococcus flavefaciens* in the presence and absence of *Methanobacterium ruminantium*. *Appl Environ Microbiol* **34**, 297–301.

Lay, C., Rigottier-Gois, L., Holmstrøm, K. & 8 other authors (2005). Colonic microbiota signatures across five northern European countries. *Appl Environ Microbiol* **71**, 4153–4155.

Lepage, P., Seksik, P., Sutren, M., de la Cochetiere, M. F., Jian, R., Marteau, P. & Doré, J. (2005). Biodiversity of the mucosa-associated microbiota is stable along the distal digestive tract in healthy individuals and patients with IBD. *Inflamm Bowel Dis* **11**, 473–480.

Leser, T. D., Amenuvor, J. Z., Jensen, T. K., Lindecrona, R. H., Boye, M. & Møller, K. (2002). Culture-independent analysis of gut bacteria: the pig gastrointestinal tract microbiota revisited. *Appl Environ Microbiol* **68**, 673–690.

Levitt, M. D., Gibson, G. R. & Christl, S. U. (1995). Gas metabolism in the large intestine. In *Human Colonic Bacteria: Role in Nutrition, Physiology and Pathology*, pp. 131–149. Edited by G. R. Gibson & G. T. Macfarlane. Boca Raton, FL: CRC Press.

Ley, R. E., Backhed, F., Turnbaugh, P., Lozupone, C. A., Knight, R. D. & Gordon, J. I. (2005). Obesity alters gut microbial ecology. *Proc Natl Acad Sci U S A* **102**, 11070–11075.

Lou, J., Dawson, K. & Strobel, H. J. (1996). Role of phosphorolytic cleavage in cellobiose and cellodextrin metabolism by the ruminal bacterium *Prevotella ruminicola*. *Appl Environ Microbiol* **62**, 1770–1773.

Louis, P., Duncan, S. H., McCrae, S. I., Millar, J., Jackson, M. S. & Flint, H. J. (2004). Restricted distribution of the butyrate kinase pathway among butyrate-producing bacteria from the human colon. *J Bacteriol* **186**, 2099–2106.

Macfarlane, G. T. & Gibson, G. R. (1995a). Carbohydrate fermentation, energy transduction and gas metabolism in the human large intestine. In *Gastrointestinal Microbiology*, vol. 1, pp. 269–318. Edited by R. I. Mackie & B. A. White. New York: Chapman & Hall.

Macfarlane, G. T. & Gibson, G. R. (1995b). Microbiological aspects of the production of short chain fatty acids in the large bowel. In *Physiological and Clinical Aspects of Short-Chain Fatty Acids*, pp. 87–105. Edited by J. H. Cummings, J. L. Rombeau & T. Sakata. Cambridge: Cambridge University Press.

Macfarlane, G. T., Gibson, G. R. & Cummings, J. H. (1992). Comparison of fermentation reactions in different regions of the human colon. *J Appl Bacteriol* **72**, 57–64.

Magee, E. A., Richardson, C. J., Hughes, R. & Cummings, J. H. (2000). Contribution of dietary protein to sulfide production in the large intestine: an in vitro and a controlled feeding study in humans. *Am J Clin Nutr* **72**, 1488–1494.

Malinen, E., Rinttila, T., Kajander, K., Matto, J., Kassinen, A., Krogius, L., Saarela, M., Korpela, R. & Palva, A. (2005). Analysis of the fecal microbiota of irritable bowel syndrome patients and healthy controls with real-time PCR. *Am J Gastroenterol* **100**, 373–382.

Mangin, I., Bonnet, R., Seksik, P. & 8 other authors (2004). Molecular inventory of faecal microflora in patients with Crohn's disease. *FEMS Microbiol Ecol* **50**, 25–36.

Manichanh, C., Rigottier-Gois, L., Bonnaud, E. & 9 other authors (2006). Reduced diversity of faecal microbiota in Crohn's disease revealed by a metagenomic approach. *Gut* **55**, 205–211. Published online ahead of print on 27 September 2005 as doi:10.1136/gut.2005.073817

Mariadason, J. M., Corner, G. A. & Augenlicht, L. H. (2000). Genetic reprogramming in pathways of colonic cell maturation induced by short chain fatty acids: comparison with trichostatin A, sulindac, and curcurmin and implications for chemoprevention of colon cancer. *Cancer Res* **60**, 4561–4572.

Marteau, P., Pochart, P., Doré, J., Béra-Maillet, C., Bernalier, A. & Corthier, G. (2001). Comparative study of bacterial groups within the human cecal and fecal microbiota. *Appl Environ Microbiol* **67**, 4939–4942.

McIntyre, A., Gibson, P. R. & Young, G. P. (1993). Butyrate production from dietary fibre and protection against large bowel cancer in a rat model. *Gut* **34**, 386–391.

McWilliam Leitch, E. C. & Stewart, C. S. (2002). *Escherichia coli* O157 and non-O157 isolates are more susceptible to L-lactate than to D-lactate. *Appl Environ Microbiol* **68**, 4676–4678.

Miller, T. L. & Wolin, M. J. (1996). Pathways of acetate, propionate, and butyrate formation by the human fecal microbial flora. *Appl Environ Microbiol* **62**, 1589–1592.

Miyazaki, K., Martin, J. C., Marinsek-Logar, R. & Flint, H. J. (1997). Degradation and utilization of xylans by the rumen anaerobe *Prevotella bryantii* (formerly *P. ruminicola* subsp. *brevis*) $B_1$4. *Anaerobe* **3**, 373–381.

Moore, W. E. C. & Moore, L. H. (1995). Intestinal floras of populations that have a high risk of colon cancer. *Appl Environ Microbiol* **61**, 3202–3207.

O'May, G. A., Reynolds, N. & Macfarlane, G. T. (2005). Effect of pH on an *in vitro* model of gastric microbiota in enteral nutrition patients. *Appl Environ Microbiol* **71**, 4777–4783.

Perrin, P., Pierre, F., Patry, Y., Champ, M., Berreur, M., Pradal, G., Bornet, F., Meflah, K. & Menanteau, J. (2001). Only fibres promoting a stable butyrate producing colonic ecosystem decrease the rate of aberrant crypt foci in rats. *Gut* **48**, 53–61.

Pryde, S. E., Richardson, A. J., Stewart, C. S. & Flint, H. J. (1999). Molecular analysis of the microbial diversity present in the colonic wall, colonic lumen, and cecal lumen of a pig. *Appl Environ Microbiol* **65**, 5372–5377.

Pryde, S. E., Duncan, S. H., Hold, G. L., Stewart, C. S. & Flint, H. J. (2002). The microbiology of butyrate formation in the human colon. *FEMS Microbiol Lett* **217**, 133–139.

Radajewski, S., Ineson, P., Parekh, N. T. R. & Murrell, J. C. (2000). Stable-isotope probing as a tool in microbial ecology. *Nature* **403**, 646–649.

Rakoff-Nahoum, S., Pagino, J., Eslami-Varzaneh, F., Edberg, S. & Medzhitov, R. (2004). Recognition of commensal microflora by toll-like receptors is required for intestinal homeostasis. *Cell* **118**, 229–241.

Ramšak, A., Peterka, M., Tajima, K., Martin, J. C., Wood, J., Johnston, M. E. A., Aminov, R. I., Flint, H. J. & Avguštin, G. (2000). Unravelling the genetic diversity of ruminal bacteria belonging to the CFB phylum. *FEMS Microbiol Ecol* **33**, 69–79.

Reeves, A. R., Wang, G. R. & Salyers, A. A. (1997). Characterization of four outer membrane proteins that play a role in utilization of starch by *Bacteroides thetaiotaomicron*. *J Bacteriol* **179**, 643–649.

Rincon, M. T., Ding, S.-Y., McCrae, S. I., Martin, J. C., Aurilia, V., Lamed, R., Shoham, Y., Bayer, E. A. & Flint, H. J. (2003). Novel organization and divergent dockerin specificities in the cellulosome system of *Ruminococcus flavefaciens*. *J Bacteriol* **185**, 703–713.

Rincon, M. T., Cepelnik, T., Martin, J. C., Lamed, R., Barak, Y., Bayer, E. A. & Flint, H. J. (2005). Unconventional mode of attachment of the *Ruminococcus flavefaciens* cellulosome to the cell surface. *J Bacteriol* **187**, 7569–7578.

Robert, C. & Bernalier-Donadille, A. (2003). The cellulolytic microflora of the human colon: evidence of microcrystalline cellulose-degrading bacteria in methane-excreting subjects. *FEMS Microbiol Ecol* **46**, 81–89.

Salyers, A. A., Vercellotti, J. R., West, S. E. H. & Wilkins, T. D. (1977a). Fermentation of mucin and plant polysaccharides by strains of *Bacteroides* from the human colon. *Appl Environ Microbiol* **33**, 319–322.

Salyers, A. A., West, S. E. H., Vercellotti, J. R. & Wilkins, T. D. (1977b). Fermentation of mucins and plant polysaccharides by anaerobic bacteria from the human colon. *Appl Environ Microbiol* **34**, 529–533.

Saunders, J. R. (1995). Population genetics of phase variable antigens. In *Population Genetics of Bacteria* (Society for General Microbiology Symposium no. 52), pp. 247–268. Edited by S. Baumberg, J. P. W. Young, E. M. H. Wellington & J. R. Saunders. Cambridge: Cambridge University Press.

Schell, M. A., Karmirantzou, M., Snel, B. & 9 other authors (2002). The genome sequence of *Bifidobacterium longum* reflects its adaptation to the human gastrointestinal tract. *Proc Natl Acad Sci U S A* **99**, 14422–14427.

Schink, B. (1992). Syntrophism among prokaryotes. In *The Prokaryotes*, 2nd edn, vol. 1, pp. 276–299. Edited by A. Balows, H. G. Trüper, M. Dworkin, W. Harder & K. H. Schleifer. New York: Springer.

Schneider, H., Schwiertz, A., Collins, M. D. & Blaut, M. (1999). Anaerobic transformation of quercetin-3-glucoside by bacteria from the human intestinal tact. *Arch Microbiol* **171**, 81–91.

Schwiertz, A., Hold, G. L., Duncan, S. H., Gruhl, B., Collins, M. D., Lawson, P. A., Flint, H. J. & Blaut, M. (2002a). *Anaerostipes caccae* gen. nov., sp. nov., a new saccharolytic, acetate-utilising, butyrate-producing bacterium from human faeces. *Syst Appl Microbiol* **25**, 46–51.

Schwiertz, A., Lehmann, U., Jacobasch, G. & Blaut, M. (2002b). Influence of resistant starch on the SCFA production and cell counts of butyrate-producing *Eubacterium* spp. in the human intestine. *J Appl Bacteriol* **93**, 157–162.

Shipman, J. A., Cho, K. H., Siegel, H. A. & Salyers, A. A. (1999). Physiological characterization of SusG, an outer membrane protein essential for starch utilization by *Bacteroides thetaiotaomicron*. *J Bacteriol* **181**, 7206–7211.

Simmering, R., Pforte, H., Jacobasch, G. & Blaut, M. (2002). The growth of the flavonoid-degrading intestinal bacterium, *Eubacterium ramulus*, is stimulated by dietary flavonoids in vivo. *FEMS Microbiol Ecol* **40**, 243–248.

Sperandio, V., Torres, A. G., Jarvis, B., Nataro, J. P. & Kaper, J. B. (2003). Bacteria-host communication: the language of hormones. *Proc Natl Acad Sci U S A* **100**, 8951–8956.

Suau, A., Bonnet, R., Sutren, M., Godon, J.-J., Gibson, G. R., Collins, M. D. & Doré, J. (1999). Direct analysis of genes encoding 16S rRNA from complex communities reveals many novel molecular species within the human gut. *Appl Environ Microbiol* **65**, 4799–4807.

Swidsinski, A., Weber, J., Loening-Baucke, V., Hale, L. P. & Lochs, H. (2005). Spatial organization and composition of the mucosal flora in patients with inflammatory bowel disease. *J Clin Microbiol* **43**, 3380–3389.

Tajima, K., Aminov, R. I., Nagamine, T., Ogata, K., Nakamura, M., Matsui, H. & Benno, Y. (1999). Rumen bacterial diversity as determined by sequence analysis of 16S rDNA libraries. *FEMS Microbiol Ecol* **29**, 159–169.

Topping, D. L. & Clifton, P. M. (2001). Short-chain fatty acids and human colonic function: roles of resistant starch and nonstarch polysaccharides. *Physiol Rev* **81**, 1031–1064.

Vernia, P., Caprilli, R., Latella, G., Barbetti, F., Magliocca, F. M. & Cittadini, M. (1988). Fecal lactate and ulcerative colitis. *Gastroenterology* **95**, 1564–1568.

Wachtershauser, A. & Stein, J. (2000). Rationale for the luminal provision of butyrate in intestinal diseases. *Eur J Nutr* **39**, 164–171.

Walker, A. W., Duncan, S. H., McWilliam Leitch, E. C., Child, M. W. & Flint, H. J. (2005). pH and peptide supply can radically alter bacterial populations and short-chain fatty acid ratios within microbial communities from the human colon. *Appl Environ Microbiol* **71**, 3692–3700.

Wang, M., Ahrne, S., Antonsson, M. & Molin, G. (2004). T-RFLP combined with principal component analysis and 16S rRNA gene sequencing: an effective strategy for comparison of fecal microbiota in infants of different ages. *J Microbiol Methods* **59**, 53–69.

Wang, M., Ahrné, S., Jeppsson, B. & Molin, G. (2005). Comparison of bacterial diversity along the human intestinal tract by direct cloning and sequencing of 16S rRNA genes. *FEMS Microbiol Ecol* **54**, 219–231.

Whitford, M. F., Forster, R. J., Beard, C. E., Gong, J. & Teather, R. M. (1998). Phylogenetic analysis of rumen bacteria by comparative sequence analysis of cloned 16S rRNA genes. *Anaerobe* **4**, 153–163.

Whitman, W. B., Coleman, D. C. & Wiebe, W. J. (1998). Prokaryotes: the unseen majority. *Proc Natl Acad Sci U S A* **95**, 6578–6583.

Wilson, K. H. & Blitchington, R. B. (1996). Human colonic biota studied by ribosomal DNA sequence analysis. *Appl Environ Microbiol* **62**, 2273–2278.

Xu, J., Bjursell, M. K., Himrod, J., Deng, S., Carmichael, L. K., Chiang, H. C., Hooper, L. V. & Gordon, J. I. (2003). A genomic view of the human-*Bacteroides thetaiotaomicron* symbiosis. *Science* **299**, 2074–2076.

Young, V. B. & Schmidt, T. M. (2004). Antibiotic-associated diarrhea accompanied by large-scale alterations in the composition of the fecal microbiota. *J Clin Microbiol* **42**, 1203–1206.

Zoetendal, E. G., Akkermans, A. D. L., Akkermans-van Vliet, W. M., de Visser, J. A. G. M. & de Vos, W. M. (2001). The host genotype affects the bacterial community in the human gastrointestinal tract. *Microb Ecol Health Dis* **13**, 129–134.

Zoetendal, E. G., von Wright, A., Vilpponen-Salmela, T., Ben-Amor, K., Akkermans, A. D. L. & de Vos, W. M. (2002a). Mucosa-associated bacteria in the human gastrointestinal tract are uniformly distributed along the colon and differ from the community recovered from feces. *Appl Environ Microbiol* **68**, 3401–3407.

Zoetendal, E. G., Ben-Amor, K., Harmsen, H. J. M., Schut, F., Akkermans, A. D. L. & de Vos, W. M. (2002b). Quantification of uncultured *Ruminococcus obeum*-like bacteria in human faecal samples by fluorescent in situ hybridization and flow cytometry using 16S rRNA-targeted probes. *Appl Environ Microbiol* **68**, 4225–4232.

Zoetendal, E. G., Collier, C. T., Koike, S., Mackie, R. I. & Gaskins, H. R. (2004). Molecular ecological analysis of the gastrointestinal microbiota: a review. *J Nutr* **134**, 465–472.

The genetics of phenotypic innovation

Hubertus J. E. Beaumont,[1] Stefanie M. Gehrig,[2] Rees Kassen,[3] Christopher G. Knight,[4] Jacob Malone,[5] Andrew J. Spiers[2] and Paul B. Rainey[1,2]

[1]School of Biological Sciences, University of Auckland, Private Bag 92019, Auckland, New Zealand

[2]Department of Plant Sciences, University of Oxford, South Parks Road, Oxford OX1 3RB, UK

[3]Department of Biology and Centre for Advanced Research in Environmental Genomics, University of Ottawa, 150 Louis Pasteur, Ottawa, ON, K1N 6N5, Canada

[4]School of Chemistry, University of Manchester, Faraday Building, Box 88, Sackville St, Manchester M60 1QD, UK

[5]Division of Molecular Microbiology, Biozentrum, University of Basel, Klingelbergstrasse 70, CH-4056 Basel, Switzerland

EVOLUTIONARY EMERGENCE OF DIVERSITY

The majority of phenotypic and ecological diversity on the planet has arisen during successive adaptive radiations, that is, periods in which a single lineage diverges rapidly to generate multiple niche-specialist types. Microbiologists tend not to think of bacteria as undergoing adaptive radiation, but there is no reason to exclude them from this general statement – in fact, rapid generation times and large population sizes suggest that bacteria may be particularly prone to bouts of rapid ecological diversification. Indeed, there is evidence from both experimental bacterial populations (Korona *et al.*, 1994; Rainey & Travisano, 1998) and natural populations (Stahl *et al.*, 2002). This being so, insight into the evolutionary emergence of diversity requires an understanding of the causes of adaptive radiation.

The causes of adaptive radiation are many and complex, but at a fundamental level there are just two: one genetic and the other ecological. Put simply, heritable phenotypic variation arises primarily by mutation, while selection working via various ecological processes shapes this variation into the patterns of phenotypic diversity evident in the world around us.

The ecological causes of adaptive radiation are embodied in theory that stems largely from Darwin's insights into the workings of evolutionary change (Darwin, 1890), but owes much to developments in the 1940s and 1950s attributable to Lack (1947), Dobzhansky (1951) and Simpson (1953). Recent work has seen a reformulation of the primary concepts (Schluter, 2000).

SGM symposium 66: Prokaryotic diversity: mechanisms and significance.
Editors N. A. Logan, H. M. Lappin-Scott & P. C. F. Oyston. Cambridge University Press. ISBN 0 521 86935 8

At its most basic, the ecological theory of adaptive radiation postulates that ecological opportunity (vacant niche space) and competition are necessary conditions for the emergence and maintenance of diversity. Imagine a single lineage, of any given species, in a pristine environment replete with ecological opportunity such that the environment provides alternative 'fitness peaks'. The population grows geometrically until the primary resource becomes limiting, at which point competition – the so-called 'engine' of adaptive radiation – becomes a significant factor. Variant types (arising by random mutation) are driven by competition to exploit new resource types where they are subject to different selective conditions. Under this scenario, the genotypes most favoured by selection are those that occupy niches different from those inhabited by the dominant type (species or genotype). Continual exposure to divergent natural selection promotes further divergence in phenotype leading, if unrestrained, to ecological speciation.

It is not our intention here to discuss the ecological causes of diversity further: this has been the subject of recent reviews (Kassen & Rainey, 2004; Rainey *et al.*, 2000, 2005) and the focus of much experimental evolutionary analysis (Rainey & Travisano, 1998; Buckling *et al.*, 2000; Kassen *et al.*, 2000, 2004; Travisano & Rainey, 2000; Buckling & Rainey, 2002a, b; Hodgson *et al.*, 2002; Rainey & Rainey, 2003). Instead, we wish to consider the genetical causes of new phenotypes – in particular, we wish to draw attention to the relationship between the molecular architecture of phenotypes (i.e. the molecular genetic and structural basis of a phenotype) and the implications of these architectures for the evolution of new phenotypes.

GENETICS OF ADAPTATION

As surprising as it may seem at first, there is no well-developed genetical theory of phenotypic evolution – at least not one that parallels the ecological theory of adaptive radiation. Some have argued that such a theory is unnecessary given our understanding of the ultimate causes of variation (mutation, recombination and migration) and the forces that determine the patterns of diversity (natural selection and genetic drift); however, there is growing awareness that understanding of these microevolutionary processes alone is insufficient to provide a comprehensive explanation for the origins of new phenotypes – the stuff of evolution. What is needed is a means of understanding how genotype, phenotype and fitness are connected through evolutionary time: in short, what is necessary is a genetical theory of evolutionary development.

At the core of the ensuing discussion is the process of adaptive evolution: the movement of a population towards a phenotype that is better suited to the prevailing environmental conditions. A good deal is known about natural selection, but we don't have a good understanding of the genetic bases of adaptation – of the kinds of mutations that generate adaptive phenotypes, whether these mutations are typically pre-existing in populations or whether they are generated *de novo*, whether adaptations arise by single or multiple mutations and whether mutations have small or large phenotypic effects. At an altogether different level, increased understanding of the molecular mechanisms that underlie life leads to questions concerning the organization of systems: whether there might be certain kinds of genes, genetic organizations or cellular architectures that increase the likelihood that mutations produce phenotypically viable solutions.

Allen Orr has been instrumental in reawakening the need to work toward a more complete theoretical understanding of adaptive evolution, and his review articles provide excellent coverage of both the history of the field and its current status (see Orr, 2005a, b). These will not be recounted here, other than to point out that the basic ideas extend back to Darwin and encompass some of biology's most influential thinkers: R. A. Fisher, M. Kimura and J. Maynard Smith. Prevalent among the basic ideas has been a strong sense of micromutationism, that is, the argument that evolutionary change through mutation and selection is a gradual process that acts on slight changes in successive generations. More recently, molecular data on the genetics of adaptation have become available – often (but not exclusively) from studies in experimental microbial evolution (see Elena & Lenski, 2003) – indicating that small mutations in 'major genes' can have major phenotypic effects. Very often these major genes have a regulatory role. In studies with virus populations, the first beneficial mutations fixed by selection as a population moves to become better adapted to its environment are often mutations with large phenotypic effects (Bull *et al.*, 1997; Burch & Chao, 1999; Wichman *et al.*, 1999; Rokyta *et al.*, 2005). All of this goes against the grain of strict micromutationism and has led to attempts to revise and extend theory so that it speaks in the same terms as the data. Progress to date has led to models that predict that the size of favourable mutations fixed by selection declines as a geometric progression (Barton, 1998; Fisher, 1930; Orr, 1998). Although these models represent a step forward in our understanding of the probabilistic nature of the phenotypic variation caused by random mutations, they say nothing about the underlying mechanisms, their effects on the dynamics of evolution and the kinds of changes expected in evolving populations. The challenge is to understand the connection between DNA sequence, phenotype and fitness and, most importantly, the relationship between the molecular architecture of an organism and the mechanisms by which changes at the DNA sequence level translate through to phenotypically useful solutions.

THE EVOLUTIONARY ORIGINS OF NEW PHENOTYPES

This is an already large but still burgeoning field: attempting to cover the entire topic – even superficially – is well beyond the scope of a single chapter. Here we focus on a single aspect, namely, the relationship between the underlying genetic architecture of an organism and the evolutionary emergence of new phenotypes. We have chosen to do so for several reasons: firstly, our own attempts to explain the evolutionary origins of new phenotypes continue to draw attention to the importance of genetic architecture; secondly, microbiologists routinely obtain data from DNA array, proteomic and metabolomic analyses that provide a glimpse of the molecular networks of organisms (genetic architecture); and thirdly, because there is a growing realization that specific wirings among the components of these cellular networks both facilitate and constrain the evolution of phenotypic novelty.

The origin of new phenotypes through evolution might, on first consideration, appear a relatively trivial issue: simply a consequence of a chance causal mutation, a specific gene acquisition event, a specific gene duplication event or similar. Although this is true in principle, this explanation for the origins of phenotypic novelty is naive in the extreme. The phenotypic consequences of random mutations in the DNA of an organism are in fact not random, but are strongly biased by the structure of organisms; certain phenotypic innovations are more likely to arise through random mutation than others. This concept is illustrated by the evolution of propanediol reductase in *Escherichia coli*. Wild-type *E. coli* is incapable of aerobic growth with propanediol as a sole carbon and energy source; however, mutation of a gene encoding lactaldehyde reductase, an enzyme normally involved in anaerobic fermentative growth on the carbon and energy source fucose, can permit growth on this novel substrate (Mortlock, 1982; Lin & Wu, 1984). In this instance, random genetic change is responsible for the evolution of a new metabolic capability, but the likelihood that this capacity emerged owed much to the fact that the genome already encoded a not too dissimilar gene (lactaldehyde reductase).

In the above example, the evolution of a novel phenotype depended largely on the gene complement of the cell. The problem for the evolution of novel phenotypes, however, is more complex and is made so by the fact that organisms are far more than a set of independently acting genes with a simple one-to-one relationship between gene and trait. Put another way, there are the interrelated problems of pleiotropy (especially antagonistic pleiotropy) and epistasis: the vast majority of potentially beneficial mutations have accompanying deleterious effects. It follows, then, that for a mutation to be favoured by natural selection the beneficial effects must outweigh the detrimental effects. Whether this is so depends largely on the network structure of connections between the components (e.g. enzymes, transcription factors, tissues, organs) that make an organism; it also depends on the number of traits under selection (Orr, 2000).

NETWORKS, POWER-LAW DISTRIBUTIONS AND ROBUSTNESS

The interconnectedness of components that make up cells – be they prokaryotic or eukaryotic – has been known in a formal sense since the early days of metabolic pathway analysis. More recently, with advances in large-scale 'omic' technologies, it has become possible to characterize protein–protein interaction networks, metabolic networks, gene regulation networks and evolutionary protein-domain networks with increasing scope and precision. This has provided a rejuvenated and more comprehensive view of living cells as an ensemble of interacting molecular components. In turn, these insights have led to *systems biology* – a 'new' research programme that seeks a deeper understanding of life by studying how higher-order phenomena result from the interactions between lower-level components (Csete & Doyle, 2002; Kitano, 2002).

At present, understanding of the functioning of networks is still relatively limited. One of the main reasons is lack of information on the functional characteristics (e.g. activity, binding coefficient, stability, etc.) of the component parts *in vivo*. Nonetheless, progress has been made in understanding how phenotypes of real organisms are realized through molecular networks and even in predicting evolutionary change on the basis of knowledge about the molecular structure of cells (Ibarra *et al*., 2002).

A typical network analysis involves plotting the basic components of the system (which might range from genes to proteins through to interacting species in a community) as nodes on a graph connected by edges (arrows) that are interactions. An understanding of the architecture of the network is then obtained using the results of graph theory, which allows a network to be quantified in a few simple statistics: for example, the number of connections per node, the average distance between nodes and the degree to which neighbours of nodes are connected. Arising from such analyses has been the surprising realization that complex biological networks are not random: in almost all cases, the interactions form a network where the frequency distribution of interactions among nodes follows that of a 'power-law' distribution, that is, there are a small number of nodes that are highly connected to other nodes (hubs) and a large number of nodes that are connected to just a single node. Such an arrangement results in a 'scale-free' topology (Proulx *et al*., 2005). The classic example of such a network is the so-called 'small world phenomenon', which hypothesizes that most people in the world are acquainted via a small number of acquaintances; the World Wide Web is a further example (Albert *et al*., 1999).

Discovery that the power-law distribution is an apparently universal feature of networks has led some to suggest that there is a universal law of life (Wolf *et al*., 2002) maintained by the 'forces of self-organization' (Barabasi, 2002) [but see Keller (2005) for an alternative view]. In the context of certain networks, e.g. protein-interaction networks, this is an appealing idea that seems to fit with intuitive expectations. For example, highly connected proteins show low rates of evolution (Fraser *et al.*, 2002), are more likely to be essential for survival (Jeong *et al.*, 2001) and less likely to be lost over evolutionary time (Krylov *et al.*, 2003) and are more pleiotropic (Promislow, 2004) than are weakly connected proteins. Similarly, for metabolic pathways and gene-regulatory networks, conformity to the power-law distribution gives the appearance of being 'fit for purpose'. For example, in *E. coli* six known central regulators including FIS, FNR and HNS regulate half of the known *E. coli* genes (Martinez-Antonio & Collado-Vides, 2003).

A consequence of networks with a power-law distribution of connections is that they are robust to mutational change. If most nodes are connected to just one other node then the random removal of a node is unlikely to have large effects (Jeong *et al.*, 2001) – of course, removal of a central hub would have a catastrophic effect and thus scale-free networks are susceptible to targeted attack. The biological significance of robustness has been much championed (see for example Wagner, 2005) and has even led some to argue that selection has favoured the evolution of scale-free networks because such networks provide a buffering against genetic and or environmental change. Indeed, many have noted the appealing connections between robustness and long-standing ideas surrounding genetic canalization and homeostasis (Waddington, 1942; Maynard Smith *et al.*, 1985).

Robustness is more than just a high-level emergent property of the networks that comprise living systems: it is apparent at different levels of biological organization, ranging from the positive and negative control systems that are commonly found in gene circuits through to redundancy of components (as found in alternative metabolic pathways of bacteria) and modularity (where subsystems are insulated so that failure in one module does not lead to a system-wide failure). From an evolutionary perspective, mechanisms that buffer against mutational change are likely to limit the capacity of a population of organisms to adapt (genetically by natural selection) to new environments. Clearly there is a tension: organisms are faced with the challenge of keeping pace with changing environments. If mutational robustness is the dominant feature of organisms, evolution would struggle to produce new phenotypes: populations would find it difficult to keep pace with the environment and it is hard to imagine how life could persist. It seems reasonable therefore to expect a balance between robustness and mechanisms that facilitate the evolution of phenotypic novelty. Indeed, as the next section shows, modular systems can be both robust and evolvable.

MODULARITY AND THE EVOLUTION OF PHENOTYPIC NOVELTY

Arguably the most important concept in terms of organizations that might facilitate evolutionary change is modularity: the bundling of features of a living system such that a change to one part should not disrupt the whole system (Barton & Partridge, 2000).

A characteristic feature of all organisms, and particularly notable in bacteria, is a modular organization: the tendency for groups of genes to interact in such a way as to limit the extent of pleiotropic effects among characters belonging to different functional complexes. This means that changes that occur within one complex have less chance of impacting negatively on others – there is robustness. Striking evidence of both modularity and robustness stems from analysis of the phenotypic consequences of individually deleting all two-component regulators from the *E. coli* genome (Oshima *et al.*, 2002; Zhou *et al.*, 2003). In that study, nearly half of the deletions had no discernible phenotypic effect even when measured across hundreds of different environmental conditions. Modular structures possess additional features that are amenable to the generation of phenotypic variability. In addition to pleiotropic robustness, the organization of organisms into parts means that the parts function as building blocks that can be reused in various combinations to increase the probability of generating viable novel (and potentially useful) phenotypic variation (Kirschner & Gerhart, 1998). Finally, although pleiotropy is often viewed as limiting the evolution of novel phenotypes, the opposite can also hold – pleiotropy can allow mutations of small genetic size to have large phenotypic effects, some of which can be beneficial (Crozat *et al.*, 2005).

Bacterial signal transduction cascades provide a good example of modular structures (Hoch & Silhavy, 1995; Reizer & Saier, 1997; Patthy, 2003). On the one hand, regulatory components usually show high fidelity and regulate a small number of structural genes, which means that changes in one regulatory module are less likely to impact negatively on other regulatory modules (Oshima *et al.*, 2002; Zhou *et al.*, 2003). In addition, the regulatory components exhibit 'genetic' modularity which confers properties relevant to evolutionary change (Kirschner & Gerhart, 1998). These properties include versatile protein elements and weak linkage among components. Versatility among protein elements is particularly common in protein kinases, where minor mutational modification can alter the specific target of activity or affect the timing of activation (Hoch & Silhavy, 1995). Weak linkage is also a feature of signal transduction pathways: components have switch-like properties and signals act to release the activity, but do not act instructively. Such regulatory organization facilitates a component's accommodation to novelty and reduces the cost of generating variation (for more information see Gerhart & Kirschner, 1997; Kirschner & Gerhart, 1998).

THE ORIGINS OF ROBUSTNESS AND MODULARITY

The evolutionary origins of robustness and modularity are of considerable interest: the central issue being the role of natural selection. Much has been written on the subject (e.g. Gerhart & Kirschner, 1997; Schlosser & Wagner, 2004; Wagner, 2005), but still the origins of robustness and modularity remain obscure (Wagner & Altenberg, 1996; Kirschner & Gerhart, 1998; Lipson *et al.*, 2002; Gardner & Zuidema, 2003; Rainey & Cooper, 2004).

Perhaps the most compelling argument in favour of an adaptive explanation for robustness and modularity comes from demand theory (Savageau, 1998), a design principle that accounts for natural selection in the evolution and optimization of modular gene circuits (see Wall *et al.*, 2004). Demand theory posits that repressors regulate frequently needed genes, whereas activators control genes needed infrequently. This configuration means that mutations that abolish regulation maintain gene expression for high-demand genes, but turn off low-demand genes. Careful analysis of the design principles of gene circuits shows that those that conform to the demand theory principle are robust to mutational change, yet evolvable, such that a large fraction of possible mutations lead to small quantitative changes in behaviour, but a smaller fraction change the wiring altogether (Wall *et al.*, 2004). Several model gene circuits have been examined in detail in order to elucidate the combined properties of robustness and evolvability. Gene circuits involving negative regulation (e.g. the *lac* operon of *E. coli* and the lambda phage lysis/lysogeny circuit) have been shown to be both robust and evolvable (Ozbudak *et al.*, 2004; Wall *et al.*, 2004). In the case of phage lambda, the circuitry is such that a wide range of simple mutations can readily tune the sensitivity and cooperativity of the regulatory switch (Little *et al.*, 1999). Similar circuitry (with similar attributes) is evident in the *sin* operon of *Bacillus subtilis* (Voigt *et al.*, 2005), leading to the suggestion that the 'dynamic plasticity' of negative regulation may be a motif that has been favoured by selection because of its inherent evolvability (Voigt *et al.*, 2005).

Computational studies have provided further insight into the evolution of modularity. A recent study showed the *de novo* evolution of modularity and network motifs, but only under conditions where the evolving system was required to adapt to a modular environment (an environment that switches regularly between two states) (Kashtan & Alon, 2005). Evident in the modular organization was a striking capacity of the system to undergo further evolution; indeed, the simplest of possible rewirings was all that was necessary for the network to optimize its response to a new environment.

STUDIES WITH EXPERIMENTAL *PSEUDOMONAS* POPULATIONS PROVIDE INSIGHT INTO THE EVOLUTION OF PHENOTYPIC NOVELTY

Our own experimental work has focused on the genetic origins of phenotypic innovations in simple populations of *Pseudomonas fluorescens* that evolve in static laboratory microcosms. The phenotypic innovation that has most preoccupied us is the 'wrinkly spreader' (WS), a niche-specialist genotype that forms a mat at the air–liquid interface of liquid cultures and grows poorly in the liquid column (Rainey & Travisano, 1998). WS genotypes arise by spontaneous mutation from the ancestral (smooth; SM) – non-mat-forming – *P. fluorescens* genotype and show a significant negative frequency-dependent fitness advantage over the ancestral strain (Rainey & Travisano, 1998). Its selective advantage is attributable to cooperation among individual WS cells: overproduction of attachment factors, while costly to individual cells, results in the interests of individuals aligning with those of the group and allows colonization of the oxygen-replete air–liquid interface (Rainey & Rainey, 2003).

WS is not a single genetic or phenotypic entity: there is a broad swathe of phenotypically diverse types of WS indicative of multiple mutational routes to the adaptive phenotype. Of central importance to all WS genotypes is the *wss* operon, a set of 10 genes that together encode the enzymes necessary for the biosynthesis of an acetylated cellulose polymer (Spiers *et al.*, 2002, 2003; Spiers & Rainey, 2005). Overproduction of the polymer is a primary cause of the WS phenotype and is brought about not by enhanced levels of transcription of the *wss* operon (Spiers *et al.*, 2002) but by overactivation of a diguanylate cyclase (DGC), which causes overproduction of cyclic-di-GMP (P. J. Goymer, S. G. Kahn, J. G. Malone, S. M. Gehrig, A. J. Spiers and P. B. Rainey, manuscript submitted) – a secondary messenger and known allosteric activator of the cellulose biosynthetic enzymes (Ross *et al.*, 1987; Tal *et al.*, 1998). Differences among WS morphs are largely attributable to differences in the activity levels of the DGC (P. J. Goymer, S. G. Kahn, J. G. Malone, S. M. Gehrig, A. J. Spiers and P. B. Rainey, manuscript submitted).

So far, two mutational routes to WS have been uncovered (P. B. Rainey, E. Bantinaki, Z. Robinson, R. Kassen, C. G. Knight and A. J. Spiers, manuscript submitted; S. M. Gehrig, A. J. Spiers and P. B. Rainey, unpublished). Both involve simple mutational changes (mostly transitions, transversions or small deletions) within components of signal transduction pathways that control the activity of different DGCs. Most significantly, in both cases, the causal mutations reside within negative regulators. In one of the negative regulators, 15 independent causal mutations have been identified – remarkably, each mutation has a different phenotypic effect.

Placing these findings in the context of the connectivities among components of the cellular system, it is probably no coincidence that the two pathways identified so far control the activities of DGCs. Given that the *P. fluorescens* genome has more than 30 DGCs (there is considerable redundancy), it is highly likely that additional causal mutations will be found within other DGC pathways. Whether these will also reside within negative regulatory components is of special interest. Although our understanding of the DGC network of cells is in its infancy (D'Argenio & Miller, 2004; Romling *et al.*, 2005), there is growing evidence that this network forms a distinct post-translational regulatory system at a high level of organization. Each of the mutational routes identified so far defines submodules within this network, and the causal mutations affect components of the signal transduction pathways that act to attenuate the output of the pathway. Loss-of-function mutations in these components therefore cause increased activity of the output modules of the systems, which results in increased levels of cyclic-di-GMP in the cell and increased activity of the enzymes that produce the cellulosic polymer and associated attachment factors. The fact that the pathway exists as an intact and operable module means that pleiotropic effects are likely to be minimal because the changes wrought by mutations in the negative regulators cause alterations in levels of output without wholesale changes in the wiring of connections (Raff, 1996; Stern, 2000).

There are additional features of the negative control systems that regulate the activity of DGCs that are likely to be relevant to understanding the causes of variation within the WS phenotype. Negative regulatory loops are highly evolvable (Csete & Doyle, 2002) such that mutations can easily tune the output status of a given pathway. For example, in the ancestral genotype, the output status (DGC activity) is controlled by an oscillating switch; however, mutations that subtly decrease (without completely abolishing) the activity of the negative regulator tune the switch to different output levels (and to a stable state). Indeed, different mutations can tune the status of the pathway to different output levels, and it is this that is responsible for variation in WS type. Similar insights into the evolutionary significance of negatively regulated systems have come from the targeted analysis of robustness and evolvability in model gene circuits subject to negative regulation [e.g. the *lac* operon of *E. coli* (Ozbudak *et al.*, 2004), the lysis/lysogeny switch in phage lambda (Little *et al.*, 1999) and the *sin* operon of *B. subtilis* (Voigt *et al.*, 2005)].

CONCLUDING COMMENTS

Explaining the origin of new phenotypes – from DNA sequence change through to phenotypic and fitness effects – is a significant problem for biologists. Crucial for progress is an understanding of how cellular systems are wired and how particular wirings act to facilitate or constrain the evolvability of the system. Advances in abilities to study the biology of organisms as complete systems and at different levels of organization are providing new insights. An emerging theme is one of robustness and modularity – the two together seeming to confer on living systems an ability to tolerate mutational and environmental noise, while at the same time increasing the likelihood that a small fraction of mutations will generate potentially useful phenotypes.

There remains much yet to unravel. A major future goal is the development of fully predictive models of phenotypic evolution: models that allow the direction of phenotypic evolution to be predicted in real organisms evolving in real environments. This requires that we understand not only the behaviour of an individual organism at a 'systems' level, but also the rules that govern the wiring of component parts.

REFERENCES

Albert, R., Jeong, H. & Barabasi, A. L. (1999). Internet: diameter of the world-wide web. *Nature* **401**, 130–131.

Barabasi, A. L. (2002). *Linked: the New Science of Networks*. Cambridge: Perseus Publishing.

Barton, N. (1998). The geometry of adaptation. *Nature* **395**, 751–752.

Barton, N. & Partridge, L. (2000). Limits to natural selection. *Bioessays* **22**, 1075–1084.

Buckling, A. & Rainey, P. B. (2002a). The role of parasites in sympatric and allopatric host diversification. *Nature* **420**, 496–499.

Buckling, A. & Rainey, P. B. (2002b). Antagonistic coevolution between a bacterium and a bacteriophage. *Proc R Soc Lond B Biol Sci* **269**, 931–936.

Buckling, A., Kassen, R., Bell, G. & Rainey, P. B. (2000). Disturbance and diversity in experimental microcosms. *Nature* **408**, 961–964.

Bull, J. J., Badgett, M. R., Wichman, H. A., Huelsenbeck, J. P., Hillis, D. M., Gulati, A., Ho, C. & Molineux, I. J. (1997). Exceptional convergent evolution in a virus. *Genetics* **147**, 1497–1507.

Burch, C. L. & Chao, L. (1999). Evolution by small steps and rugged landscapes in the RNA virus ϕ6. *Genetics* **151**, 921–927.

Crozat, E., Philippe, N., Lenski, R. E., Geiselmann, J. & Schneider, D. (2005). Long-term experimental evolution in *Escherichia coli*. XII. DNA topology as a key target of selection. *Genetics* **169**, 523–532.

Csete, M. E. & Doyle, J. C. (2002). Reverse engineering of biological complexity. *Science* **295**, 1664–1669.

D'Argenio, D. A. & Miller, S. I. (2004). Cyclic di-GMP as a bacterial second messenger. *Microbiology* **150**, 2497–2502.

Darwin, C. (1890). *The Origin of Species*, 6th edn. London: John Murray.

Dobzhansky, T. (1951). *Genetics and the Origin of Species*, 3rd edn. New York: Columbia University Press.

Elena, S. F. & Lenski, R. E. (2003). Evolution experiments with microorganisms: the dynamics and genetic bases of adaptation. *Nat Rev Genet* **4**, 457–469.

Fisher, R. A. (1930). *The Genetical Theory of Natural Selection*. Oxford: Oxford University Press.

Fraser, H. B., Hirsh, A. E., Steinmetz, L. M., Scharfe, C. & Feldman, M. W. (2002). Evolutionary rate in the protein interaction network. *Science* **296**, 750–752.

Gardner, A. & Zuidema, W. (2003). Is evolvability involved in the origin of modular variation? *Evolution* **57**, 1448–1450.

Gerhart, J. & Kirschner, M. (1997). *Cells, Embryos, and Evolution*. Malden: Blackwell.

Hoch, J. A. & Silhavy, T. J. (1995). *Two-Component Signal Transduction*. Washington, DC: American Society for Microbiology.

Hodgson, D. J., Rainey, P. B. & Buckling, A. (2002). Mechanisms linking diversity, productivity and invasibility in experimental bacterial communities. *Proc R Soc Lond B Biol Sci* **269**, 2277–2283.

Ibarra, R. U., Edwards, J. S. & Palsson, B. O. (2002). *Escherichia coli* K-12 undergoes adaptive evolution to achieve *in silico* predicted optimal growth. *Nature* **420**, 186–189.

Jeong, H., Mason, S. P., Barabasi, A. L. & Oltvai, Z. N. (2001). Lethality and centrality in protein networks. *Nature* **411**, 41–42.

Kashtan, N. & Alon, U. (2005). Spontaneous evolution of modularity and network motifs. *Proc Natl Acad Sci U S A* **102**, 13773–13778.

Kassen, R. & Rainey, P. B. (2004). The ecology and genetics of microbial diversity. *Annu Rev Microbiol* **58**, 207–231.

Kassen, R., Buckling, A., Bell, G. & Rainey, P. B. (2000). Diversity peaks at intermediate productivity in a laboratory microcosm. *Nature* **406**, 508–512.

Kassen, R., Llewellyn, M. & Rainey, P. B. (2004). Ecological constraints on diversification in a model adaptive radiation. *Nature* **431**, 984–988.

Keller, E. F. (2005). Revisiting "scale-free" networks. *Bioessays* **27**, 1060–1068.

Kirschner, M. & Gerhart, J. (1998). Evolvability. *Proc Natl Acad Sci U S A* **95**, 8420–8427.

Kitano, H. (2002). Systems biology: a brief overview. *Science* **295**, 1662–1664.

Korona, R., Nakatsu, C. H., Forney, L. J. & Lenski, R. E. (1994). Evidence for multiple adaptive peaks from populations of bacteria evolving in a structured habitat. *Proc Natl Acad Sci U S A* **91**, 9037–9041.

Krylov, D. M., Wolf, Y. I., Rogozin, I. B. & Koonin, E. V. (2003). Gene loss, protein sequence divergence, gene dispensability, expression level, and interactivity are correlated in eukaryotic evolution. *Genome Res* **13**, 2229–2235.

Lack, D. (1947). *Darwin's Finches*. Cambridge: Cambridge University Press.

Lin, E. C. C. & Wu, T. T. (1984). Functional divergence of the L-fucose system in mutants of *Escherichia coli*. In *Microorganisms as Model Systems for Studying Evolution*, pp. 135–164. Edited by R. P. Mortlock. New York: Plenum.

Lipson, H., Pollack, J. B. & Suh, N. P. (2002). On the origin of modular variation. *Evolution* **56**, 1549–1556.

Little, J. W., Shepley, D. P. & Wert, D. W. (1999). Robustness of a gene regulatory circuit. *EMBO J* **18**, 4299–4307.

Martinez-Antonio, A. & Collado-Vides, J. (2003). Identifying global regulators in transcriptional regulatory networks in bacteria. *Curr Opin Microbiol* **6**, 482–489.

Maynard Smith, J., Burian, R., Kauffman, S., Alberch, P., Campbell, J., Goodwin, B., Lande, R., Raup, D. & Wolpert, L. (1985). Developmental constraints and evolution. *Q Rev Biol* **60**, 265–287.

Mortlock, R. P. (1982). Metabolic acquisitions through laboratory selection. *Annu Rev Microbiol* **36**, 259–284.

Orr, H. A. (1998). The population genetics of adaptation: the distribution of factors fixed during adaptive evolution. *Evolution* **52**, 935–949.

Orr, H. A. (2000). Adaptation and the cost of complexity. *Evolution* **54**, 13–20.

Orr, H. A. (2005a). Theories of adaptation: what they do and don't say. *Genetica* **123**, 3–13.

Orr, H. A. (2005b). The genetic theory of adaptation: a brief history. *Nat Rev Genet* **6**, 119–127.

Oshima, T., Aiba, H., Masuda, Y., Kanaya, S., Sugiura, M., Wanner, B. L., Mori, H. & Mizuno, T. (2002). Transcriptome analysis of all two-component regulatory system mutants of *Escherichia coli* K-12. *Mol Microbiol* **46**, 281–291.

Ozbudak, E. M., Thattai, M., Lim, H. N., Shraiman, B. I. & van Oudenaarden, A. (2004). Multistability in the lactose utilization network of *Escherichia coli*. *Nature* **427**, 737–740.

Patthy, L. (2003). Modular assembly of genes and the evolution of new functions. *Genetica* **118**, 217–231.

Promislow, D. E. L. (2004). Protein networks, pleiotropy and the evolution of senescence. *Proc R Soc Lond B Biol Sci* **271**, 1225–1234.

Proulx, S. R., Promislow, D. E. & Phillips, P. C. (2005). Network thinking in ecology and evolution. *Trends Ecol Evol* **20**, 345–353.

Raff, R. A. (1996). *The Shape of Life: Genes, Development, and the Evolution of Animal Form*. Chicago: University of Chicago Press.

Rainey, P. B. & Cooper, T. F. (2004). Evolution of bacterial diversity and the origins of modularity. *Res Microbiol* **155**, 370–375.

Rainey, P. B. & Rainey, K. (2003). Evolution of cooperation and conflict in experimental bacterial populations. *Nature* **425**, 72–74.

Rainey, P. B. & Travisano, M. (1998). Adaptive radiation in a heterogeneous environment. *Nature* **394**, 69–72.

Rainey, P. B., Buckling, A., Kassen, R. & Travisano, M. (2000). The emergence and maintenance of diversity: insights from experimental bacterial populations. *Trends Ecol Evol* **15**, 243–247.

Rainey, P. B., Brockhurst, M., Buckling, A., Hodgson, D. J. & Kassen, R. (2005). The use of model *Pseudomonas fluorescens* populations to study the causes and consequences of microbial diversity. In *Biological Diversity and Function in Soils*, pp. 83–99. Edited by R. D. Bardgett, M. B. Usher & D. W. Hopkins. Cambridge: Cambridge University Press.

Reizer, J. & Saier, M. H., Jr (1997). Modular multidomain phosphoryl transfer proteins of bacteria. *Curr Opin Struct Biol* **7**, 407–415.

Rokyta, D. R., Joyce, P., Caudle, S. B. & Wichman, H. A. (2005). An empirical test of the mutational landscape model of adaptation using a single-stranded DNA virus. *Nat Genet* **37**, 441–443.

Romling, U., Gomelsky, M. & Galperin, M. Y. (2005). C-di-GMP: the dawning of a novel bacterial signalling system. *Mol Microbiol* **57**, 629–639.

Ross, P., Weinhouse, H., Aloni, Y. & 8 other authors (1987). Regulation of cellulose synthesis in *Acetobacter xylinum* by cyclic diguanylic acid. *Nature* **325**, 279–281.

Savageau, M. A. (1998). Demand theory of gene regulation. I. Quantitative development of the theory. *Genetics* **149**, 1665–1676.

Schlosser, G. & Wagner, G. (2004). *Modularity in Development and Evolution*. Chicago: University of Chicago Press.

Schluter, D. (2000). *The Ecology of Adaptive Radiation*. Oxford: Oxford University Press.

Simpson, G. G. (1953). *The Major Features of Evolution*. New York: Columbia University Press.

Spiers, A. J. & Rainey, P. B. (2005). The *Pseudomonas fluorescens* SBW25 wrinkly spreader biofilm requires attachment factor, cellulose fibre and LPS interactions to maintain strength and integrity. *Microbiology* **151**, 2829–2839.

Spiers, A. J., Kahn, S. G., Bohannon, J., Travisano, M. & Rainey, P. B. (2002). Adaptive divergence in experimental populations of *Pseudomonas fluorescens*. I. Genetic and phenotypic bases of wrinkly spreader fitness. *Genetics* **161**, 33–46.

Spiers, A. J., Bohannon, J., Gehrig, S. M. & Rainey, P. B. (2003). Biofilm formation at the air–liquid interface by the *Pseudomonas fluorescens* SBW25 wrinkly spreader requires an acetylated form of cellulose. *Mol Microbiol* **50**, 15–27.

Stahl, D. A., Fishbain, S., Klein, M., Baker, B. J. & Wagner, M. (2002). Origins and diversification of sulfate-respiring microorganisms. *Antonie van Leeuwenhoek* **81**, 189–195.

Stern, D. L. (2000). Evolutionary developmental biology and the problem of variation. *Evolution* **54**, 1079–1091.

Tal, R., Wong, H. C., Calhoon, R. & 11 other authors (1998). Three *cdg* operons control cellular turnover of cyclic di-GMP in *Acetobacter xylinum*: genetic organization and occurrence of conserved domains in isoenzymes. *J Bacteriol* **180**, 4416–4425.

Travisano, M. & Rainey, P. B. (2000). Studies of adaptive radiation using model microbial systems. *Am Nat* **156**, S35–S44.

Voigt, C. A., Wolf, D. M. & Arkin, A. P. (2005). The *Bacillus subtilis sin* operon: an evolvable network motif. *Genetics* **169**, 1187–1202.

Waddington, C. H. (1942). The canalization of development and genetic assimilation of acquired characters. *Nature* **150**, 1008–1023.

Wagner, A. (2005). *Robustness and Evolvability in Living Systems*. Princeton: Princeton University Press.

Wagner, G. P. & Altenberg, L. (1996). Complex adaptations and the evolution of evolvability. *Evolution* **50**, 967–976.

Wall, M. E., Hlavacek, W. S. & Savageau, M. A. (2004). Design of gene circuits: lessons from bacteria. *Nat Rev Genet* **5**, 34–42.

Wichman, H. A., Badgett, M. R., Scott, L. A., Boulianne, C. M. & Bull, J. J. (1999). Different trajectories of parallel evolution during viral adaptation. *Science* **285**, 422–424.

Wolf, Y. I., Karev, G. & Koonin, E. V. (2002). Scale-free networks in biology: new insights into the fundamentals of evolution? *Bioessays* **24**, 105–109.

Zhou, L., Lei, X. H., Bochner, B. R. & Wanner, B. L. (2003). Phenotype microarray analysis of *Escherichia coli* K-12 mutants with deletions of all two-component systems. *J Bacteriol* **185**, 4956–4972.

Minimal genomes required for life

Rosario Gil, Vicente Pérez-Brocal, Amparo Latorre and Andrés Moya

Insituto Cavanilles de Biodiversidad y Biología Evolutiva, Universitat de València, Apartado Postal 22085, 46071 València, Spain

INTRODUCTION

Even the simplest unicellular organisms on Earth display an amazing degree of complexity. The question is whether such complexity is a necessary attribute of cellular life or whether, instead, cellular life could also be possible with a much smaller number of molecular components, in the form of what has been called a minimal cell (Luisi *et al.*, 2002; Islas *et al.*, 2004). The first step to envisage such a minimal cell implies the identification of the necessary and sufficient features of life, leading to a clear definition of what 'life' is in this context. This is an extremely complex question with a long tradition in theoretical biology because, in addition to the understanding of the essential properties of a living system, the definition of cellular life is also related to the debated issue of the origin of life, since in the early stages of cellular evolution cells must have been close to the simplest possible life forms. Nowadays, there is a reasonable degree of consensus in defining life as the property of a system that displays simultaneously three features: homeostasis, self-reproduction and evolution (Luisi *et al.*, 2002).

Life can be considered an emergent property: it is a quality that arises from the assembly of non-living elements, properly arranged in space and time (Luisi, 2002). Therefore, to understand life it is necessary to understand first the main non-living components. Metabolism and genetics can be considered as the two central pillars that sustain life (Peretó, 2005), since every living being has some form of metabolism and genetic replication from a template, both of them taking place within a boundary that in modern cells is a phospholipidic membrane. Considering that the functional parts

SGM symposium 66: Prokaryotic diversity: mechanisms and significance.
Editors N. A. Logan, H. M. Lappin-Scott & P. C. F. Oyston. Cambridge University Press. ISBN 0 521 86935 8

of a living cell are proteins and RNA molecules and that the instructions for making these parts are encoded by genes, we can come up with a definition of the necessary elements to keep a minimal cell alive by knowing its complete gene set, what has been called a minimal genome (Mushegian, 1999). Most studies aimed at defining such a minimal genome limit their scope to minimal proteomes, ignoring functional RNA molecules, as well as the regulatory and non-coding sequences on chromosomes. Therefore, this review will refer only to the minimal set of protein-coding genes.

Experimental work is increasingly being complemented by computational research in this field. However, trying to define a general model of all kinds of living beings would be fruitless. Instead, models to define a minimal genome have to be tied to particular levels of biological organization. The increasing knowledge on complete genomes from bacteria makes these prokaryotes a suitable model to try to define what a modern minimal genome should be like. The ultimate goal in this area of research would be to construct a minimal living cell, an aspiration that is also related, as stated by Mushegian (1999), with the challenge of modelling the early history of life using knowledge of prebiotic chemistry.

THE MINIMAL GENOME CONCEPT: BACKGROUND AND HISTORY

Bacterial genomes differ vastly in their sizes and gene repertoires. The number of protein-coding genes in well-characterized bacterial genomes ranges from more than 9300 in the actinomycete *Rhodococcus* sp. RHA1 to as few as 362 in the endosymbiotic bacterium *Buchnera aphidicola* BCc (unpublished results by our group). However, no matter how small, cells have to rely on their own gene products to perform all essential functions for life, since they can usually import metabolites but not functional proteins.

A minimal genome must contain the smallest number of genetic elements sufficient to build a modern-type free-living cellular organism (Mushegian, 1999). However, such a minimal gene-set has no clear meaning by itself, but needs to be associated with a defined set of environmental conditions. Taking that consideration into account, the absolute minimal genome will contain the smallest possible group of protein-coding genes that would be sufficient to sustain cellular life in the most favourable conditions, that is, in a rich environment in which all essential nutrients are provided, and in the absence of any adverse factors (Koonin, 2000).

In recent years, several theoretical and experimental studies have attempted to outline the minimal gene-set for bacterial life by different experimental and computational methods that will be reviewed in the following sections.

Comparative genomics

Estimation of the minimal set of functions conserved in all living cells had been attempted even before complete genome sequences become available, by exhaustive analysis of all identified protein families (Gonnet *et al.*, 1992). This pioneering study revealed that it was possible to apply computer methods in a systematic way in order to draw conclusions about essential cellular functions. With the increasing knowledge of completely characterized genomes, comparative genomics has flourished as a powerful discipline, revealing that most bacterial proteins are highly conserved in evolution. For that reason, it has been possible to predict the functions of genes present in uncharacterized genomes based on genetic and biochemical information obtained from model organisms. Consequently, for many years, the reconstruction of the minimal genome has become the domain of computational comparative genomics. Comparative genomic analyses have an evolutionary basis, since they are primarily based on alignment of DNA sequences in order to identify orthologous genes in genomes of distantly related species, that is, homologous genes that are present in different species and derive from the same ancestral gene in the last common ancestor of the compared species. Although the definition of a minimal genome based on the comparative analysis of known genomes is not independent of the number of genomes used to define it, and these analyses might give a disproportionate weight to taxa favoured, for whatever reason, by sequencers (Charlebois & Doolittle, 2004), these kinds of studies have proven to be very useful in understanding which are the essential functions that define a living cell.

The first attempt to define a minimal genome based on comparative genomics (Mushegian & Koonin, 1996) was made soon after the two first bacterial genomes, from *Haemophilus influenzae* (Fleischmann *et al.*, 1995) and *Mycoplasma genitalium* (Fraser *et al.*, 1995), were completely sequenced. Due to their parasitic lifestyle, these two bacteria present reduced genomes compared with other phylogenetically related free-living species. This is a common trait in bacteria with a host-associated lifestyle, associated with the loss of a great number of genes that are not required for survival in their protected environments, where many molecules don't need to be synthesized because they can be obtained from the host. In fact, *M. genitalium* possesses what has been for a decade the smallest fully characterized bacterial genome, with only 480 protein-coding genes in a 540 kb genome (Fraser *et al.*, 1995). In addition, *M. genitalium* and *H. influenzae* are Gram-positive and Gram-negative bacteria, respectively, separated from their last common ancestor by at least 1·5 billion years of evolution, so genes conserved across such phylogenetic distance were good candidates to be considered essential. The reconstruction of this first minimal gene-set was performed in three steps. As a first step, orthologous genes between the two genomes were identified. Secondly, non-orthologous gene displacements (i.e. non-orthologous genes that encode proteins with a similar function) were considered, in order to fill the gaps in biochemical pathways that were assumed to be essential. Finally, genes that appeared to be functionally redundant or parasite-specific were removed from the list, leading to a minimal gene-set composed of only 256 genes. This hypothetical minimal set is enriched in universally conserved proteins, since 71 % of them show significant similarity to eukaryotic or archaeal proteins. More than half of the genes are involved in genetic information storage and processing, including more or less complete systems for replication, translation and transcription and a surprisingly large set of molecular chaperones. The rest of the genes encode proteins necessary to sustain a simplified metabolism (glycolysis and substrate phosphorylation, salvage pathways for nucleotide biosynthesis and incomplete lipid and cofactor biosynthesis), as well as a limited repertoire for transport systems and protein export, plus 18 apparently essential genes of uncharacterized function.

Although this first version of the minimal genome was just a crude approximation based on a very limited amount of information, it appeared to correspond to a plausible minimalist bacterium. However, a massive transposon-insertion experiment performed on *M. genitalium* and *Mycoplasma pneumoniae* (Hutchison *et al.*, 1999) showed that some of the genes included in this first hypothetical minimal genome can be disrupted, and therefore couldn't be considered as essential. It is reasonable to consider that enlarging the number of compared genomes will significantly reduce the number of shared genes and, therefore, the number of genes included in a minimal genome. The computational comparative analysis of 21 complete genomes of bacteria, archaea and eukaryotes (Koonin, 2000) suggested that a set of about 150 genes would be sufficient to maintain a living cell possessing basal systems for replication, transcription and translation, a reduced repair machinery, a small set of molecular chaperones, an intermediate metabolism reduced to glycolysis, a primitive transport system and no cell wall.

A more recent approach to the minimal gene-set has included the small genomes of insect endosymbionts in these computational comparisons. Mutualistic and obligate insect–bacteria symbioses have been intensively studied in recent years, particularly those involving the gammaproteobacteria *Buchnera aphidicola* (three different strains), *Wigglesworthia glossinidia* and two species of '*Candidatus* Blochmannia', primary endosymbionts of aphids, tsetse flies and carpenter ants, respectively, whose genomes have been fully sequenced (Shigenobu *et al.*, 2000; Tamas *et al.*, 2002; van Ham *et al.*, 2003; Akman *et al.*, 2002; Gil *et al.*, 2003; Degnan *et al.*, 2005). Insects confer to their partners a stable environment, inside specialized cells called bacteriocytes, while bacteria provide the host with new metabolic capabilities, complementing the insects' nutritional requirements and contributing to the exploitation of new ecological niches.

Analysis of these endosymbiont genomes revealed that a massive genome reduction occurred after the establishment of their respective symbioses. Therefore, it can be considered that genes that have been preserved in all cases are good candidates to be essential for endosymbiotic life. A comparative analysis performed among the first five sequenced genomes of endosymbionts showed that they share only 277 protein-coding genes (281 if non-orthologous gene displacement is taken into account), corresponding to half of their coding capacity (Gil *et al.*, 2003). Some of these shared genes must be involved in endosymbiotic processes, while the rest should be essential for any kind of cellular life. All these genomes are relatively similar in size and they encode quite similar numbers of genes in each functional COG (cluster of orthologous genes) category (Tatusov *et al.*, 2001). Interestingly, as was already observed in the first computational approach to the minimal genome, about one-third of the genes in all five genomes are devoted to information storage and processing, most of these genes being shared by all five genomes, and also a considerable number of molecular chaperones have been preserved. Remarkable differences were only found among the genes that encode proteins involved in cell envelope biogenesis, flagellar biosynthesis and metabolism of amino acids, nucleotides and coenzymes, suggesting that the molecular mechanisms necessary for survival in an intracellular environment may be quite similar for any endosymbiotic association, while about a third of the coding capacity of each endosymbiont seems to be dedicated to specific requirements of the corresponding symbiosis, mainly reflecting differences in host lifestyle, nutritional needs and location within the host cell. To identify the subset of genes that can be considered essential for cellular life, the complete set of genes shared by all five endosymbionts was compared with the reduced genome of the epicellular parasite *M. genitalium*. The study showed that these six genomes share only 180 housekeeping protein-coding genes (Gil *et al.*, 2003). Once more, the analysis unveiled the essentiality of the genes involved in informational processes, since about half of these genes belong to this category. The number of shared genes among endosymbiotic and parasitic bacteria was further reduced to 156 when the intracellular parasites *Rickettsia prowazekii* and *Chlamydia trachomatis* were added to the comparison (Klasson & Andersson, 2004).

However, computational approaches to minimal gene-set analysis have important limitations that lead them to underestimate the number of candidate genes to be included in a minimal genome. First of all, identifying shared (orthologous) genes in genomes of distantly related species is not a trivial matter, since orthologous genes have been often duplicated after the divergence of the species being compared, leading to more complex situations in which two or more paralogous genes (i.e. homologous genes present in one genome that originate by duplication of an ancestral gene) can be present simultaneously in a given genome, making it difficult to identify the true pair of orthologues between two compared genomes. Furthermore, when distantly related genes are compared, the low degree of sequence conservation might make their identification as true orthologous genes difficult, which again may lead to an underestimation of shared genes. In addition, genes with a different ancestral origin (i.e. non-orthologous genes) might be responsible for a similar function in different organisms. Therefore, when trying to identify the minimal gene-set by comparative genomics, cases of non-orthologous gene displacement must be taken into account. This last statement also implies that, even under the same environmental and living conditions, numerous versions of minimal genomes can be envisaged, as will be discussed below.

Experimental approaches

Some other estimates of the minimal genome are based on indirect evidence from random mutagenesis or systematic gene disruption. The first experimental attempt to make such an estimate was performed even before the advent of comparative genomics, based on the analysis of the viability of a limited number of randomly generated gene knockouts in *Bacillus subtilis* (Itaya, 1995). In this pioneering work, a number of approximately 300 essential genes was inferred, although the identification of such genes was not possible at that time. Remarkably, this number is fairly close to the number of genes that was later estimated by Mushegian & Koonin (1996) by comparative genomics. Several genome-wide analyses to identify genes that are essential under particular growth conditions have been performed since then, using three different experimental approaches: massive transposon mutagenesis (the most widely used approach, reviewed by Judson & Mekalanos, 2000), the use of antisense RNA to inhibit gene expression (Ji *et al.*, 2001; Forsyth *et al.*, 2002) and the systematic inactivation of each individual gene present in a genome (Mori *et al.*, 2000; Gerdes *et al.*, 2003; Kobayashi *et al.*, 2003; Kang *et al.*, 2004). All of these approaches yielded minimal gene-sets that are compatible with the comparative-genomics inferences. Very few genes included in the computationally derived minimal genomes were found to be dispensable and, among these, it remains to be determined whether their dispensability reflects artefacts of the experimental strategies or unexpected functional redundancy (Koonin, 2003).

Although the systematic targeted approaches will be powerful tools for future studies, they require more substantial organization and financial commitment. And, once again, these experimental approaches have their limitations and also tend to underestimate the number of genes included in the minimal genome. Transposon mutagenesis might overestimate the set by misclassification of dispensable genes that slow down growth without arresting it, but can also miss essential genes that tolerate transposon insertions, while the use of antisense RNA is limited to genes for which adequate expression of the inhibitory RNA can be obtained in the organism under study. In addition to a possible significant error rate (dispensable genes considered as essential or essential genes that are missed), inactivation of single genes does not detect essential functions encoded by redundant genes. The essential gene-set is not the same as the minimal genome, since genes that are individually non-essential may not be simultaneously dispensable.

A combined approach

All the above-mentioned experimental and computational approaches to the minimal genome gave sets of essential genes with similar functional features, which are distinct from those of the general population of conserved bacterial genes (the latter represented in the database of protein COGs). They are substantially enriched in genes that encode components of genetic-information processing systems, mainly genes of the transcriptional apparatus, and contain relatively few genes for metabolic enzymes plus a very small fraction of functionally uncharacterized genes. This fact can be explained because, since no alternative genes involved in information processes have been found in living cells, such genes are not exchangeable in the way that genes for operational processes such as metabolism are. However, in order to be considered alive, a cell must also be able to maintain metabolic homeostasis, and therefore the minimal genome must include the necessary genes to allow the cell to perform all the necessary reactions to maintain a minimal and coherent metabolic functionality.

Trying to get closer to the minimal bacterial genome, a combined investigation of all previously used computational and experimental strategies for addressing this issue was performed, ensuring that all genes involved in metabolic pathways that could be considered as essential to maintain a reasonable metabolic homeostasis for any living cell were included (Gil *et al.*, 2004). The analysis rendered a minimal genome containing 206 protein-coding genes (Table 1, Fig. 1). Two-thirds of these genes are involved in genetic-information storage and processing, including a virtually complete DNA replication machinery, a very rudimentary system for DNA repair, a virtually complete transcriptional machinery without any transcriptional regulator and a nearly complete translational system, plus protein processing, folding, secretion and degradation functions. Among cellular functions, cell division can be driven by FtsZ only, considering that, in a protected environment, the cell wall might not be necessary for cellular structure. The genes necessary to perform all these functions are very well conserved in all analysed organisms, and most of them are also present in all other approaches to the minimal genome already described, which means that any proposed minimal genome would probably include more or less the same genes for these purposes.

The remaining third of genes in this hypothetical minimal genome encode proteins implicated in the transport and use of nutrients from the environment to obtain energy

Table 1. Classification of genes included in the minimal genome for bacterial life proposed by Gil *et al.* (2004)

Functional category	Number of genes
DNA metabolism	16
Basic replication machinery	13
DNA repair, restriction and modification	3
RNA metabolism	106
Basic transcription machinery	8
Translation: aminoacyl-tRNA synthesis	21
Translation: tRNA maturation and modification	6
Translation: ribosomal proteins	50
Translation: ribosome function, maturation and modification	7
Translation factors	12
RNA degradation	2
Protein processing, folding and secretion	15
Protein post-translational modification	2
Protein folding	5
Protein translocation and secretion	5
Protein turnover	3
Cellular processes	5
Cell division	1
Transport	4
Energetic and intermediary metabolism	56
Glycolysis	10
Proton motive force generation	9
Pentose phosphate pathway	3
Lipid metabolism	7
Biosynthesis of nucleotides	15
Biosynthesis of cofactors	12
Poorly characterized	8
Total	206

and basic molecules to maintain cellular structures. Eight genes of unknown function have also been included in this minimal genome, because they have been preserved in all analysed reduced genomes and are essential in free-living bacteria. Further studies are needed to determine whether they are really necessary to maintain a living cell. The list of genes involved in nutrient transport and metabolic functions varies slightly depending on the different minimal genome proposals. However, this is not surprising,

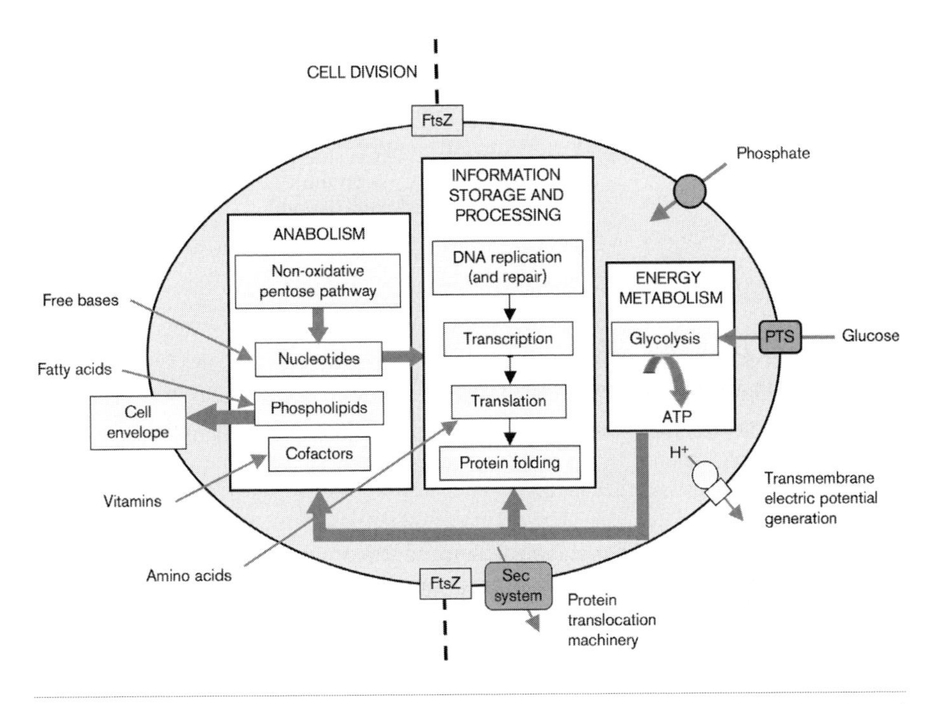

Fig. 1. Simplified integrated overview of the essential cellular functions implemented by a hypothetical minimal genome of 206 protein-coding genes derived by an integrated approach, taking into account genome-wide computational, experimental and metabolic studies on completely sequenced bacterial genomes.

since there is no conceptual or experimental support for the existence of *one* form of minimal bacterial cell, at least from a metabolic point of view. Different essential functions can be defined depending on the environmental conditions, and numerous versions of minimal genomes can be conceived to fulfil such functions even for the same set of conditions. Although it still remains questionable whether such a minimalist cell (or any of those previously mentioned) could survive under any realistic conditions, there is no doubt that future studies will highlight a diversity of minimal ecologically dependent metabolic charts that support a universal genetic machinery.

GENOME MINIMIZATION

Even though we still have a long way to go before we understand the specific functions that are essential for different bacteria in their natural environments, it appears that we are already aware of most essential cellular functions and, clearly, free-living bacteria possess many dispensable functions that would not be needed in an ideally controlled environment. A minimal strain consisting of just the backbone genes involved in essential functions would be interesting for laboratory studies on bacterial physiology, genetics and ecology, as well as for biotechnological purposes, where a simplified

Table 2. Gene content in recently sequenced naturally reduced genomes and comparison with the hypothetical minimal genome containing 206 protein-coding genes

Species/strain	Genome size (Mb)	Protein-coding genes	Genes shared with the minimal genome	Reference
Buchnera aphidicola BCc	0·42	363	161	Our group (unpublished)
'*Blochmannia pennsylvanicus*'*	0·79	658	193	Degnan *et al.* (2005)
Wolbachia wBm	1·08	806	170	Foster *et al.* (2005)
Prochlorococcus marinus MED4	1·66	1716	180	Rocap *et al.* (2003)
Prochlorococcus marinus SS120	1·75	1882	183	Dufresne *et al.* (2003)

*The name '*Blochmannia pennsylvanicus*' has not been validly published and is enclosed in quotes simply to indicate this fact.

bacterium, free of unwanted metabolic side-products, would increase the efficiency of obtaining the desired product and reduce the cost of its purification. While some experimental attempts at genome minimization are being performed, using our current knowledge of dispensable cellular functions in model free-living organisms, scientists can also learn from nature about the functions that can be removed without impairing cellular life, in order to improve functional knowledge of what a minimal cell should be like.

Experimental genome minimization

Several attempts at experimental genome minimization in bacteria have already been reported using different strategies for large-scale genome reduction, most of which involve the insertion of modified transposons into the genome under study and the excision of the flanked genomic segments. Yu *et al.* (2002) developed the first rapid and efficient genome-engineering method. By cumulatively combining multiple random deletions in the chromosome, they were able to eliminate about 10 % of the *Escherichia coli* genome without detectable phenotypic effects in a rich growth medium. A similar strategy has recently been used to engineer the genome of *Corynebacterium glutamicum*, an important micro-organism for enhanced production of biochemicals, leading to a strain with a 7·5 % reduction of its genome (Suzuki *et al.*, 2005). A second method for large-scale genome reduction using modified transposons has been designed by Goryshin *et al.* (2003); after 20 serial repeats of the procedure, they were able to obtain four different *E. coli* strains that had lost around 250 kb (about 4·5 %) of the parental genome. The advantages of this method include that it can be used in many bacterial species, it allows the selection for fast-growing cells among several possible deletion strains and it allows the deleted material to be saved as a conditional plasmid, available for further studies of the deleted genes. A different deletion strategy for precise genomic surgery, involving the use of the λ red recombination system, has been designed by Kolisnychenko *et al.* (2002) to delete large regions identified by comparative genomics as recent horizontal acquisitions, such as K-islands, resulting in an *E. coli* strain with 8·1 % reduced genome size. However, this reductive method relies on previous knowledge of the genome sequence and the dispensability of the regions to be removed. Nevertheless, a similar approach has been applied for the removal of dispensable regions in *Bacillus subtilis*, using a plasmid-based chromosomal integration–excision system, leading to a 7·7 % reduction in genome size in this bacterium (Westers *et al.*, 2003).

It seems that 'playing' with minimalized genomes might become an interesting avenue for research, as important technical advances in genome engineering are expected in the near future, allowing exploration of the combinatorial space of possible minimal genomes and the compatibility (or incompatibility) of essential genes from different, phylogenetically distinct sources.

Genome minimization in nature

All the attempts to approach a minimal genome using computational and experimental methods mentioned above resemble those already performed by nature (Table 2), where genome reduction is associated mainly with a transition from free-living to a host-dependent lifestyle. Many genomic sequences of bacteria that can replicate only within eukaryotic host cells have been published in the last few years, revealing that these obligately host-associated bacteria (parasites and mutualistic symbionts) possess the smallest known genomes in nature, often below 1 Mb in size. The genome of a bacterium that lives in such a close relationship with a eukaryotic host is referred to in the literature as a resident genome (Anderson & Kurland, 1998). Among the completely sequenced resident genomes, many correspond to obligate pathogens of the classes *Mollicutes* (several species of the genus *Mycoplasma*, *Ureaplasma urealyticum*, '*Candidatus* Phytoplasma asteris' and *Mesoplasma florum*), *Chlamydiae* (several strains of *Chlamydia trachomatis* and *Chlamydia pneumoniae*), *Spirochaetes* (*Treponema pallidum* and two species of the genus *Borrelia*), *Actinobacteria* (two strains of *Tropheryma whipplei*) and *Alphaproteobacteria* (several species of the genus *Rickettsia* and *Anaplasma marginale*). All genomes of mutualistic endosymbionts sequenced to date correspond to two classes of the phylum *Proteobacteria*: the six gammaproteobacterial endosymbiont of insects already mentioned and two species of *Wolbachia* (alphaproteobacteria), endosymbionts of a fruit-fly and a nematode, respectively.

In bacteria, gene content is basically correlated with genome size (Casjens, 1998) and, therefore, the reduced size of resident genomes reflects the presence of a smaller number of genes, when compared with their free-living relatives, due to relaxed selection on the maintenance of genes that are rendered unnecessary in the protected environment provided by the host. In addition, these reduced genomes have lost most regulatory elements. Because of their obligately intracellular lifestyles, the transmission of these bacteria from one host to the next frequently involves bottlenecks and replication in populations of small effective size, with little opportunity for recombination between variants. For this reason, the resident genomes accumulate deleterious mutations by genetic drift at a rate that is higher than that for free-living organisms, a phenomenon known as Muller's ratchet (Moran, 1996). Furthermore, the extremely reduced genomes of endosymbiotic bacteria and some pathogens have lost most of the genes involved in recombination processes and, consequently, the genome size can not be increased by acquisition of foreign DNA (Silva *et al.*, 2003).

Recently, it has been shown that natural genome reduction can also take place in free-living bacteria. Three species of *Prochlorococcus*, free-living cyanobacteria that represent the smallest and most abundant photosynthetic organism in the ocean, have been sequenced (Dufresne *et al.*, 2003; Rocap *et al.*, 2003). Two of these species present genomes that have been reduced in size by about 30 % in comparison with other close relatives, representing nearly minimal oxyphototrophic genomes. These genomes also share some features with those of host-dependent prokaryotes: bias toward a low G+C content, acceleration of the rate of evolution of protein-coding genes and loss of DNA-repair genes. In contrast, however, these changes in *Prochlorococcus* genomes appear to be a consequence of a selective process favouring the adaptation of this organism to its environment, allowing these organisms to make substantial economies in energy and material for cell maintenance (Dufresne *et al.*, 2005).

The most extreme case of natural genome minimization known in bacteria corresponds to *Buchnera aphidicola* BCc, the primary endosymbiont of the aphid *Cinara cedri*. Its genome of 422 kb contains only 362 protein-coding genes (unpublished results). Comparative analysis of two selected regions from this genome with the same regions in the other three sequenced *Buchnera aphidicola* genomes (Pérez-Brocal *et al.*, 2005) indicated that, at least in these two regions, the reduction process is mainly due to the loss of genes by a process of gene disintegration (Silva *et al.*, 2001). Gene loss has not been even all along the genome, but it is highly dependent on the nature of the gene function, since more divergent genes are significantly more prone to loss than more conserved ones. The genome reduction in this bacterium appears to be due mainly to the loss of genes belonging to metabolic and cellular processes, which turned out to be non-essential in the protected environment provided by the insect host. Therefore, the size of the *Buchnera aphidicola* genome appears to be controlled by restrictions on the loss of function of crucial genes, because, when any gene is inactivated, its nucleotides are progressively removed from the genome in a relatively short period of time (Gómez-Valero *et al.*, 2004). Consequently, the genes that have been retained in this reduced genome must be a closer estimate of a minimal genome. Nevertheless, this assumption must be taken with caution. It remains to be determined whether progressive genome reduction by gene loss is, instead, driving this bacterium towards extinction and replacement by another, healthier bacterium able to perform the necessary functions to retain host fitness. In this last scenario, some essential genes might have already been lost, meaning that this genome would not contain all the information necessary to fulfil the requisites to be considered a minimal genome.

MINIMAL GENE-SET AND THE LAST UNIVERSAL COMMON ANCESTOR (LUCA)

A minimal genome is a purely functional concept that does not explicitly incorporate evolution, and it may appear that we learn relatively little about the fundamentals of life from a computationally derived or experimentally defined minimal gene-set. In practice, however, analysis of the evolutionary conservation of minimal gene-sets indicates that they are relevant for the reconstruction of the gene-sets of ancestral life-forms (Harris *et al.*, 2003; Koonin, 2003). The comparative-genomic approach is based on the key evolutionary notion of orthology, and the resulting sets of genes should approximate those of ancestral life-forms. Regarding experimental approaches to identifying essential genes, although they do not have an evolutionary basis, essential genes tend to be highly evolutionarily conserved, in terms of wide phyletic spread and rate of sequence evolution. Some authors consider that the notion of a single common ancestor for a group of genomes might be a simplification, and other alternative hypotheses in the form of a population of diverse organisms that exchanged their genes at an extremely high rate have been proposed (Woese, 1998; Doolittle, 2000). Regardless of whether LUCA was a single organism or a community of organisms, a pool of ancient genes must have existed and can be traced back by analysis of minimal genomes. One of the main findings from the analysis of computationally and experimentally derived minimal genomes is the remarkable evolutionary plasticity of even the essential biological functions, since only a small group of genes (most of them involved in translation and transcription) is truly ubiquitous among living beings. Minimal genomes can also help scientists to determine the optimal evolutionary parameters for ancestral gene-sets by examining the most parsimonious reconstructions of different gene categories included in the minimal set (Koonin, 2003). Thus, depending on selective rates of gene loss and horizontal gene transfer, the size and composition of LUCA would be quite different.

CONCLUDING REMARKS

The notion of a minimal genome is, at this point, mostly a conceptual tool for discussion of the minimal requisites for cellular life and the possible experimental design and construction of simpler cellular models. The field of the minimal cell is anchored partly in solid experimental facts and still belongs in part to the sphere of theoretical research. Projects designed to scan entire microbial genomes for essential genes have revealed a remarkably compact and conserved, but not universal, set of genes whose functions are necessary for survival or reproduction (Stephens & Laub, 2003). A minimal cell cannot be sharply defined, given that different essential functions can be defined depending on the environmental conditions. Nevertheless, it is possible to try to delineate which functions should be performed in any modern living cell and list the genes that would be necessary to maintain such functions. It should be noticed that numerous alternative minimal genomes can be conceived to fulfil such functions even for the same set of conditions, as two or more unrelated or distantly related proteins have evolved for most essential cellular functions (Koonin, 2003).

The relationship between 'minimal cell' and 'primitive life' is one of the focuses of research in the field, since the early cells adhere to the notion of the minimal cell. However, one must keep in mind that it is impossible to identify one of the above-mentioned diverse solutions with the one adopted by the more primitive cells (Morange, 2003). Any attempt to universalize the conclusions would necessarily include comparison with archaeal genomes, more specifically the smallest ones (Waters *et al.*, 2003). Remarkably, the different approaches used to define a minimal genome converge on the conclusion that genes dealing with RNA biosynthesis are common to all cellular life, while some of the main components of the DNA-replication machinery are not universal. These facts may be linked to the idea of a primitive 'RNA world', although such an assumption is far from immediate.

In a more technological vein, the development of more sophisticated techniques for genomic engineering, together with continued efforts to define the minimal genome, will help to achieve the exciting goal of experimentally constructing a modern-type minimal living cell in perhaps the not-so-distant future.

ACKNOWLEDGEMENTS

This research was funded by grants Grupos03/204 from Govern Valencià and BFM2003-00305 from Ministerio de Ciencia y Tecnología (MCyT) to A. M. R. G. is a recipient of a contract in the Ramon y Cajal Program from the MCyT, Spain.

REFERENCES

Akman, L., Yamashita, A., Watanabe, H., Oshima, K., Shiba, T., Hattori, M. & Aksoy, S. (2002). Genome sequence of the endocellular obligate symbiont of tsetse flies, *Wigglesworthia glossinidia*. *Nat Genet* **32**, 402–407.

Anderson, S. G. E. & Kurland, C. G. (1998). Reductive evolution of resident genomes. *Trends Microbiol* **6**, 263–268.

Casjens, S. (1998). The diverse and dynamic structure of bacterial genomes. *Annu Rev Genet* **32**, 339–377.

Charlebois, R. L. & Doolittle, W. F. (2004). Computing prokaryotic gene ubiquity: rescuing the core from extinction. *Genome Res* **14**, 2469–2477.

Degnan, P. H., Lazarus, A. B. & Wernegreen, J. J. (2005). Genome sequence of *Blochmannia pennsylvanicus* indicates parallel evolutionary trends among bacterial mutualists of insects. *Genome Res* **15**, 1023–1033.

Doolittle, W. F. (2000). The nature of the universal ancestor and the evolution of the proteome. *Curr Opin Struct Biol* **10**, 355–358.

Dufresne, A., Salanoubat, M., Partensky, F. & 18 other authors (2003). Genome sequence of the cyanobacterium *Prochlorococcus marinus* SS120, a nearly minimal oxyphototrophic genome. *Proc Natl Acad Sci U S A* **100**, 10020–10025.

Dufresne, A., Garczarek, L. & Partensky, F. (2005). Accelerated evolution associated with genome reduction in a free-living prokaryote. *Genome Biol* **6**, R14. doi:10.1186/gb-2005-6-2-r14

Fleischmann, R. D., Adams, M. D., White, O. & 37 other authors (1995). Whole-genome random sequencing and assembly of *Haemophilus influenzae* Rd. *Science* **269**, 496–512.

Forsyth, R. A., Haselbeck, R. J., Ohlsen, K. L. & 20 other authors (2002). A genome-wide strategy for the identification of essential genes in *Staphylococcus aureus*. *Mol Microbiol* **43**, 1387–1400.

Foster, J., Ganatra, M., Kamal, I. & 23 other authors (2005). The *Wolbachia* genome of *Brugia malayi*: endosymbiont evolution within a human pathogenic nematode. *PLoS Biol* **3**, e121.

Fraser, C. M., Gocayne, J. D., White, O. & 26 other authors (1995). The minimal gene complement of *Mycoplasma genitalium*. *Science* **270**, 397–403.

Gerdes, S. Y., Scholle, M. D., Campbell, J. W. & 18 other authors (2003). Experimental determination and system level analysis of essential genes in *Escherichia coli* MG1655. *J Bacteriol* **185**, 5673–5684.

Gil, R., Silva, F. J., Zientz, E. & 10 other authors (2003). The genome sequence of *Blochmannia floridanus*: comparative analysis of reduced genomes. *Proc Natl Acad Sci U S A* **100**, 9388–9393.

Gil, R., Silva, F. J., Peretó, J. & Moya, A. (2004). Determination of the core of a minimal bacterial gene set. *Microbiol Mol Biol Rev* **68**, 518–537.

Gómez-Valero, L., Latorre, A. & Silva, F. J. (2004). The evolutionary fate of nonfunctional DNA in the bacterial endosymbiont *Buchnera aphidicola*. *Mol Biol Evol* **21**, 2172–2181.

Gonnet, G. H., Cohen, M. A. & Benner, S. A. (1992). Exhaustive matching of the entire protein sequence database. *Science* **256**, 1443–1445.

Goryshin, I. Y., Naumann, T. A., Apodaca, J. & Reznikoff, W. S. (2003). Chromosomal deletion formation system based on Tn*5* double transposition: use for making minimal genomes and essential gene analysis. *Genome Res* **13**, 644–653.

Harris, J. K., Kelley, S. T., Spiegelman, G. B. & Pace, N. R. (2003). The genetic core of the universal ancestor. *Genome Res* **13**, 407–412.

Hutchison, C. A., III, Peterson, S. N., Gill, S. R., Cline, R. T., White, O., Fraser, C. M., Smith, H. O. & Venter, J. C. (1999). Global transposon mutagenesis and a minimal mycoplasma genome. *Science* **286**, 2165–2169.

Islas, S., Becerra, A., Luisi, P. L. & Lazcano, A. (2004). Comparative genomics and the gene complement of a minimal cell. *Orig Life Evol Biosph* **34**, 243–256.

Itaya, M. (1995). An estimation of minimal genome size required for life. *FEBS Lett* **362**, 257–260.

Ji, Y., Zhang, B., van Horn, S. F., Warren, P., Woodnutt, G., Burnham, M. K. R. & Rosemberg, M. (2001). Identification of critical staphylococcal genes using conditional phenotypes generated by antisense RNA. *Science* **293**, 2266–2269.

Judson, N. & Mekalanos, J. J. (2000). Transposon-based approaches to identify essential bacterial genes. *Trends Microbiol* **8**, 521–526.

Kang, Y., Durfee, T., Glasner, J. D., Qiu, Y., Frisch, D., Winterberg, K. M. & Blattner, F. R. (2004). Systematic mutagenesis of the *Escherichia coli* genome. *J Bacteriol* **186**, 4921–4930.

Klasson, L. & Andersson, S. G. (2004). Evolution of minimal-gene-sets in host-dependent bacteria. *Trends Microbiol* **12**, 37–43.

Kobayashi, K., Ehrlich, S. D., Albertini, A. & 96 other authors (2003). Essential *Bacillus subtilis* genes. *Proc Natl Acad Sci U S A* **100**, 4678–4683.

Kolisnychenko, V., Plunkett, G., III, Herring, C. D., Fehér, T., Pósfai, J., Blattner, F. R. & Pósfai, G. (2002). Engineering a reduced *Escherichia coli* genome. *Genome Res* **12**, 640–647.

Koonin, E. V. (2000). How many genes can make a cell: the minimal-gene-set concept. *Annu Rev Genomics Hum Genet* **1**, 99–116.

Koonin, E. V. (2003). Comparative genomics, minimal gene-sets and the last universal common ancestor. *Nat Rev Microbiol* **1**, 127–136.

Luisi, P. L. (2002). Toward the engineering of minimal living cells. *Anat Rec* **268**, 208–214.

Luisi, P. L., Oberholzer, T. & Lazcano, A. (2002). The notion of a DNA minimal cell: a general discourse and some guidelines for an experimental approach. *Helv Chim Acta* **85**, 1759–1777.

Moran, N. A. (1996). Accelerated evolution and Muller's ratchet in endosymbiotic bacteria. *Proc Natl Acad Sci U S A* **93**, 2873–2878.

Morange, M. (2003). *La Vie Expliquée? 50 Ans Après la Double Helix*. Paris: Odile Jacob (in French).

Mori, H., Isono, K., Horiuchi, T. & Miki, T. (2000). Functional genomics of *Escherichia coli* in Japan. *Res Microbiol* **151**, 121–128.

Mushegian, A. (1999). The minimal genome concept. *Curr Opin Genet Dev* **9**, 709–714.

Mushegian, A. R. & Koonin, E. V. (1996). A minimal gene set for cellular life derived by comparison of complete bacterial genomes. *Proc Natl Acad Sci U S A* **93**, 10268–10273.

Peretó, J. (2005). Controversies on the origin of life. *Int Microbiol* **8**, 23–31.

Pérez-Brocal, V., Latorre, A., Gil, R. & Moya, A. (2005). Comparative analysis of two genomic regions among four strains of *Buchnera aphidicola*, primary endosymbiont of aphids. *Gene* **345**, 73–80.

Rocap, G., Larimer, F. W., Lamerdin, J. & 21 other authors (2003). Genome divergence in two *Prochlorococcus* ecotypes reflects oceanic niche differentiation. *Nature* **424**, 1042–1047.

Shigenobu, S., Watanabe, H., Hattori, M., Sakaki, Y. & Ishikawa, H. (2000). Genome sequence of the endocellular bacterial symbiont of aphids *Buchnera* sp. APS. *Nature* **407**, 81–86.

Silva, F. J., Latorre, A. & Moya, A. (2001). Genome size reduction through multiple events of gene disintegration in *Buchnera* APS. *Trends Genet* **17**, 615–618.

Silva, F. J., Latorre, A. & Moya, A. (2003). Why are the genomes of endosymbiotic bacteria so stable? *Trends Genet* **19**, 176–180.

Stephens, C. M. & Laub, M. T. (2003). Microbial genomics: all that you can't leave behind. *Curr Biol* **13**, R571–R573.

Suzuki, N., Okayama, S., Nonaka, H., Tsuge, Y., Inui, M. & Yukawa, H. (2005). Large-scale engineering of the *Corynebacterium glutamicum* genome. *Appl Environ Microbiol* **71**, 3369–3372.

Tamas, I., Klasson, L., Canback, B., Naslund, A. K., Eriksson, A. S., Wernegreen, J. J., Sandström, J. P., Moran, N. A. & Andersson, S. G. E. (2002). 50 million years of genomic stasis in endosymbiotic bacteria. *Science* **296**, 2376–2379.

Tatusov, R. L., Natale, D. A., Garkavtsev, I. V. & 7 other authors (2001). The COG database: new developments in phylogenetic classification of protein from complete genomes. *Nucleic Acids Res* **29**, 22–28.

van Ham, R. C. H. J., Kamerbeek, J., Palacios, C. & 13 other authors (2003). Reductive genome evolution in *Buchnera aphidicola*. *Proc Natl Acad Sci U S A* **100**, 581–586.

Waters, E., Hohn, M. J., Ahel, I. & 19 other authors (2003). The genome of *Nanoarchaeum equitans*: insights into early archaeal evolution and derived parasitism. *Proc Natl Acad Sci U S A* **100**, 12984–12988.

Westers, H., Dorenbos, R., van Dijl, J. M. & 21 other authors (2003). Genome engineering reveals large dispensable regions in *Bacillus subtilis*. *Mol Biol Evol* **20**, 2076–2090.

Woese, C. (1998). The universal ancestor. *Proc Natl Acad Sci U S A* **95**, 6854–5859.

Yu, B. J., Sung, B. H., Koob, M. D., Lee, C. H., Lee, J. H., Lee, W. S., Kim, M. S. & Kim, S. C. (2002). Minimization of the *Escherichia coli* genome using a Tn5-targeted Cre/loxP excision system. *Nat Biotechnol* **20**, 1018–1023.

Evolution of the core of genes

Vincent Daubin and Emmanuelle Lerat

Laboratoire de Biométrie et Biologie Evolutive, 43 Bld du 11 Novembre 1918, Université Lyon 1, 69622 Villeurbanne cedex, France

INTRODUCTION

The extravagant diversity of microbes has only been fully appreciated with the development of comparative genomics. Comparisons of gene repertoires among prokaryotes have revealed striking differences among species and even among strains of the same species. For example, the genomes of three *Escherichia coli* strains have been shown to share only 40 % of their genes, with most of the remaining genes being strain-specific (Welch *et al.*, 2002). More generally, although most prokaryotic genomes contain thousands of genes, only a handful can be identified as truly ubiquitous in modern organisms. This so-called 'core' of universal genes has received much interest from evolutionary biologists because it probably represents a relic of the last universal common ancestor (LUCA) and provides valuable information for reconstructing the tree of life. It has also been viewed as the *sine qua non* condition of life, since no living organism seems able to survive without it. However, perhaps more interesting is the paucity of these ubiquitous genes, as it shows the formidable evolutionary plasticity of biological systems and points to the mechanisms necessary for acquiring and generating new genes.

WHAT'S IN A GENOME?

An inventory

All cellular organisms have in common the use of DNA as the support of genetic information, RNA as an intermediate of protein expression and the same genetic code (with only a few exceptions) as well as catabolism and metabolism based on a limited number of amino acids and sugars. These and other shared fundamental characteristics

SGM symposium 66: Prokaryotic diversity: mechanisms and significance.
Editors N. A. Logan, H. M. Lappin-Scott & P. C. F. Oyston. Cambridge University Press. ISBN 0 521 86935 8 ©SGM 2006

suggest that the basal functioning of a living cell has been roughly conserved through evolution and predicts that genomes provide evidence of such conservation. Therefore, the genetic repertoire of an organism can be schematically divided into three parts: (i) those genes that fulfil the elementary functions of cell metabolism, including replication, transcription, translation and the synthesis of the precursors needed for these operations (these functions being universal, the genes involved are expected to be found in all cellular organisms), (ii) those genes responsible, in contrast, for the uniqueness of an organism, its specific ecology and phenotype and (iii) genes without any beneficial effect on the organism, such as parasitic elements.

The core of universal genes

Because of the complexity of the machinery necessary for replicating and expressing a genome, the definition of the set of universal genes, based on comparisons of complete genomes, has come as a surprise. In a recent review, Koonin (2003) estimated that all of the around 100 genomes available at the time shared only 63 genes. Although most of these genes are, as expected, implicated in the fundamental processes of translation (mostly), transcription and replication, they are clearly insufficient to fulfil these functions. Furthermore, as more genomes are added to the comparison, the core is only expected to melt away. Gene inactivation assays have shown that this core is included in the set of genes that are essential for a given cell (Gerdes *et al.*, 2003; Akerley *et al.*, 2002; Kobayashi *et al.*, 2003), which is consistent with their conservation, but also that it only represents a small fraction of them. The paucity of truly universal genes is likely to be the result of several combined factors: (i) the last ancestor of the genes considered here goes back several billion years, and homologous genes might have accumulated so many differences that we are no longer able to recognize their homology; (ii) since their common ancestor, each lineage may have invented new genes and adopted new molecular strategies to accomplish these fundamental tasks; and (iii) the definition of the minimal core of genes is usually based on comparisons that include genomes from parasites, which rely on their hosts for certain essential functions and can no longer be considered as autonomous forms of life.

Several studies have stressed the importance of gene replacement in the evolution of genomes (Daubin & Ochman, 2004a; Koonin, 2003; Lerat *et al.*, 2005), particularly because the failure to recognize homology can not explain the difference of gene repertoires in closely related organisms. In this view, the core of genes may be considered as the set of true evolutionarily essential genes, in the sense that evolution has apparently never found stable alternatives to achieve these tasks. The corollary is that most ancestrally essential genes have been replaced during the course of evolution, which suggests that genetic innovation has played an important role, not only in the adaptation of organisms to new environments, but also in continually refining the most central cellular processes. The shared basic characteristics of all life forms disguise a tremendous diversity of mechanisms.

Why, then has any gene been conserved since LUCA? The 'complexity hypothesis' (Jain *et al.*, 1999) proposes that genes which are part of protein complexes and have consequently developed tight interactions with each other are unlikely to be replaced. This is consistent with the presence of about 30 ribosomal proteins (~50 %) in the core of universal genes (Koonin, 2003). However, the core also contains many proteins that are probably not part of any protein complexes, such as tRNA synthetases. Although there is no doubt that they are essential, the reasons for the conservation of such genes remain to be elucidated (but see later).

THE EVOLUTION OF GENETIC REPERTOIRES

Horizontal gene transfer

The processes by which new genes are produced are not fully known; however, the most broadly held view is that regions of the genome are duplicated and that subsequent functional diversification can produce genes that confer novel properties (Lynch & Conery, 2000). Because the invention of useful genes through the modification of existing sequences is a slow and tentative process, many organisms have drawn on an alternative means, namely, the enlistment of established traits from unrelated organisms. Various mechanisms allow bacteria to integrate genetic material into their genomes, such as conjugation via a plasmid, transduction via a bacteriophage and the direct transformation of competent bacteria. This phenomenon of gene acquisition has been named horizontal (or lateral) gene transfer (HGT), as opposed to the vertical transmission of genetic material from parent to daughter cell. There is strong evidence that HGT has occurred among very distantly related organisms, from different genera, groups, phyla and even domains of life. In this case, HGT offers a novel and powerful mechanism of adaptation. There are obvious advantages to this strategy in that the genes from other organisms have already been refined by selection and the benefits can be instantaneous. Numerous lineages have successfully exploited novel or previously unsuitable environments by such gene-acquisition events, ranging from those leading to antibiotic resistance (Carattoli *et al.*, 2002; Deng *et al.*, 2003) and thermophily (Forterre *et al.*, 2000) in some bacterial species to photosynthesis (Archibald *et al.*, 2003; Douglas, 1998; Moreira *et al.*, 2000) and aerobiosis (Andersson & Kurland, 1999) in eukaryotes.

The ever-changing core

Because of the difficulty of resolving deep relationships in the tree of life and the possibility that, over billions of years, homologous genes may have lost all traces of similarity, it is an arduous task to study the renewal of genetic repertoires since LUCA.

However, the evolution of gene content in more recent groups can be monitored to highlight the processes of genetic innovation. A core of genes can be described for every natural group of organisms and any clade of a phylogenetic tree. For example, by comparing strains of *E. coli*, it is possible to define a set of genes that is ubiquitous in this species. This core in *E. coli* encompasses about 3000 genes (Welch *et al.*, 2002). In comparison, the core of the *Gammaproteobacteria* (a more ancient group of bacteria which includes *E. coli* and relatively distantly related species such as *Vibrio cholerae* or *Haemophilus influenzae*) contains only about 300 genes (Lerat *et al.*, 2003). Just like the universal core, this set of genes is probably not sufficient to support the entire metabolism of the cell. This suggests a high turnover of genes in this lineage, ancestral genes being regularly replaced, in one lineage or another, by new ones. The nature of these new genes has been studied and they tend to display interesting characteristics that highlight their origins (Daubin & Ochman, 2004a); the composition of most of the species- or clade-specific genes is strongly biased toward A+T nucleotides relative to the rest of the genome, which supports a relationship with bacteriophages and other parasitic elements. In addition, these genes are often found in the vicinity of mobile genes in the genome. These characteristics tend to attenuate with time, with genes specific to deeper clades being less biased in composition and often more dispersed in the genome. Thus, there is strong evidence that the process of gene replacement is facilitated by HGT and particularly by selfish elements such as bacteriophages. In addition, when one aims to identify the species which potentially donated these genes, many of them appear to be unrelated to any other known protein families present in other sequenced genomes. For this reason, these genes have been called ORFans (i.e. open reading frames that are apparently the sole members of their protein families). It has been proposed that these genes originate, or at least evolve their new function, in bacteriophages before being enlisted in the genome of their new host (Daubin & Ochman, 2004b). In this view, bacteriophages and other 'selfish' elements, by continually experimenting with new ways of interacting with their various hosts, would sometimes come up with genetic innovations that benefit both parts and be major actors in the evolution of their host genome.

Antique genes: authentic or forgery?

The approaches used to study the evolution of gene repertoires generally focus on the occurrence of homologous genes in genomes and thus only document events of non-homologous replacement (i.e. acquisition of genes that do not have homologues in the genome). However, if HGT is frequent enough to generate such a turnover of genes, one might expect that replacement by a homologue from a distant species occurs just as frequently. Pushing this reasoning to the limit, Charlebois & Doolittle (2004) have proposed that the ubiquity of core genes is the result of frequent exchanges of these genes among species rather than common ancestry. Indeed, cases of HGT in the core of genes have been described, particularly for tRNA synthetases (Brochier *et al.*, 2002; Brown *et al.*, 2001; Koonin, 2003). As mentioned earlier, those proteins may be a peculiar case in the universal core because they are generally not part of a protein complex and have very conserved substrates (i.e. an amino acid and a tRNA), which may be factors that favour successful transfers among distantly related organisms. However, whether the frequency of such exchanges among species is sufficient to confer universal distribution on a gene is unclear. A number of recent studies, based on careful phylogenetic analyses of protein families, though confirming the importance of HGT, support the view that the transmission of genes in microbial evolution is nevertheless mostly vertical (Beiko *et al.*, 2005; Ge *et al.*, 2005; Gu & Zhang, 2004; Lerat *et al.*, 2005). Several possibilities exist to evaluate the frequency of homologous replacements in genomes. One is to measure the congruence of gene phylogenies, either with each other (when nothing is known of the species phylogeny) or with a reference tree. Another approach is to study possible intermediates of homologous replacement. Indeed, the complete replacement of a gene by a homologue is predicted to occur via several different paths. Firstly, it may occur by homologous recombination. However, such events can take place only among very closely related species, because homologous recombination is strongly constrained by the identity of the sequences involved (Matic *et al.*, 1996). Replacement by this means therefore has very small if not undetectable effects on the gene phylogeny. Secondly, the genome can lose a gene and later replace it with a homologous gene. These events can be documented in certain cases by mapping gene presence on a phylogeny using parsimony. Thirdly and finally, a replacement can take place via an intermediate state where two homologous genes are present in the genome. Lerat *et al.* (2005) have proposed the term 'synologue' to designate genes with several representatives in a genome, because, in the presence of HGT, it is impossible to tell a priori whether such genes have originated from duplication ('paralogue') or from HGT ('xenologue'). With a phylogenetic approach, these authors have estimated the proportion of synology accounted for by each of these two mechanisms in the group of the *Gammaproteobacteria*. They found relatively few gene families with synologues, compared with families represented by one copy per genome, but HGT appears to account for a large part of the observed synology. Interestingly, this high degree of phylogenetic conflict associated with synology is not found in families with one gene per species, which suggests that functional redundancy in a genome is only rarely resolved by the loss of the native gene, but more generally to the disadvantage of the alien gene.

The phylogenetic core

Although vertical transmission of genes predominates in bacterial evolution, HGT can, over long periods of time, affect many gene families, and it is possible that no phylogeny derived from a single gene may faithfully describe the phylogeny of species.

Indeed, cases of HGT in the universal core of genes have been described. As discussed before, in the core defined by Koonin (2003), 15 genes are tRNA synthetases, which are known to illustrate some of the most clear-cut cases of HGT. Therefore, the concept of the phylogenetic core has been introduced to represent those genes which produce concordant phylogenies and have potentially been transferred strictly vertically since the common ancestor. Depending on the group of species considered, this phylogenetic core encompasses more or fewer genes (Brochier *et al.*, 2005; Koonin, 2003; Lerat *et al.*, 2003). From the universal core, no gene can be safely considered a priori as exempt from HGT (Daubin *et al.*, 2002). Several approaches have been proposed to circumvent this problem and to infer a tree of life without restricting the analysis to ubiquitous genes. These approaches either combine the phylogenetic information of several non-ubiquitous genes (Daubin *et al.*, 2002; Clarke *et al.*, 2002; Beiko *et al.*, 2005) or consider the presence and absence of genes in genomes as phylogenetic characters (Snel *et al.*, 1999; Wolf *et al.*, 2001). Interestingly, these methods, though being largely independent of each other, support very similar topologies of the tree of life. These results suggest in particular that both gene phylogenies and genetic repertoires contain congruent and useful phylogenetic information on the history of life. Incidentally, this congruence further supports the idea that HGT in prokaryotes consists mainly of acquisition of novel genes and to a lesser extent of gene exchange among species.

CONCLUSIONS

Bacteria, archaea and eukaryotes all share the same support of genetic information, genetic code and intermediate of protein synthesis, as well as fundamental mechanisms such as recombination, but only a handful of genes are truly ubiquitous in these organisms. This suggests that evolution has produced various ways of accomplishing the same tasks and that the history of life is a long story of gene replacement. In prokaryotes, it appears that HGT has played a significant role in the evolution of genetic repertoires, not only by favouring the exchange of genes among species, but more importantly by allowing the incorporation of novel genes, elaborated in selfish elements. The growing number of completely sequenced genomes and the development of comparative genomics approaches will certainly bring new insight into these mechanisms.

REFERENCES

Akerley, B. J., Rubin, E. J., Novick, V. L., Amaya, K., Judson, N. & Mekalanos, J. J. (2002). A genome-scale analysis for identification of genes required for growth or survival of *Haemophilus influenzae*. *Proc Natl Acad Sci U S A* **99**, 966–971.

Andersson, S. G. & Kurland, C. G. (1999). Origins of mitochondria and hydrogenosomes. *Curr Opin Microbiol* **2**, 535–541.

Archibald, J. M., Rogers, M. B., Toop, M., Ishida, K. & Keeling, P. J. (2003). Lateral gene transfer and the evolution of plastid-targeted proteins in the secondary plastid-containing alga *Bigelowiella natans*. *Proc Natl Acad Sci U S A* **100**, 7678–7683.

Beiko, R. G., Harlow, T. J. & Ragan, M. A. (2005). Highways of gene sharing in prokaryotes. *Proc Natl Acad Sci U S A* **102**, 14332–14337.

Brochier, C., Bapteste, E., Moreira, D. & Philippe, H. (2002). Eubacterial phylogeny based on translational apparatus proteins. *Trends Genet* **18**, 1–5.

Brochier, C., Forterre, P. & Gribaldo, S. (2005). An emerging phylogenetic core of Archaea: phylogenies of transcription and translation machineries converge following addition of new genome sequences. *BMC Evol Biol* **5**, 36.

Brown, J. R., Douady, C. J., Italia, M. J., Marshall, W. E. & Stanhope, M. J. (2001). Universal trees based on large combined protein sequence data sets. *Nat Genet* **28**, 281–285.

Carattoli, A., Filetici, E., Villa, L., Dionisi, A. M., Ricci, A. & Luzzi, I. (2002). Antibiotic resistance genes and *Salmonella* genomic island 1 in *Salmonella enterica* serovar Typhimurium isolated in Italy. *Antimicrob Agents Chemother* **46**, 2821–2828.

Charlebois, R. L. & Doolittle, W. F. (2004). Computing prokaryotic gene ubiquity: rescuing the core from extinction. *Genome Res* **14**, 2469–2477.

Clarke, G. D., Beiko, R. G., Ragan, M. A. & Charlebois, R. L. (2002). Inferring genome trees by using a filter to eliminate phylogenetically discordant sequences and a distance matrix based on mean normalized BLASTP scores. *J Bacteriol* **184**, 2072–2080.

Daubin, V. & Ochman, H. (2004a). Bacterial genomes as new gene homes: the genealogy of ORFans in *E. coli*. *Genome Res* **14**, 1036–1042.

Daubin, V. & Ochman, H. (2004b). Start-up entities in the origin of new genes. *Curr Opin Genet Dev* **14**, 616–619.

Daubin, V., Gouy, M. & Perriere, G. (2002). A phylogenomic approach to bacterial phylogeny: evidence of a core of genes sharing a common history. *Genome Res* **12**, 1080–1090.

Deng, W., Liou, S.-R., Plunkett, G., III, Mayhew, G. F., Rose, D. J., Burland, V., Kodoyianni, V., Schwartz, D. C. & Blattner, F. R. (2003). Comparative genomics of *Salmonella enterica* serovar Typhi strains Ty2 and CT18. *J Bacteriol* **185**, 2330–2337.

Douglas, S. E. (1998). Plastid evolution: origins, diversity, trends. *Curr Opin Genet Dev* **8**, 655–661.

Forterre, P., Bouthier De La Tour, C., Philippe, H. & Duguet, M. (2000). Reverse gyrase from hyperthermophiles: probable transfer of a thermoadaptation trait from archaea to bacteria. *Trends Genet* **16**, 152–154.

Ge, F., Wang, L. S. & Kim, J. (2005). The cobweb of life revealed by genome-scale estimates of horizontal gene transfer. *PLoS Biol* **3**, e316.

Gerdes, S. Y., Scholle, M. D., Campbell, J. W. & 18 other authors (2003). Experimental determination and system level analysis of essential genes in *Escherichia coli* MG1655. *J Bacteriol* **185**, 5673–5684.

Gu, X. & Zhang, H. (2004). Genome phylogenetic analysis based on extended gene contents. *Mol Biol Evol* **21**, 1401–1408.

Jain, R., Rivera, M. C. & Lake, J. A. (1999). Horizontal gene transfer among genomes: the complexity hypothesis. *Proc Natl Acad Sci U S A* **96**, 3801–3806.

Kobayashi, K., Ehrlich, S. D., Albertini, A. & 96 other authors (2003). Essential *Bacillus subtilis* genes. *Proc Natl Acad Sci U S A* **100**, 4678–4683.

Koonin, E. V. (2003). Comparative genomics, minimal gene-sets and the last universal common ancestor. *Nat Rev Microbiol* **1**, 127–136.

Lerat, E., Daubin, V. & Moran, N. A. (2003). From gene trees to organismal phylogeny in prokaryotes: the case of the γ-Proteobacteria. *PLoS Biol* **1**, E19.

Lerat, E., Daubin, V., Ochman, H. & Moran, N. A. (2005). Evolutionary origins of genomic repertoires in bacteria. *PLoS Biol* **3**, e130.

Lynch, M. & Conery, J. S. (2000). The evolutionary fate and consequences of duplicate genes. *Science* **290**, 1151–1155.

Matic, I., Taddei, F. & Radman, M. (1996). Genetic barriers among bacteria. *Trends Microbiol* **4**, 69–72.

Moreira, D., Le Guyader, H. & Philippe, H. (2000). The origin of red algae and the evolution of chloroplasts. *Nature* **405**, 69–72.

Snel, B., Bork, P. & Huynen, M. A. (1999). Genome phylogeny based on gene content. *Nat Genet* **21**, 108–110.

Welch, R. A., Burland, V., Plunkett, G., III & 16 other authors (2002). Extensive mosaic structure revealed by the complete genome sequence of uropathogenic *Escherichia coli*. *Proc Natl Acad Sci U S A* **99**, 17020–17024.

Wolf, Y. I., Rogozin, I. B., Grishin, N. V., Tatusov, R. L. & Koonin, E. V. (2001). Genome trees constructed using five different approaches suggest new major bacterial clades. *BMC Evol Biol* **1**, 8.

Biogeographical diversity of archaeal viruses

Kenneth M. Stedman,[1] Adam Clore[1] and Yannick Combet-Blanc[2]

[1]Department of Biology, Center for Life in Extreme Environments, Portland State University, PO Box 751, Portland, OR 97207-0751, USA

[2]Laboratoire de Microbiologie IRD, Université de Provence, CESB/ESIL case 925, 163 avenue de Luminy, F-13288 Marseille Cedex 9, France

INTRODUCTION

Biogeography, or the spatial distribution of biological diversity, has been studied since Darwin and Wallace in the 1800s. Their studies, and most later studies, concentrated on macroscopic organisms, mostly animals and plants, and many differences between species were observed, often correlated with geographical isolation. The theoretical basis for these differences was established later and is still being refined. The theories of island biogeography have been extremely influential in many fields of biology (Bell *et al*., 2005). Critical to biogeographical studies are comparable organisms from different locations with quantifiable diversity, often sequence diversity.

Microbial biogeography

More recently, micro-organisms have been studied (Finlay, 2002), especially with the advent of molecular tools. Studies using enrichment cultures indicated that identical micro-organisms were present wherever they were collected (Smith *et al*., 1991); however, this is clearly biased due to the relatively small number of micro-organisms that can be cultivated (Pace, 1997). The advent of small-subunit (SSU) rRNA gene sequence analysis indicated that 'everything is everywhere', particularly for spore-forming bacteria (Roberts & Cohan, 1995). It was unclear whether this indicated that there was so much dispersal of these spore-forming organisms that they were identical throughout the world or whether it was general for bacteria due to their extremely large population sizes. For the most part, however, only one gene, generally the SSU rRNA gene, was investigated. Extremophiles are thought to have more barriers to dispersal than mesophilic organisms. Studies in the last few years using multilocus sequence typing

SGM symposium 66: Prokaryotic diversity: mechanisms and significance.
Editors N. A. Logan, H. M. Lappin-Scott & P. C. F. Oyston. Cambridge University Press. ISBN 0 521 86935 8 ©SGM 2006

and variable regions in hyperthermophilic archaea and thermophilic cyanobacteria indicate that there is considerable endemism, i.e. geographically isolated populations that have speciated or are on their way to speciation (Papke *et al.*, 2003; Whitaker *et al.*, 2003).

Virus biogeography

Viruses have been studied considerably less well than either metazoans or bacteria. The mosquito and tick-borne flaviviruses, many of which are pathogenic to humans, have recently been subjected to biogeographical study (Gaunt *et al.*, 2001). These viruses were apparently brought unsuspectingly from Africa to America with the slave trade (Gould *et al.*, 2001). Generally, their genome sequence diversity parallels the epidemiology of disease and the intermediate hosts for the virus, particularly mosquitoes, rather than their geographical dispersal (Gaunt *et al.*, 2001). The biogeography of human immunodeficiency virus 1 has been extremely well documented, but it is also clearly transmitted by geographically mobile humans (Rambaut *et al.*, 2004).

Some plant viruses have also been studied, for obvious agricultural reasons. There is 80–97 % nucleotide sequence identity in complete genomes of isolates of *Wheat streak mosaic virus* (WSMV) from Mexico, Turkey, the Czech Republic and the USA. Unusually, there is a closer relationship between WSMV genomes from Turkey and the USA than between isolates from the USA and Mexico and European isolates. This is thought to be due to human transport, generally through mites. Among other viruses, partial genome sequences within the family *Potyviridae* are 44–99 % identical, with slight correlation to geographical distance, again with potentially confounding anthropogenic transport (Rabenstein *et al.*, 2002).

Bacteriophage diversity and abundance

Viruses that do not cause human or plant disease have been studied even less well, despite animal- and plant-pathogenic viruses being a vanishingly small proportion of total viruses in the environment. A number of groups have shown that viruses of bacteria and archaea greatly outnumber all other organisms; some estimates are as high as 10^{31} in ocean surface waters (Hendrix, 2002; Rohwer & Edwards, 2002; Rohwer, 2003). Bacteriophage have been shown to be highly diverse in sequence, with some intriguing exceptions (see below). In a group of whole genomes of bacteriophage of *Mycobacterium smegmatis*, Hatfull and Hendrix and colleagues found only 50 % of the genes to be similar to any known genes (Pedulla *et al.*, 2003), prompting the proposition that phage sequence diversity is ‘the dark matter of sequence space’ as a result of their extremely large populations and sequence diversity (Pedulla *et al.*, 2003). In parallel, Rohwer and colleagues have been randomly sequencing concentrated liquid samples that should contain the vast majority of virus particles. Very little overlap was observed in their clone libraries, leading them to propose that there are 800 species of virus in 1 kg of soil and up to 10 000 in the same amount of marine sediment (Breitbart *et al.*, 2002, 2004a).

Bacteriophage biogeography

Breitbart *et al.* (2004b) also found identical sequences while using PCR to amplify a T7-like DNA polymerase gene from locations throughout the world. Short & Suttle (2005) have also reported finding identical bacteriophage sequences in samples from throughout the world. However, none of the latter studies studied whole genomes, just portions thereof. This has led to a model of environmental viruses including a relatively small pool of 'active' viruses and a very large pool of 'latent' viruses (Breitbart & Rohwer, 2005).

VIRUSES OF ARCHAEA

Archaeal viruses, particularly those of thermophiles, have been shown to be extremely diverse both morphologically and in their genome sequences. The structures range from turreted icosahedra with previously unknown triangulation numbers (Rice *et al.*, 2004) to bottle-shaped viruses (Häring *et al.*, 2005). Their genomes rarely contain sequences that are similar to sequences from outside their own virus families; in fact, in the extreme case of the Pyrobaculum spherical virus (PSV), none of the 48 ORFs showed significant matches to any other known sequence (Häring *et al.*, 2004; Prangishvili & Garrett, 2005). This has led to the unprecedented assignment of five new virus families by the International Committee on Taxonomy of Viruses (ICTV) to account for these new viruses. Two new proposed families are pending review by the ICTV.

Rudiviruses

One of these families is the *Rudiviridae*. Rudiviruses are rod-shaped viruses, originally found in Iceland and subsequently discovered in Yellowstone National Park (YNP) in the USA and in Italy (Prangishvili *et al.*, 1999; Rice *et al.*, 2001; Vestergaard *et al.*, 2005). The rudiviruses have quite a wide host range. *Sulfolobus virus SIRV-1* and *Sulfolobus virus SIRV-2* (SIRV-1 and SIRV-2) were originally isolated from hyperthermophilic and acidophilic *Sulfolobus islandicus* isolates from Iceland (Prangishvili *et al.*, 1999). *Thermoproteus virus 1* (TTV-1; recently reclassified in the *Lipothrixviridae*) was isolated from an Icelandic *Thermoproteus tenax* isolate (Janekovic *et al.*, 1983) and Acidianus rod-shaped virus 1 (ARV-1) from an Italian *Acidianus* isolate (Vestergaard *et al.*, 2005). Rudivirus-like particles were observed in samples from YNP (Rice *et al.*, 2001). SIRV-1 was shown to have an extremely high rate of mutation for a DNA virus, whereas SIRV-2 seemed much more stable (Prangishvili *et al.*, 1999). These mutations were found to cluster in small parts of the genome flanked by short 12 bp sequences (Peng *et al.*, 2004). SIRV-1 from strain KVEM10/H3 and SIRV-2 from strain HVE10/2

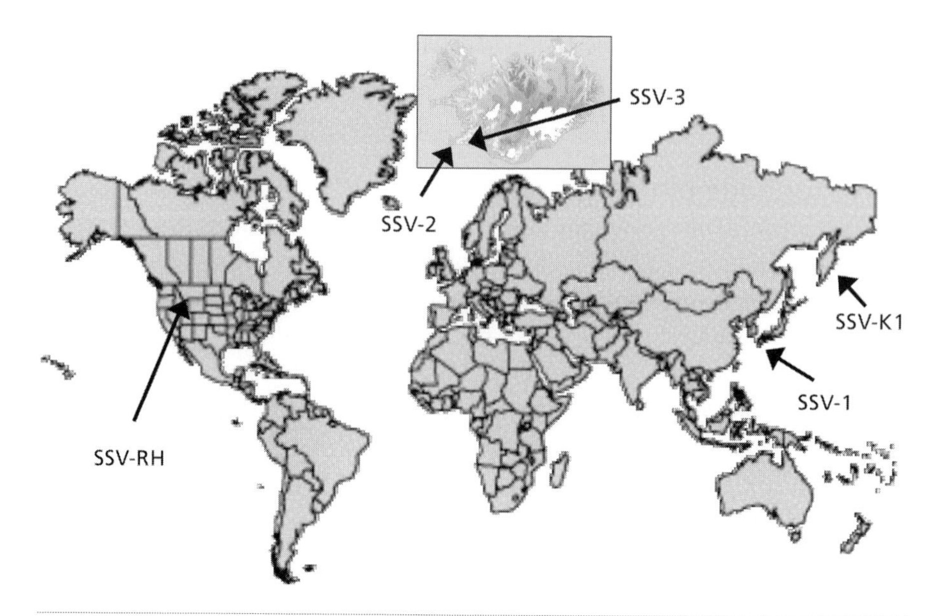

Fig. 1. Map of the world, with Iceland enlarged, indicating *Sulfolobus* fusellovirus isolation sites. SSV-1 was isolated from hot springs at Beppu Onsen on the island of Kyushu in Japan; SSV-2 was isolated from the Reykjanes thermal area of Iceland; SSV-3 was isolated from the Krisovik thermal area of Iceland; SSV-RH was isolated from the Ragged Hills thermal area of the Norris Geyser Basin in YNP; and SSV-K1 was isolated from the Valley of the Geysers in the Kronotsky Natural Reserve in Kamchatka, Russia. World map reproduced with permission from http://www.worldatlas.com.

are very closely related to each other, with blocks of nucleotide sequence that range from 74 to 94 % identical with gaps and insertions of less-similar sequences. Some of these sequences are very similar to sequences from another Icelandic *Sulfolobus* virus, from the family *Lipothrixviridae*, *Sulfolobus islandicus filamentous virus* (SIFV), indicating potential horizontal gene transfer between the two virus families (Peng *et al.*, 2001). There is considerable sequence diversity between *Acidianus filamentous virus 1* (*Lipothrixviridae*) and the *Sulfolobus* SIRVs. Some of this may be due to their different hosts, so they are not directly comparable, making biogeographical studies difficult. The high mutation rate in SIRV-1 and the availability of genome sequences of only two isolates of *Sulfolobus* rudiviruses currently precludes analysis of these viruses. Culture-independent studies of SIRV-like sequences in YNP are currently under way (J. Snyder and M. Young, personal communication).

Fuselloviruses

The best studied and most abundant of archaeal viruses are the *Sulfolobus* fuselloviruses, of the family *Fuselloviridae*, found in acidic hot springs throughout the world (Fig. 1). The first of these, *Sulfolobus spindle-shaped virus 1* (SSV-1), was discovered by Wolfram Zillig and collaborators in an isolate from Beppu, Kyushu, Japan (Martin

Table 1. Approximate distances (km) between springs harbouring fuselloviruses

GPS co-ordinates are not available for the sites of isolation of SSV-1 and SSV-2.

Virus	SSV-1	SSV-2	SSV-RH
SSV-2	9000	–	
SSV-RH	9400	5500	–
SSV-K1	3400	6800	6100

et al., 1984; Yeats *et al.*, 1982). Like most archaeal viruses, SSV-1 has a novel morphology, that of a 60 × 90 nm spindle with a short tail at one end. The complete genome sequence of SSV-1 was determined, and only one of 34 ORFs, that of the viral integrase, had sequence similarity to known proteins (Palm *et al.*, 1991). The genes for three structural proteins, VP1 and VP3 coat proteins and VP2, a DNA-binding protein, were also identified (Reiter *et al.*, 1987). Zillig and colleagues have since discovered many fuselloviruses in samples from throughout Iceland (Zillig *et al.*, 1998); about 8 % of samples collected contained an isolatable virus. One of these, the REY 15/4 *Sulfolobus* isolate, from Reykjanes, Iceland, was found to contain not only the SSV-2 fusellovirus, but also a virus–plasmid hybrid pSSVx that could be transmitted only in the presence of a full-sized fusellovirus (Arnold *et al.*, 1999). Under laboratory conditions, SSVs have a much lower mutation rate than that observed in SIRV-1 (Prangishvili *et al.*, 1999).

***Sulfolobus spindle-shaped virus 2* (SSV-2), a fusellovirus from Iceland.** The complete sequence of SSV-2 was recently determined and analysed (Stedman *et al.*, 2003). Surprisingly, compared with bacteriophage and other *Sulfolobus* viruses, the sequences of SSV-1 and SSV-2 were only 55 % identical to each other. Moreover, one of the structural genes, that for VP2, was missing from the SSV-2 genome. Nonetheless, the genomic gene order or synteny was well conserved and about half of the ORFs were clearly homologous. The question then arose whether the sequences were different as a result of their geographical separation (southern Japan to Iceland is approximately 9000 km) or because of other factors such as the chemistry of the spring where the virus was isolated.

Fuselloviruses worldwide. In addition to Iceland, SSV-like particles were observed in samples from YNP in the USA and Kamchatka, Russia (Rice *et al.*, 2001). Fuselloviruses were isolated from both locations, *Sulfolobus spindle-shaped virus Ragged Hills* (SSV-RH) from YNP and *Sulfolobus spindle-shaped virus Kamchatka 1* (SSV-K1) from Kamchatka. The complete genome sequence was determined and analysed for each of them (Wiedenheft *et al.*, 2004). Approximate geographical distances that separate the sites of isolation are shown in Table 1; the distances varied from a high of about 9500 km (Japan to YNP) to a low of 3400 km (Japan to Kamchatka), with Iceland to YNP and Kamchatka being intermediate. Despite this difference in geographical

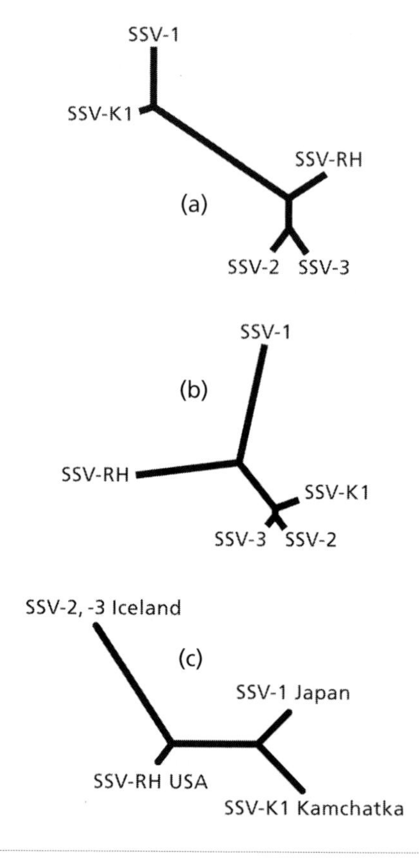

Fig. 2. (a)–(b) Phylogenetic trees of ORFs from complete fusellovirus genomes using maximum-likelihood for the largest ORF (approx. 800 amino acids) (a) and for the integrase gene (b). Trees were determined by maximum-likelihood analysis carried out on manually adjusted alignments of predicted protein sequences using ProML in PHYLIP version 4 using program defaults. All trees were bootstrapped for 100 iterations and used to generate a consensus tree. Bootstrap values were between 70 and 100. (c) Geographical distance tree generated from the data in Table 1. Tree graphics were constructed using Drawtree in PHYLIP.

separation, the sequences of the SSV genomes were equally different, about 55 % identical, similar to the identity between SSV-1 and SSV-2 (Wiedenheft *et al.*, 2004). This is in distinct contrast to their *Sulfolobus* hosts, which showed geographical endemism (Whitaker *et al.*, 2003). These new SSV genomes also appeared to be arranged similarly, with the exception of a large inverted region in SSV-K1. None of the SSVs other than SSV-1 contain a VP2 gene, and it remains to be determined whether any of them contain a DNA-binding protein. In addition, the laboratory host for all of these SSVs, *Sulfolobus solfataricus* strain P2, does not contain an obvious homologue of the VP2 protein in its genome, making virus genome packaging a mystery (She *et al.*, 2001). Eighteen of the ORFs in the genomes of these viruses are conserved, and they map to approximately half of the virus genome. Attempts to make phylogenetic trees using individual ORFs led to conflicting trees (e.g. Fig. 2a, b). These trees did not agree with a distance tree of geographical distance (Fig. 2c), indicating no correlation between genetic and geographical distance. Trees from concatenated sequences also did not show correlation (not shown).

Potentially, many SSV genotypes co-exist in the environment, and the selection of a single virus clone for sequencing has obscured their diversity. All of the complete genome sequences determined and viruses isolated to date were isolated through enrichment cultures, which have been shown to select strongly for a few virus genotypes (Snyder *et al.*, 2004).

Two complementary approaches were taken to address the origin of this sequence diversity. First, the complete sequence of Sulfolobus spindle-shaped virus 3 (SSV-3), a fusellovirus isolated from a sample taken near to the source of SSV-2, was determined. In parallel, culture-independent techniques were used to determine fusellovirus sequence diversity in three isolated but nearby springs in YNP and samples from Lassen Volcanic National Park and elsewhere in California in the USA.

A new fusellovirus from Iceland: SSV-3. The complete sequence of SSV-3 from Krisovik, Iceland, was found to be approximately 70 % identical in complete nucleotide sequence to SSV-2, from Reykjanes, Iceland, about 30 km distant (Y. Combet-Blanc and K. M. Stedman, unpublished; Fig. 1 inset). This is considerably higher than the approximately 55 % overall identity found between SSVs from Japan, Kamchatka, YNP and SSV-2 from Iceland, but much less than that between the Icelandic rudiviruses SIRV-1 and SIRV-2. This indicates that there is a geographical component to fusellovirus sequence diversity, similar to that shown by the *Sulfolobus* hosts (Whitaker *et al.*, 2003). However, some ORFs in the SSV-3 genome are more similar to viruses from other parts of the world than to SSV-2. There are three ORFs that show no similarity whatsoever to ORFs in SSV-2 but are clearly homologous to sequences from SSV-K1 from Kamchatka (Fig. 3). Other ORFs are more similar to the Japanese and USA viruses than to SSV-2 from Iceland (Fig. 3). Some of these ORFs may be different as a result of different functions in SSV-2 and SSV-3, such as the integrase genes and putative packaging ORFs possibly important for competition. Nevertheless, this indicates overall that individual genes in the fuselloviruses have different phylogenies, apparently like the 'morons' described in bacteriophage (Hendrix *et al.*, 1999), and indicating that there is horizontal gene transfer between viruses, as in their hosts (Nelson *et al.*, 1999) and SIRV and SIFV. It is also possible that this is an indication of the lack of sufficient sequence data for complete fusellovirus genomes. Work to remedy this omission is ongoing.

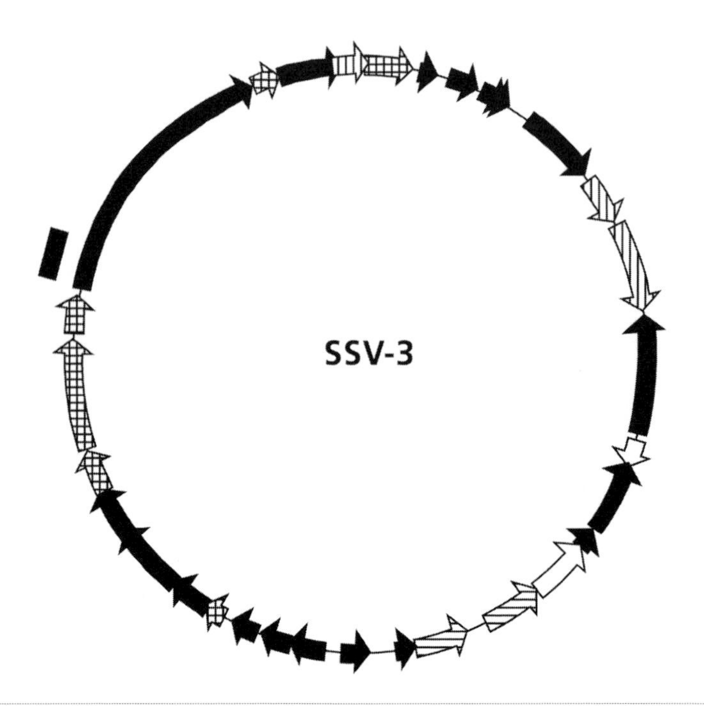

Fig. 3. Genome map of SSV-3. ORFs are shown as arrows. Filled arrows indicate ORFs most similar to corresponding ORFs in SSV-2, from Iceland. Cross-hatched arrows indicate ORFs most similar to corresponding ORFs in SSV-K1 from Kamchatka. Arrows filled with vertical bars indicate ORFs most similar to corresponding ORFs in SSV-1 from Japan. Arrows filled with horizontal bars indicate ORFs most similar to corresponding ORFs in SSV-RH from the USA. Open arrows indicate ORFs with no known similar sequences. The bar outside the main genome indicates the region amplified by conserved primers used on environmental samples.

Culture-independent surveys of virus diversity

In parallel to the determination of complete genomes of isolated viruses, techniques were developed by Snyder *et al.* (2004) to isolate fusellovirus sequences from the environment without the requirement for cell culture. These PCR-based techniques allowed the determination of many fusellovirus-like sequences in samples collected from a single spring (Snyder *et al.*, 2004). Based on the four complete genome sequences known at the time, degenerate primers were designed to amplify a portion of each of these SSV genomes (Fig. 3). These primers were then used to amplify DNA isolated from three different hot springs in YNP on a monthly basis and from a number of different hot springs in Lassen Volcanic National Park and other hot springs in California (Snyder, 2005; Snyder *et al.*, 2004; K. M. Stedman, M. Young, J. Boone and A. Daugherty, unpublished).

Surprisingly, very large amounts of sequence diversity were observed, even in a single spring (Snyder *et al.*, 2004). This diversity was reflected in samples from California

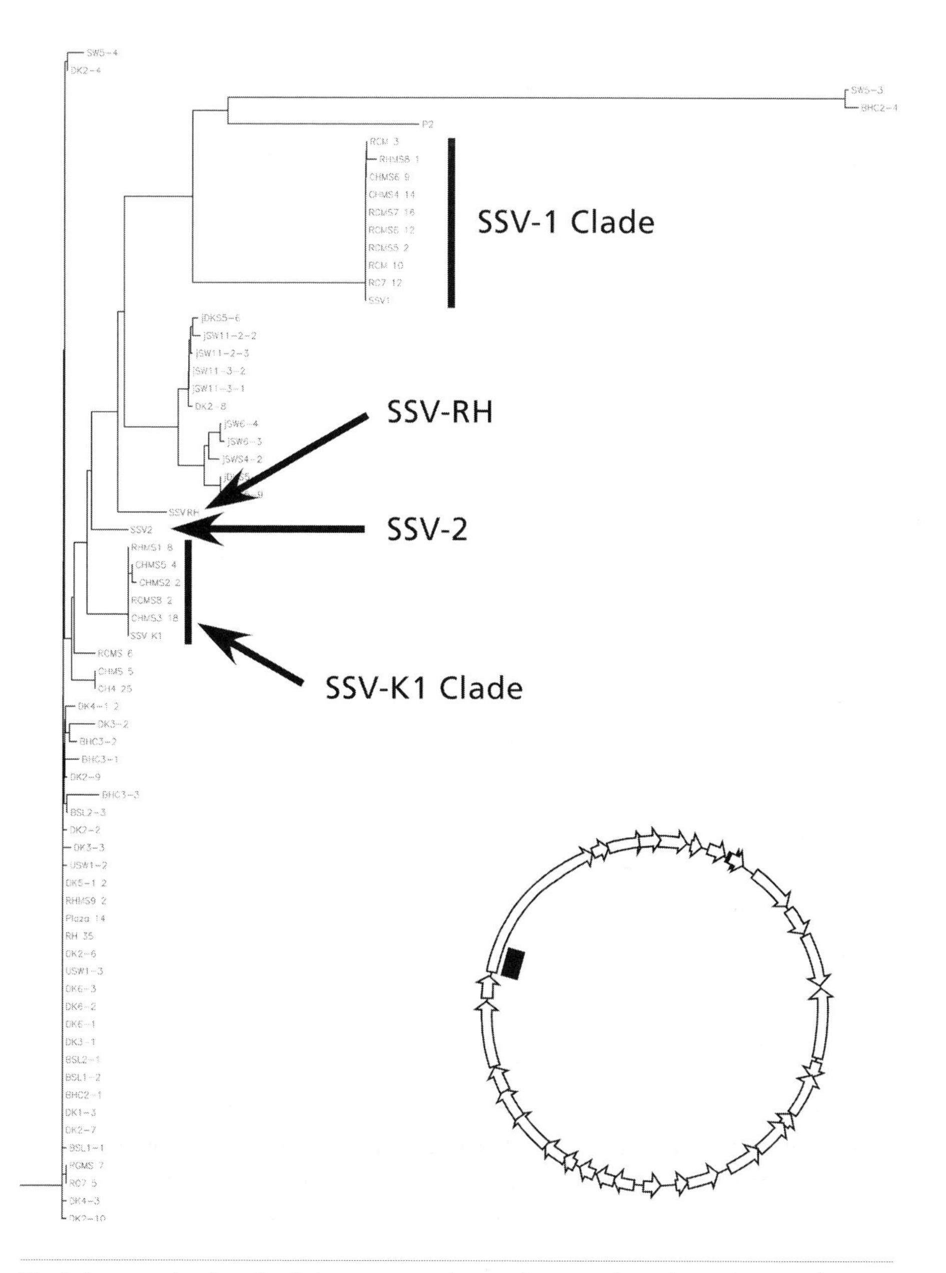

Fig. 4. Sequence tree from fusellovirus sequences from environmental samples. Environmental DNA was amplified with conserved primers (Snyder *et al.*, 2004) to the most conserved sequences in known SSV genomes (dark bar in inset). Approximately 250 nucleotides of sequence was determined from cloned PCR products. Sequences were aligned using CLUSTAL W and a tree was calculated from that alignment (http://workbench.sdsc.edu). Sequences most similar to the sequenced SSV genomes are labelled. Environmental samples were from Lassen Volcanic National Park (SW, DK, USW and BSL), the Long Valley Caldera in eastern California, USA (BHC), and YNP (RC, RH, CH and Plaza). A defective provirus found in the *S. solfataricus* P2 genome is labelled as 'P2'.

(Fig. 4). A few sequences appeared to be related to the sequences of the complete virus genomes from throughout the world, indicating that there is more fusellovirus sequence diversity in a single spring in YNP or Lassen Volcanic National Park than was found in sequences from Iceland, Japan and Kamchatka (Snyder *et al.*, 2004).

The puzzle of fusellovirus biogeography

This poses a conundrum; if there is as much sequence diversity in a single spring at a single time-point as there is between viruses isolated approximately 10 000 km apart, the sequences of viruses isolated close together, e.g. SSV-2 and SSV-3 from Iceland, should not be more similar to each other than to geographically isolated viruses. It is also not clear how the sequence diversity is generated, as fuselloviruses seem to have relatively stable genomes under laboratory conditions. The culture-independent studies were done on small (250–1000 nucleotide) segments of the fusellovirus genomes, which may have different histories than the rest of the genome. There was also no selection for free viruses versus integrated proviruses, which may be under different selective pressure. It is also not clear whether all of the sequences collected are from *Sulfolobus* viruses. Most of the microbes found in the sampled springs by culture-independent screening of SSU rRNA gene clones are not *Sulfolobus* or known hosts for SSVs (Snyder, 2005). Fuselloviruses are known to infect the closely related *Sulfolobus shibatae*, '*Sulfolobus islandicus*' and *S. solfataricus*, but not *Sulfolobus acidocaldarius* or other members of the family *Sulfolobales* (Schleper *et al.*, 1992; Stedman *et al.*, 2003). Some of the observed diversity may be the result of virus dispersal. Fuselloviruses are not very stable at their ambient temperature (80 °C) and pH (3) (Schleper *et al.*, 1992), but are stable for weeks at lower temperatures, potentially allowing distribution (S. Morris and K. M. Stedman, unpublished results).

SUMMARY AND CONCLUDING REMARKS

The fuselloviruses of the hyperthermophilic archaeon *Sulfolobus* are excellent model systems to study virus biogeography. They are not animal pathogens and are apparently not transported by artificially mobile hosts. The SSV hosts seem to be isolated by their growth conditions and show geographical differentiation (Whitaker *et al.*, 2003). They have relatively small (15 kbp) genomes but are quite diverse (Wiedenheft *et al.*, 2004). Preliminary analysis of complete virus genomes indicates that there is some contribution of geographical separation leading to virus divergence. However, culture-independent techniques indicate that there is very high diversity in single springs and that this changes relatively rapidly. To resolve this apparent discrepancy, more complete virus genomes are being sequenced and larger fragments of DNAs from the environment are being analysed.

ACKNOWLEDGEMENTS

Research on virus diversity in the Stedman lab is supported by an NSF-Microbial Observatories grant MCB-0132156. Thanks to Jamie Snyder, Mark Young, Steven Morris, Jane Boone and Aaron Daughtery for sharing their data before publication and for stimulating discussions.

REFERENCES

Arnold, H. P., She, Q., Phan, H., Stedman, K., Prangishvili, D., Holz, I., Kristjansson, J. K., Garrett, R. A. & Zillig, W. (1999). The genetic element pSSVx of the extremely thermophilic crenarchaeon *Sulfolobus* is a hybrid between a plasmid and a virus. *Mol Microbiol* **34**, 217–226.

Bell, T., Ager, D., Song, J. I., Newman, J. A., Thompson, I. P., Lilley, A. K. & van der Gast, C. J. (2005). Larger islands house more bacterial taxa. *Science* **308**, 1884.

Breitbart, M. & Rohwer, F. (2005). Here a virus, there a virus, everywhere the same virus? *Trends Microbiol* **13**, 278–284.

Breitbart, M., Salamon, P., Andresen, B., Mahaffy, J. M., Segall, A. M., Mead, D., Azam, F. & Rohwer, F. (2002). Genomic analysis of uncultured marine viral communities. *Proc Natl Acad Sci U S A* **99**, 14250–14255.

Breitbart, M., Felts, B., Kelley, S., Mahaffy, J. M., Nulton, J., Salamon, P. & Rohwer, F. (2004a). Diversity and population structure of a near-shore marine-sediment viral community. *Proc Biol Sci* **271**, 565–574.

Breitbart, M., Miyake, J. H. & Rohwer, F. (2004b). Global distribution of nearly identical phage-encoded DNA sequences. *FEMS Microbiol Lett* **236**, 249–256.

Finlay, B. J. (2002). Global dispersal of free-living microbial eukaryote species. *Science* **296**, 1061–1063.

Gaunt, M. W., Sall, A. A., de Lamballerie, X., Falconar, A. K., Dzhivanian, T. I. & Gould, E. A. (2001). Phylogenetic relationships of flaviviruses correlate with their epidemiology, disease association and biogeography. *J Gen Virol* **82**, 1867–1876.

Gould, E. A., de Lamballerie, X., Zanotto, P. M. & Holmes, E. C. (2001). Evolution, epidemiology, and dispersal of flaviviruses revealed by molecular phylogenies. *Adv Virus Res* **57**, 71–103.

Häring, M., Peng, X., Brugger, K., Rachel, R., Stetter, K. O., Garrett, R. A. & Prangishvili, D. (2004). Morphology and genome organization of the virus PSV of the hyperthermophilic archaeal genera *Pyrobaculum* and *Thermoproteus*: a novel virus family, the *Globuloviridae*. *Virology* **323**, 233–242.

Häring, M., Rachel, R., Peng, X., Garrett, R. A. & Prangishvili, D. (2005). Viral diversity in hot springs of Pozzuoli, Italy, and characterization of a unique archaeal virus, Acidianus bottle-shaped virus, from a new family, the *Ampullaviridae*. *J Virol* **79**, 9904–9911.

Hendrix, R. W. (2002). Bacteriophages: evolution of the majority. *Theor Popul Biol* **61**, 471–480.

Hendrix, R. W., Smith, M. C., Burns, R. N., Ford, M. E. & Hatfull, G. F. (1999). Evolutionary relationships among diverse bacteriophages and prophages: all the world's a phage. *Proc Natl Acad Sci U S A* **96**, 2192–2197.

Janekovic, D., Wunderl, S., Holz, I., Zillig, W., Gierl, A. & Neumann, H. (1983). TTV-1, TTV-2 and TTV-3: a family of viruses of the extremely thermophilic anaerobic sulfur reducing archaebacterium *Thermoproteus tenax*. *Mol Gen Genet* **192**, 39–45.

Martin, A., Yeats, S., Janekovic, D., Reiter, W. D., Aicher, W. & Zillig, W. (1984). SAV-1, a temperate UV inducible DNA virus-like particle from the archaebacterium *Sulfolobus acidocaldarius* isolate B-12. *EMBO J* **3**, 2165–2168.

Nelson, K. E., Clayton, R. A., Gill, S. R. & 26 other authors (1999). Evidence for lateral gene transfer between archaea and bacteria from genome sequence of *Thermotoga maritima*. *Nature* **399**, 323–329.

Pace, N. R. (1997). A molecular view of microbial diversity and the biosphere. *Science* **276**, 734–740.

Palm, P., Schleper, C., Grampp, B., Yeats, S., McWilliam, P., Reiter, W. D. & Zillig, W. (1991). Complete nucleotide sequence of the virus SSV1 of the archaebacterium *Sulfolobus shibatae*. *Virology* **185**, 242–250.

Papke, R. T., Ramsing, N. B., Bateson, M. M. & Ward, D. M. (2003). Geographical isolation in hot spring cyanobacteria. *Environ Microbiol* **5**, 650–659.

Pedulla, M. L., Ford, M. E., Houtz, J. M. & 17 other authors (2003). Origins of highly mosaic mycobacteriophage genomes. *Cell* **113**, 171–182.

Peng, X., Blum, H., She, Q., Mallok, S., Brugger, K., Garrett, R. A., Zillig, W. & Prangishvili, D. (2001). Sequences and replication of genomes of the archaeal rudiviruses SIRV1 and SIRV2: relationships to the archaeal lipothrixvirus SIFV and some eukaryal viruses. *Virology* **291**, 226–234.

Peng, X., Kessler, A., Phan, H., Garrett, R. A. & Prangishvili, D. (2004). Multiple variants of the archaeal DNA rudivirus SIRV1 in a single host and a novel mechanism of genomic variation. *Mol Microbiol* **54**, 366–375.

Prangishvili, D. & Garrett, R. A. (2005). Viruses of hyperthermophilic crenarchaea. *Trends Microbiol* **13**, 535–542.

Prangishvili, D., Arnold, H. P., Götz, D., Ziese, U., Holz, I., Kristjansson, J. K. & Zillig, W. (1999). A novel virus family, the *Rudiviridae*: structure, virus-host interactions and genome variability of the Sulfolobus viruses SIRV1 and SIRV2. *Genetics* **152**, 1387–1396.

Rabenstein, F., Seifers, D. L., Schubert, J., French, R. & Stenger, D. C. (2002). Phylogenetic relationships, strain diversity and biogeography of tritimoviruses. *J Gen Virol* **83**, 895–906.

Rambaut, A., Posada, D., Crandall, K. A. & Holmes, E. C. (2004). The causes and consequences of HIV evolution. *Nat Rev Genet* **5**, 52–61.

Reiter, W. D., Palm, P., Henschen, A., Lottspeich, F., Zillig, W. & Grampp, B. (1987). Identification and characterization of the genes encoding three structural proteins of the Sulfolobus virus-like particle SSV1. *Mol Gen Genet* **206**, 144–153.

Rice, G., Stedman, K., Snyder, J., Wiedenheft, B., Willits, D., Brumfield, S., McDermott, T. & Young, M. J. (2001). Viruses from extreme thermal environments. *Proc Natl Acad Sci U S A* **98**, 13341–13345.

Rice, G., Tang, L., Stedman, K., Roberto, F., Spuhler, J., Gillitzer, E., Johnson, J. E., Douglas, T. & Young, M. (2004). The structure of a thermophilic archaeal virus shows a double-stranded DNA viral capsid type that spans all domains of life. *Proc Natl Acad Sci U S A* **101**, 7716–7720.

Roberts, M. S. & Cohan, F. M. (1995). Recombination and migration rates in natural populations of *Bacillus subtilis* and *Bacillus mojavensis*. *Evolution* **49**, 1081–1094.

Rohwer, F. (2003). Global phage diversity. *Cell* **113**, 141.

Rohwer, F. & Edwards, R. (2002). The phage proteomic tree: a genome-based taxonomy for phage. *J Bacteriol* **184**, 4529–4535.

Schleper, C., Kubo, K. & Zillig, W. (1992). The particle SSV1 from the extremely thermophilic archaeon *Sulfolobus* is a virus: demonstration of infectivity and of transfection with viral DNA. *Proc Natl Acad Sci U S A* **89**, 7645–7649.

She, Q., Singh, R. K., Confalonieri, F. & 28 other authors (2001). The complete genome of the crenarchaeon *Sulfolobus solfataricus* P2. *Proc Natl Acad Sci U S A* **98**, 7835–7840.

Short, C. M. & Suttle, C. A. (2005). Nearly identical bacteriophage structural gene sequences are widely distributed in both marine and freshwater environments. *Appl Environ Microbiol* **71**, 480–486.

Smith, J. M., Dowson, C. G. & Spratt, B. G. (1991). Localized sex in bacteria. *Nature* **349**, 29–31.

Snyder, J. (2005). *Virus dynamics, archaeal populations, and water chemistry of three acidic hot springs in Yellowstone National Park*. PhD thesis, Montana State University, Bozeman, MT, USA.

Snyder, J. C., Spuhler, J., Wiedenheft, B., Roberto, F. F., Douglas, T. & Young, M. J. (2004). Effects of culturing on the population structure of a hyperthermophilic virus. *Microb Ecol* **48**, 561–566.

Stedman, K. M., She, Q., Phan, H., Arnold, H. P., Holz, I., Garrett, R. A. & Zillig, W. (2003). Relationships between fuselloviruses infecting the extremely thermophilic archaeon *Sulfolobus*: SSV1 and SSV2. *Res Microbiol* **154**, 295–302.

Vestergaard, G., Haring, M., Peng, X., Rachel, R., Garrett, R. A. & Prangishvili, D. (2005). A novel rudivirus, ARV1, of the hyperthermophilic archaeal genus *Acidianus*. *Virology* **336**, 83–92.

Whitaker, R. J., Grogan, D. W. & Taylor, J. W. (2003). Geographic barriers isolate endemic populations of hyperthermophilic archaea. *Science* **301**, 976–978.

Wiedenheft, B., Stedman, K., Roberto, F., Willits, D., Gleske, A. K., Zoeller, L., Snyder, J., Douglas, T. & Young, M. (2004). Comparative genomic analysis of hyperthermophilic archaeal *Fuselloviridae* viruses. *J Virol* **78**, 1954–1961.

Yeats, S., McWilliam, P. & Zillig, W. (1982). A plasmid in the archaebacterium *Sulfolobus solfataricus*. *EMBO J* **1**, 1035–1038.

Zillig, W., Arnold, H. P., Holz, I. & 7 other authors (1998). Genetic elements in the extremely thermophilic archaeon *Sulfolobus*. *Extremophiles* **2**, 131–140.

Is there a link between *Chlamydia* and heart disease?

Lee Ann Campbell and Cho-chou Kuo

Department of Pathobiology, University of Washington, Seattle, WA 98195, USA

CHLAMYDIAE

Chlamydiae are Gram-negative bacteria that are obligately intracellular parasites (Moulder *et al.*, 1984). *Chlamydia trachomatis* and *Chlamydia pneumoniae* are human pathogens. *C. trachomatis* is a leading cause of preventable blindness in developing nations and sexually transmitted disease in the Western world (Moulder *et al.*, 1984). *C. pneumoniae* is an aetiological agent of acute respiratory diseases (Grayston *et al.*, 1998) and causes approximately 10 % of pneumonia cases and 5 % of bronchitis and sinusitis cases in adults (Grayston *et al.*, 1993). *Chlamydia psittaci* is an avian pathogen. Humans can be infected with *C. psittaci*, through direct contact with an infected bird or by inhalation of dust contaminated with excreta of infected birds. This disease, known as psittacosis, is a severe pneumonia with systemic symptoms (Moulder *et al.*, 1984; Everett *et al.*, 1999). Other chlamydial species infect animals (Fukushi & Hirai, 1992; Everett *et al.*, 1999). Although rare, humans may be infected through contact with lower mammals.

Chlamydial infection often becomes chronic and induces chronic inflammatory responses leading to tissue destruction, fibrosis and scarring and, thus, has been termed a disease of immunopathology (Grayston *et al.*, 1985). For example, repeated/chronic ocular infection with *C. trachomatis* can result in blindness and genital tract infection can result in obstruction of the fallopian tube and infertility. Considerable recent attention has been focused on the association of *C. pneumoniae* with atherosclerosis, a disease of chronic inflammation and the leading cause of morbidity and mortality in the Western world. Although *C. trachomatis* and *C. psittaci* have been associated with

SGM symposium 66: Prokaryotic diversity: mechanisms and significance.
Editors N. A. Logan, H. M. Lappin-Scott & P. C. F. Oyston. Cambridge University Press. ISBN 0 521 86935 8 ©SGM 2006

rare cases of acute cardiac diseases, such as myocarditis, endocarditis and pericarditis, only *C. pneumoniae* has been associated with chronic cardiovascular disease, coronary artery disease and other atherosclerotic syndromes including carotid stenosis, stroke, aortic aneurysm and occlusion of the artery of the lower extremity (Kuo *et al.*, 1995; Kuo & Campbell, 2003). This review will cover the observational studies on the association of serum antibody against *C. pneumoniae* and atherosclerosis and detection of the organism in atherosclerotic lesions and *in vitro* and animal model studies on elucidation of putative mechanisms by which *C. pneumoniae* contributes to the disease process.

SEROLOGICAL EVIDENCE OF AN ASSOCIATION OF *C. PNEUMONIAE* WITH ATHEROSCLEROSIS

Key to the recognition of the association of *C. pneumoniae* with respiratory disease and, subsequently, atherosclerosis and other chronic infectious disease was the development of the micro-immunofluorescence test for detection of *C. pneumoniae*-specific antibodies and differentiation of acute infection from chronic or repeated infection (Wang & Grayston, 1986). This test was also instrumental in determination of the prevalence, incidence and geographical distribution of infection. Cumulatively, these studies have demonstrated that infection is ubiquitous and everybody is infected and reinfected in their lifetime (Grayston, 2000). Early studies in Seattle, WA, USA, showed that infection was rare in children under the age of 5, but 75 % of children were infected between the ages of 5 and 14 years (Grayston, 2005). The antibody prevalence in young adults is 50 % and continues to increase throughout adulthood to reach 80 % in the elderly (Wang & Grayston, 1998). Coincidentally, atherosclerosis begins in childhood and develops throughout life.

The first report of an association of *C. pneumoniae* infection and atherosclerosis was by Saikku *et al.* (1988), who found that Finnish men with IgG antibody against *C. pneumoniae* were at increased risk for myocardial infarction and coronary heart disease in comparison with population controls. The most recent study to demonstrate an association of *C. pneumoniae* with coronary heart disease is novel because the population studied was young adults (30- to 50-year-old men in the military), hospitalized for the first time for acute myocardial infarction, and pre-hospitalization sera, obtained 1 year or more before the occurrence, were available. Control subjects were from the same cohort and individually matched by age, race and time of serum collection. In this study, a significant association was observed between high titres to *C. pneumoniae* IgG and IgA antibody and acute myocardial infarction. The authors concluded that recent or chronic infection was associated with an increased risk of heart attacks in this cohort of young men (Arcari *et al.*, 2005). Over 100 papers have been published to address the association of *C. pneumoniae* antibody and atherosclerotic disease (Grayston, 2005). Many of these studies have found an association between antibody against *C. pneumoniae* and atherosclerosis, while a few well-designed studies did not find an association. Overall, these serological studies are difficult to compare and interpret because different study designs were used, different classes of antibodies were measured, different titres of antibodies were considered positive, in some studies, genus-reactive antibodies that could not differentiate *C. pneumoniae* infection from other chlamydial infection were measured and, in some studies, confounding risk factors were not considered while, in those that did, different adjustment methods were used (Grayston, 2005). Although most of these studies used the micro-immunofluorescence test, there is interlaboratory variation in interpretation of results even among investigators with experience with this method (Peeling *et al.*, 2000). Regardless of these issues, seroepidemiological studies provided the impetus for further studies in demonstrating the presence of the organism in atherosclerotic tissues.

EVIDENCE OF *C. PNEUMONIAE* IN ATHEROSCLEROTIC LESIONS

Chlamydiae have a unique developmental cycle and two morphological forms, the elementary body (EB) and the reticulate body (RB). The EB is infectious but not metabolically active, while the RB is metabolically active but not infectious. One of the characteristics of *C. pneumoniae* that was key to the discovery of the organism in the atherosclerotic lesion is the unique pear-shaped ultrastructure of the EB and a large periplasmic space in comparison with the round and dense structure of the EB (Grayston *et al.*, 1988). The first report of *C. pneumoniae* in atherosclerotic tissue was based on finding these ultrastructures in electron micrographs of atheromas and by immunocytochemical staining using *C. pneumoniae*-specific monoclonal antibodies (Shor *et al.*, 1992). An expanded study, using PCR as an additional method, confirmed these findings (Kuo *et al.*, 1993a). In a series of studies carried out at the University of Washington, *C. pneumoniae* was detected in 54 % of atherosclerotic tissue samples examined ($n=272$) and the organism and was not detected in tissue from normal arteries ($n=52$) (Kuo & Campbell, 2000). There have been over 60 studies conducted by investigators worldwide that have found the organism in atherosclerotic lesions throughout the arterial tree using various methodologies including immunocytochemical staining, PCR, *in situ* hybridization and culture (Ramirez, 1996; Jackson *et al.*, 1997a; Maass *et al.*, 1998), although isolation of *Chlamydia* from chronic infections is rare. In the atheroma, *C. pneumoniae* has been identified in foam cells derived from macrophages and smooth-muscle cells and in endothelial cells (Kuo *et al.*, 1993b; Ouchi *et al.*, 2000). A few reports have not found the organism in atherosclerotic lesions. Regardless of the differences in detection rates, methodologies used and inter-laboratory variation, the findings from these cumulative studies leave no question that *C. pneumoniae* exists in atherosclerotic tissues, and any more such observational studies will not further advance the field.

BASIC SCIENCE STUDIES SUPPORT A ROLE FOR *C. PNEUMONIAE* IN ATHEROSCLEROTIC PROCESSES

The most significant question is whether *C. pneumoniae* contributes to the chronic inflammatory processes and progression of atherosclerosis. In the past few years, both *in vitro* studies in cell culture and *in vivo* studies in animal models in four different species (mice, rabbits, rats and pigs) have demonstrated putative mechanisms by which *C. pneumoniae* could contribute to the pathogenesis of atherosclerotic disease (Fig. 1). In view of the lack of success of recent large-scale clinical trials of antibiotic treatment of patients with advanced atherosclerosis, the importance of identifying alternative targets and strategies for intervention or prevention can not be ignored if chronic infection contributes to immunopathology of cardiovascular disease. As discussed below, while the treatment failures are disappointing, treating chronic chlamydial infection has proven difficult, especially late in the stage of inflammatory processes, as exemplified by treatment of chronic *C. trachomatis* infection of the eye and fallopian tube, which may lead to blindness and tubal infertility, respectively. This underscores the need for continuing basic science approaches to elucidate the role of *C. pneumoniae* infection in atherosclerosis.

Atherosclerosis

In order to ascertain whether *C. pneumoniae* or other infectious agents contribute to atherogenesis, it is critical to determine whether the interaction of the organism with the host elicits responses that are characteristic of the disease process. Atherosclerosis results from a chronic, fibroproliferative inflammatory stimulation in which oxidative damage to the subendothelial layer of the artery is important in its pathogenesis (Ross, 1999). An early event in lesion development is endothelial activation, which can be triggered by multiple traditional risk factors, such as cigarette smoking, hyperglycaemia, hyperlipidaemia and circulating immune complexes. Subsequently, monocytes/macrophages are recruited and adhere to the injured endothelium and migrate into the intima. The expression of adhesion molecules on monocytes/macrophages promotes their adherence to their corresponding receptors on the endothelial cells (Hogg & Berlin, 1995; Iiyama *et al.*, 1999). In the intima, macrophages take up oxidized low-density lipoproteins (LDL). These foam cells release cytokines, which further upregulate endothelial cell adhesion molecule expression and result in additional recruitment of leukocytes. Subsequently, smooth-muscle cells migrate into the intima, proliferate and secrete collagen, elastin and proteoglycans to form a fibrous matrix. These processes can cause narrowing of the arterial lumen. The mature atherosclerotic plaque consists of a fibrous cap, which encapsulates an acellular, lipid-rich necrotic core derived from dead foam cells. Destabilization of the advanced atherosclerotic lesion is characterized by expansion of the necrotic core and erosion of the media and fibrous cap. This can result in plaque rupture and thrombosis, which may lead to myocardial infarction.

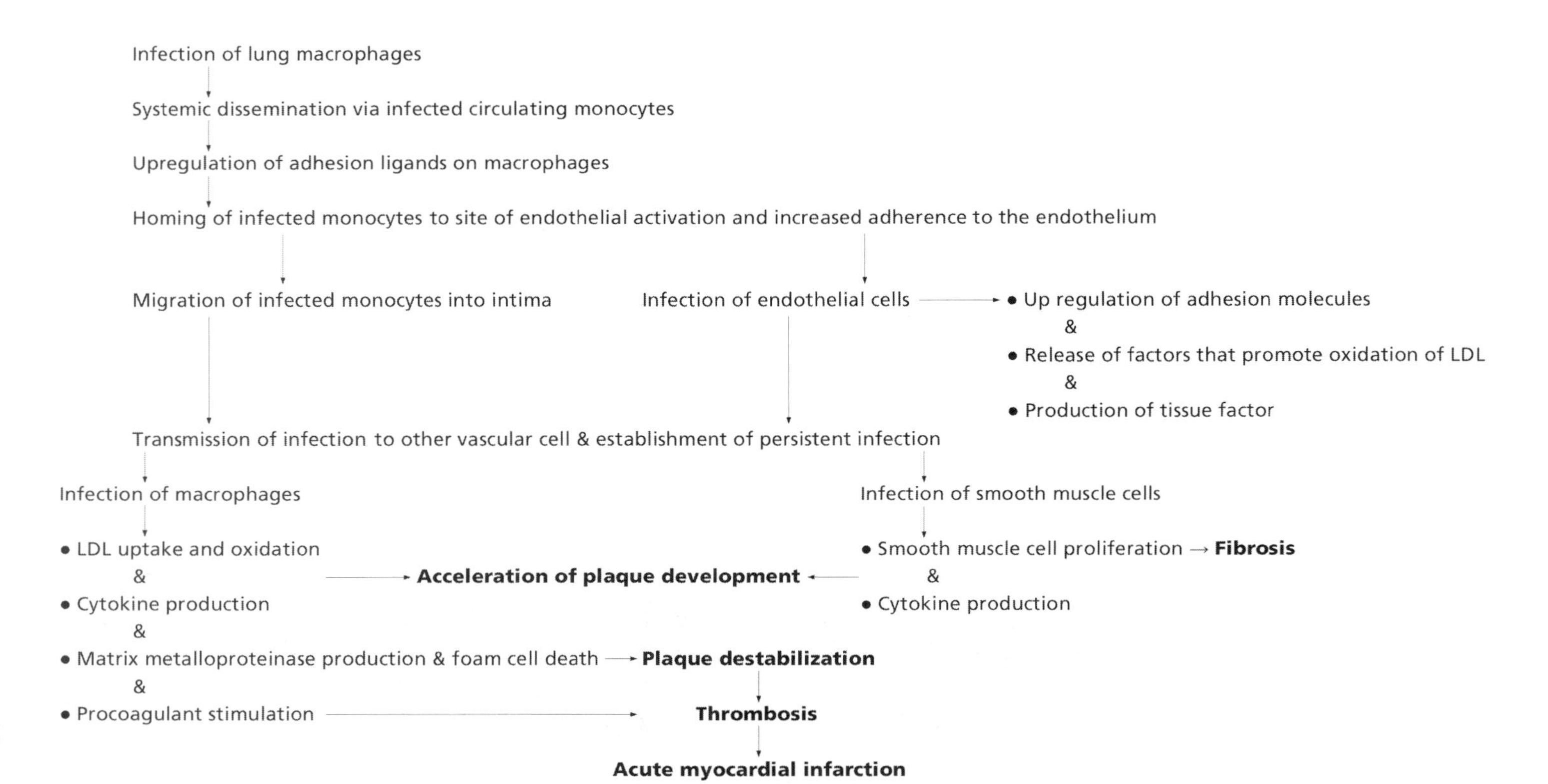

Fig. 1. *Chlamydia pneumoniae* pathogenesis and atherosclerosis.

In vitro studies

In vitro studies on how *C. pneumoniae* infection could contribute to the chronic inflammatory processes of atherosclerosis have demonstrated effects of infection consistent with these processes. *C. pneumoniae* can infect macrophages, smooth-muscle cells and endothelial cells, which are cellular components of the atherosclerotic lesion (Godzik *et al*., 1995; Gaydos *et al*., 1996). Infected monocytes/macrophages can transmit infection to smooth-muscle cells and endothelial cells and interaction of monocytes/macrophages with smooth-muscle cells and endothelial cells enhances susceptibility of these cells to *C. pneumoniae* infection (Lin *et al*., 2000; Puolakkainen *et al*., 2003). *C. pneumoniae* infection of macrophages enhances monocyte adhesion to the endothelium by upregulating expression of $\beta2$ integrin on the surface of macrophages (Kalaygolu *et al*., 2001). Infection of endothelial cells upregulates expression of adhesion molecules on endothelial cells, which results in increased leukocyte adherence to the endothelium and migration to the intima (Kaukoranta-Tolvanen *et al*., 1996; Coombes & Mahony, 2001). *C. pneumoniae* infection of endothelial cells induces platelet-derived growth factor (PDGF), a strong mitogen, which results in smooth-muscle cell proliferation (Coombes *et al*., 2002). Infection of these vascular cells results in the release of proinflammatory cytokines, which could enhance chronic inflammatory responses that accelerate lesion development (Coombes & Mahony, 2001; Heinemann *et al*., 1996; Summersgill *et al*., 2000; Kol *et al*., 1999). Co-culture of infected macrophages with vascular smooth-muscle cells results in activation of a signal transduction profile that is pro-atherosclerotic, including prenylation of Rac1 and RhoA, activation of nuclear transcription factor κB and production of reactive oxygen species and chemokines (Dechend *et al*., 2003). Statins, which lower cholesterol and affect cellular immune responses, decrease transmission of *C. pneumoniae* infection from infected macrophages to smooth-muscle cells, induction of cellular signalling and production of chemokines by smooth-muscle cells and macrophages (Dechend *et al*., 2003; Kothe *et al*., 2000). *C. pneumoniae* can also contribute to lesion formation by its ability to induce foam cell formation in the presence of LDL and oxidation of LDL and promote production of proinflammatory cytokines through contact with foam cells (Kalayoglu & Byrne, 1998; Kalayoglu *et al*., 2000). *C. pneumoniae* can survive and replicate in foam cells (Blessing *et al*., 2002a), which is consistent with the demonstration of the organism in foam cells in the human atherosclerotic lesion (Kuo *et al*., 1993b). Interestingly, *C. pneumoniae* infection of endothelial cells results in the release of acellular components leading to oxidation of LDL (Dittrich *et al*., 2004). *C. pneumoniae* also induces effects that can lead to plaque destabilization, including induction of matrix metalloproteinase (Choi *et al*., 2002; Kol *et al*., 1998), production of tissue factor (TF), which can result in procoagulant activities (Bea *et al*., 2003; Fryer *et al*., 1997; Dechend *et al*., 1999), and promotion of foam cell death through a non-caspase-dependent mechanism (Yaraei *et al*., 2005). TF production in *C. pneumoniae*-

infected macrophages occurs through induction of early growth response factor-1 (Egr-1) (Bea *et al.*, 2003), a proatherogenic factor that contributes to the expression of growth factors such as PDGF-B, cytokines such as TNF-α, adhesion molecules such as ICAM-1 and factors involved in coagulation such as TF (McCaffrey *et al.*, 2000; Silverman & Collins, 1999). Infection of smooth-muscle cells with *C. pneumoniae* also induces Egr-1, resulting in smooth-muscle cell proliferation (Rupp *et al.*, 2005). Injection of *C. pneumoniae*-infected peripheral blood mononuclear cells (PBMCs) into mice caused upregulation of Egr-1 mRNA (Rupp *et al.*, 2005). Interestingly, expression of Egr-1 is elevated in human and mouse atherosclerotic lesions (McCaffrey *et al.*, 2000). Overall, *C. pneumoniae* affects the expression of multiple cellular pathways which contribute to proatherogenic processes, which could result in accelerated lesion formation and plaque destabilization.

Animal models of *C. pneumoniae* infection and atherosclerosis

The association of *C. pneumoniae* and cardiovascular disease has been further strengthened by aetiological studies that demonstrate that infection accelerates atherosclerotic lesion development in mouse, rabbit and rat models of *C. pneumoniae* infection and atherosclerosis. In mice, *C. pneumoniae* disseminates from the lungs to the arteries via monocytes/macrophages and establishes persistent infection in the arteries of hyperlipidaemic mice, in which the organism was detected in foam cells within the atherosclerotic lesion (Moazed *et al.*, 1997, 1998). In contrast, persistent infection was not established in normolipidaemic mice (Moazed *et al.*, 1997). These studies suggested that *C. pneumoniae* has a tropism for the atherosclerotic lesion. The mice used in these studies were apolipoprotein E-deficient (ApoE$^{-/-}$) mice, which are hypercholesterolaemic and develop atherosclerosis spontaneously, and C57BL6/J mice, which do not develop atherosclerosis unless fed a high fat/high cholesterol diet (Reddick *et al.*, 1994; Nakashima *et al.*, 1994). These mouse models mimic what has been reported in humans, as *C. pneumoniae* has been detected in PBMCs (Boman *et al.*, 1998; Smieja *et al.*, 2002) and in foam cells in atherosclerotic tissue (Kuo *et al.*, 1993b), but rarely in normal arterial tissue (Kuo & Campbell, 2000), and is more frequently found in cardiovascular tissue in comparison with non-cardiovascular tissue in autopsy tissue from different sites of the same individual (Jackson *et al.*, 1997b). In ApoE$^{-/-}$ mice (Moazed *et al.*, 1999), C57BL/6J mice fed an atherogenic diet (Blessing *et al.*, 2001), LDL receptor knockout mice (LDLR$^{-/-}$) fed a cholesterol diet (Hu *et al.*, 1999), New Zealand white rabbits fed a diet supplemented with cholesterol (Muhlestein *et al.*, 1998; Fong *et al.*, 2002) and Tg53 male rats, a transgenic hyperlipidaemia–genetic hypertension model (Herrera *et al.*, 2003), *C. pneumoniae* infection has been shown to accelerate atherosclerotic lesion development. In contrast, in the ApoE$^{-/-}$ and LDLR$^{-/-}$ models, *C. trachomatis* had no effect on plaque development (Hu *et al.*, 1999; Blessing *et al.*, 2000) and there was no effect on atherosclerosis in the LDLR$^{-/-}$ model inoculated with heat-inactivated *C. pneumoniae* organisms (Sharma *et al.*, 2004). Likewise, inoculation of the rabbit model with *Mycoplasma pneumoniae*, another respiratory pathogen, had no effects on atherosclerotic lesion development (Fong *et al.*, 2002). In LDLR$^{-/-}$ mice or C57BL/6J mice fed a regular diet not supplemented with cholesterol, infection did not induce atherosclerosis (Hu *et al.*, 1999; Blessing *et al.*, 2002b), suggesting that atherogenic effects of *C. pneumoniae* were dependent on cholesterol. Interestingly, infection accelerates lesion formation if C57BL/6J mice are fed an atherogenic diet at the time of infection, simulating infection of adults, but not if infection occurs prior to feeding a hypercholesterolaemic diet, simulating infection of young children (Blessing *et al.*, 2002b). Another interesting finding from the studies with the rat model was the demonstration that intraperitoneal infection of *C. pneumoniae* induced recruitment of macrophages and accelerated foam cell formation in hyperlipidaemic rats, but not in control normolipidaemic rats on the same genetic background (Herrera *et al.*, 2003), demonstrating what has been observed previously *in vitro*, that *C. pneumoniae* can also induce foam cell formation *in vivo*. In another approach, the direct impact of *C. pneumoniae* infection of the arterial wall was addressed by delivery of the organism to the carotid artery in LDLR-deficient mice and measuring the effects on collar-induced atherosclerosis and plaque composition. This study demonstrated that delivery of *C. pneumoniae*, but not *M. pneumoniae*, resulted in enhanced collar-induced atherosclerosis and a lesion morphology indicative of a more vulnerable plaque (Hauer *et al.*, 2006). Cumulatively, these studies suggest that *C. pneumoniae* is a co-risk factor with hyperlipidaemia and that the atherogenic effects of *C. pneumoniae* infection are contingent on the vascular responses to hyperlipidaemia.

These models of *C. pneumoniae*-accelerated atherosclerosis have provided the opportunity to identify factors that lead to acceleration. In the first study, the role of TNF-α, a proinflammatory cytokine, was addressed by using knockout mice that were deficient in the p55 receptor, which mediates the majority of its effects on inflammatory processes. These studies showed that *C. pneumoniae*-accelerated atherosclerosis was ablated in p55 receptor-knockout mice, suggesting a role of TNF-α in this process (Campbell *et al.*, 2005). In the second study, mice that were deficient in inducible nitric oxide synthase, an enzyme involved in an anti-chlamydial defence mechanism, were fed an atherogenic diet and found to have further acceleration in comparison with control infected mice fed a high fat/high cholesterol diet (Chesebro *et al.*, 2003). These findings suggest that, under conditions of an impaired innate immunity, there was an increase in persistent infection resulting in enhancement of inflammatory processes by infection.

In addition to hyperlipidaemia, another co-factor, vascular injury, has been identified for *C. pneumoniae* and atherosclerosis in the pig model, which is the animal model of atherosclerosis that most closely resembles atherosclerosis in humans. In this model, the coronary artery was injured by a balloon catheter and *C. pneumoniae* was inoculated directly into the pulmonary or coronary artery. An increase in maximal intimal thickening, indicative of smooth-muscle cell proliferation, was observed in the damaged coronary artery, but not in arteries not injured with the catheter. This observation suggests that pre-existing coronary lesions were a pre-requisite for neointima formation by *C. pneumoniae* (Pislaru *et al.*, 2003).

Another mechanism that has been identified by which *C. pneumoniae* could contribute to atherosclerosis is the promotion of endothelial dysfunction. The endothelium plays a role in arterial tone, which is mediated in part by endothelial nitric oxide. In atherosclerotic arteries, endothelial function is impaired by decreased availability of endothelial nitric oxide, which is a vasodilator, or increased production of vasoconstricting factors. Studies in both the ApoE knockout model and the pig model showed that *C. pneumoniae* infection caused endothelium dysfunction (Liuba *et al.*, 2000, 2003). This impairment of endothelial relaxation was through decreased availability of nitric oxide, which, in the pig model, was due to increased nitric oxide synthase activity (Liuba *et al.*, 2003). *C. pneumoniae* infection of the pig also resulted in increased risk of vascular thrombosis by increasing the systemic levels of fibrinogen, to induce a procoagulant state. These authors concluded that the effects of *C. pneumoniae* on impairment of endothelial relaxation and increased fibrinogen levels support a role of infection in atherosclerosis development and may explain the association of *C. pneumoniae* infection with acute coronary syndrome (Liuba *et al.*, 2003).

To address a role of *C. pneumoniae* infection in acute coronary syndromes further, Ezzahiri *et al.* (2003) investigated whether *C. pneumoniae* infection contributed to plaque instability. Consistent with *in vitro* studies, these investigators demonstrated that infection increases production of MMP-2 and MMP-9 in LDLR/ApoE double-knockout mice and reduces the fibrous cap area, suggesting that *C. pneumoniae* could contribute to plaque erosion (Ezzahiri *et al.*, 2003).

Animal models of *C. pneumoniae*-accelerated atherosclerosis have also been used to determine whether treatment with anti-chlamydial antibiotics can prevent the consequences of infection on atherosclerosis. The first study conducted in a rabbit model demonstrated that azithromycin treatment, administered immediately after the last inoculation, decreased *C. pneumoniae*-accelerated plaque development; however, the organism was not eradicated from the aorta (Muhlestein *et al.*, 1998). Another study examined effects of the length of time after *C. pneumoniae* inoculation that treatment was initiated on *C. pneumoniae*-accelerated lesion formation. Initiation of treatment within 5 days after infection with azithromycin, clarithromycin, moxifloxacin and doxycyline largely prevented the aortic lesions (Fong, 2000). However, azithromycin did not reduce atherosclerotic lesion size in *C. pneumoniae*-infected rabbits, while clarithromycin was still somewhat effective. In the ApoE mouse model, if azithromycin treatment was initiated as early as 3 days (Blessing *et al.*, 2005) or as late as 2 weeks after infection (Rothstein *et al.*, 2001), no effect was observed on the acceleration of lesion development induced by *C. pneumoniae*. Cumulatively, these studies suggest that early treatment of infection is necessary to prevent atherogenic effects of infection.

In summary, in multiple animal models of *C. pneumoniae* infection and atherosclerosis, *C. pneumoniae* has been shown to contribute to atherosclerotic processes. These studies provide strong evidence that *C. pneumoniae* is not an innocent bystander but plays a role in the progression of atherogenesis in conjunction with other risk factors. Although animal models have limitations in that they may not mimic human disease processes exactly, these models should be exploited to determine specific mechanisms, provide guidance for development of intervention methods and permit testing of alternative strategies.

CLINICAL TRIALS

Eight early small-scale clinical treatment trials with macrolide antibiotics for prevention of cardiac events in humans yielded mixed results, three showing positive results, four negative results and one an equivocal result (Grayston, 2003). Of the three large trials of secondary prevention in the United States, none have demonstrated a prolonged benefit of antibiotic treatment. In the first study, WIZARD [Weekly Intervention with Zithromax (azithromycin) for Atherosclerosis and Its Related Disorders], over 7000 patients that had a documented myocardial infarction that occurred at least 6 weeks previously and serological evidence of *C. pneumoniae* were enrolled and treated with azithromycin or placebo for 3 months (O'Connor *et al.*, 2003). Overall, the trial was considered negative because, although a favourable trend was noted during the treatment, long-term benefits were not observed. The ACES (Azithromycin and Coronary Events Study) study enrolled 4000 patients with stable coronary heart disease, disregarding *C. pneumoniae* serology, in a randomized, double-blind placebo control trial. Patients were treated with azithromycin or placebo, weekly, for 1 year. The primary end point, a composite of death due to coronary heart disease, non-fatal myocardial infarction, coronary revascularization or hospitalization for unstable angina, did not differ in the azithromycin-treated group in comparison with the patient group receiving placebo. There were also no differences with regard to any of the components of the primary end point, death from any cause or stroke (Grayston *et al.*, 2005). The PROVE IT (Prevastin or atorvastatin evaluation and infection therapy) trial, a double-blind, randomized trial, enrolled 4000 patients with unstable angina.

Patients received a 2 year treatment course of gatifloxacin, in which the antibiotic was given for the first 10 days of each month throughout the trial. The primary end point was a composite of death from any cause, myocardial infarction, unstable angina requiring hospitalization, revascularization performed at least 30 days after randomization or stroke. No benefits were noted in reduction of events in those patients treated with gatifloxacin in comparison with patients given placebo (Cannon *et al.*, 2005). Overall, in a meta-analysis of 11 prospective, randomized placebo-controlled trials using anti-chlamydial antibiotics, including the WIZARD, ACES and PROVE-IT studies, no overall benefit was observed for antibiotic treatment in reducing mortality or cardiovascular events in patients with coronary artery disease (Andraws *et al.*, 2005).

The lack of success of the large-scale clinical trials in secondary prevention of coronary events has stimulated discussion as to whether these failures indicate that chronic infection with *C. pneumoniae* is not a major cause of coronary heart disease (Danesh, 2005) or whether *C. pneumoniae* is indeed an important contributor with other risk factors to the disease process, but that treating chronic infection in patients with advanced, irreversible disease is too late in the inflammatory process and, therefore, antibiotics are unlikely to have an effect (Grayston, 2005; Taylor-Robinson & Boman, 2005; Anderson, 2005). In general, treatment of chronic chlamydial infections has proven difficult and successful treatment often requires long-term therapy. Two factors contributing to this problem are that the infectious EB is non-replicating and therefore not susceptible to antibiotics and that it has been shown *in vitro* that a persistent state can be induced (Beatty *et al.*, 1994), arresting the developmental cycle and rendering the organism unsusceptible to antibiotics. A recent study found that levels of azithromycin or rifampicin that were effective in eradication of *C. pneumoniae* infection of fibroblasts had no effect on infected human PBMCs that were infected with *C. pneumoniae* (Gieffers *et al.*, 2001), supporting the notion that treatment of cardiovascular infection may be problematic as suggested by the inability of antibiotics to eradicate infection in cultured PBMCs. Thus, the negative results of these clinical studies do not rule out a role for infection in atherosclerosis, but underscore the need for basic science studies aimed at the development of targets for earlier intervention or prevention.

REFERENCES

Anderson, J. L. (2005). Infection, antibiotics, and atherothrombosis – end of the road or new beginnings? *N Engl J Med* **352**, 1706–1709.

Andraws, R., Berger, J. S. & Brown, D. L. (2005). Effects of antibiotic therapy on outcomes of patients with coronary artery disease: a meta-analysis of randomized controlled trials. *JAMA* **293**, 2641–2647.

Arcari, C. M., Gaydos, C. A., Nieto, F. J., Krauss, M. & Nelson, K. E. (2005). Association between *Chlamydia pneumoniae* and acute myocardial infarction in young men in the United States military: the importance of timing of exposure measurement. *Clin Infect Dis* **40**, 1123–1130.

Bea, F., Puolakkainen, M. H., McMillen, T., Hudson, F. N., Mackman, N., Kuo, C.-C., Campbell, L. A. & Rosenfeld, M. E. (2003). *Chlamydia pneumoniae* induces tissue factor expression in mouse macrophages via activation of Egr-1 and the MEK-ERK1/2 pathway. *Circ Res* **92**, 394–401.

Beatty, W., Morrison, R. P. & Byrne, G. I. (1994). Persistent chlamydiae: from cell culture to a paradigm for chlamydial pathogenesis. *Microbiol Rev* **58**, 686–699.

Blessing, E., Nagano, S., Campbell, L. A., Rosenfeld, M. E. & Kuo, C.-C. (2000). Effect of *Chlamydia trachomatis* infection on atherosclerosis in apolipoprotein E-deficient mice. *Infect Immun* **68**, 7195–7197.

Blessing, E., Campbell, L. A., Rosenfeld, M. E., Chough, N. & Kuo, C.-C. (2001). *Chlamydia pneumoniae* infection accelerates hyperlipidemia induced atherosclerotic lesion development in C57BL/6J mice. *Atherosclerosis* **158**, 13–17.

Blessing, E., Kuo, C.-C., Lin, T.-M., Campbell, L. A., Bea, F., Chesebro, B. & Rosenfeld, M. E. (2002a). Foam cell formation inhibits growth of *Chlamydia pneumoniae* but does not attenuate *Chlamydia pneumoniae*-induced secretion of proinflammatory cytokines. *Circulation* **105**, 1976–1982.

Blessing, E., Campbell, L. A., Rosenfeld, M. E. & Kuo, C.-C. (2002b). *Chlamydia pneumoniae* and hyperlipidemia are co-risk factors for atherosclerosis: infection prior to induction of hyperlipidemia does not accelerate development of atherosclerotic lesions in C57BL/6J mice. *Infect Immun* **70**, 5332–5334.

Blessing, E., Campbell, L. A., Rosenfeld, M. E., Chesebro, B. & Kuo, C.-C. (2005). A 6 week course of azithromycin treatment has no beneficial effect on atherosclerotic lesion development in apolipoprotein E-deficient mice chronically infected with *Chlamydia pneumoniae*. *J Antimicrob Chemother* **55**, 1037–1040.

Boman, J., Soderberg, S., Forsberg, J. & 8 other authors (1998). High prevalence of *Chlamydia pneumoniae* DNA in peripheral blood mononuclear cells in patients with cardiovascular disease and in middle-aged blood donors. *J Infect Dis* **178**, 274–277.

Campbell, L. A., Nosaka, T., Rosenfeld, M. E., Yaraei, K. & Kuo, C.-C. (2005). Tumor necrosis factor alpha plays a role in the acceleration of atherosclerosis by *Chlamydia pneumoniae* in mice. *Infect Immun* **73**, 3164–3165.

Cannon, C. P., Braunwald, E., McCabe, C. H., Grayston, J. T., Muhlestein, B., Giugliano, R. P., Cairns, R. & Skene, A. M. (2005). Antibiotic treatment of *Chlamydia pneumoniae* after acute coronary syndrome. Pravastatin or Atorvastatin Evaluation and Infection Therapy: Thrombolysis in Myocardial Infarction 22 Investigators. *N Engl J Med* **352**, 1646–1654.

Chesebro, B. B., Blessing, E., Kuo, C.-C., Rosenfeld, M. E., Puolakkainen, M. & Campbell, L. A. (2003). Nitric oxide synthase plays a role in *Chlamydia pneumoniae*-induced atherosclerosis. *Cardiovasc Res* **60**, 170–174.

Choi, E. Y., Kim, D., Hong, B. K., Kwon, H. M., Song, Y. G., Byun, K. H., Park, H. Y., Whang, K. C. & Kim, H. S._(2002). Upregulation of extracellular matrix metalloproteinase inducer (EMMPRIN) and gelatinases in human atherosclerosis infected with *Chlamydia pneumoniae*: the potential role of *Chlamydia pneumoniae* infection in the progression of atherosclerosis. *Exp Mol Med* **34**, 391–400.

Coombes, B. K. & Mahony, J. B. (2001). cDNA array analysis of altered gene expression in human endothelial cells in response to *Chlamydia pneumoniae* infection. *Infect Immun* **69**, 1420–1427.

Coombes, B. K., Chiu, B., Fong, I. W. & Mahony, J. B. (2002). *Chlamydia pneumoniae* infection of endothelial cells induces transcriptional activation of platelet-derived growth factor-B: a potential link to intimal thickening in a rabbit model of atherosclerosis. *J Infect Dis* **185**, 1621–1630.

Danesh, J. (2005). Antibiotics in the prevention of heart attacks. *Lancet* **365**, 365–367.

Dechend, R., Maass, M., Gieffers, J., Dietz, R., Scheidereit, C., Leutz, A. & Gulba, D. C. (1999). *Chlamydia pneumoniae* infection of vascular smooth muscle and endothelial cells activates NF-kappaB and induces tissue factor and PAI-1 expression: a potential link to accelerated arteriosclerosis. *Circulation* **100**, 1369–1373.

Dechend, R., Gieffers, J., Dietz, R., Joerres, A., Rupp, J., Luft, F. C. & Maass, M. (2003). Hydroxymethylglutaryl coenzyme A reductase inhibition reduces *Chlamydia pneumoniae*-induced cell interaction and activation. *Circulation* **108**, 261–265.

Dittrich, R., Dragonas, C., Mueller, A., Maltaris, T., Rupp, J., Beckmann, M. W. & Maass, M. (2004). Endothelial *Chlamydia pneumoniae* infection promotes oxidation of LDL. *Biochem Biophys Res Commun* **319**, 501–505.

Everett, K. D., Bush, R. M. & Andersen, A. A. (1999). Emended description of the order *Chlamydiales*, proposal of *Parachlamydiaceae* fam. nov. and *Simkaniaceae* fam. nov., each containing one monotypic genus, revised taxonomy of the family *Chlamydiaceae*, including a new genus and five new species, and standards for the identification of organisms. *Int J Syst Bacteriol* **49**, 415–440.

Ezzahiri, R., Stassen, F. R., Kurvers, H. A., van Pul, M. M., Kitslaar, P. J. & Bruggeman, C. A. (2003). *Chlamydia pneumoniae* infection induces an unstable atherosclerotic plaque phenotype in LDL-receptor, ApoE double knockout mice. *Eur J Vasc Endovasc Surg* **26**, 88–95.

Fong, I. W. (2000). Antibiotics effects in a rabbit model of *Chlamydia pneumoniae*-induced atherosclerosis. *J Infect Dis* **181** (Suppl. 3), S514–S518.

Fong, I. W., Chiu, B., Viira, E., Jang, D. & Mahony, J. B. (2002). Influence of clarithromycin on early atherosclerotic lesions after *Chlamydia pneumoniae* infection in a rabbit model. *Antimicrob Agents Chemother* **46**, 2321–2326.

Fryer, R. H., Schwobe, E. P., Woods, M. L. & Rodgers, G. M. (1997). *Chlamydia* species infect human vascular endothelial cells and induce procoagulant activity. *J Investig Med* **45**, 168–174.

Fukushi, H. & Hirai, K. (1992). Proposal of *Chlamydia pecorum* sp. nov. for *Chlamydia* strains derived from ruminants. *Int J Syst Bacteriol* **42**, 306–308.

Gaydos, C. A., Summersgill, J. T., Sahney, N. N., Ramirez, J. A. & Quinn, T. C. (1996). Replication of *Chlamydia pneumoniae in vitro* in human macrophages, endothelial cells, and aortic artery smooth muscle cells. *Infect Immun* **64**, 1614–1620.

Gieffers, J., Fullgraf, H., Jahn, J., Klinger, M., Dalhoff, K., Katus, H. A., Solbach, W. & Maass, M. (2001). *Chlamydia pneumoniae* infection in circulating human monocytes is refractory to antibiotic treatment. *Circulation* **103**, 351–356.

Godzik, K., O'Brien, E. R., Wang, S.-P. & Kuo, C. C. (1995). *In vitro* susceptibility of human vascular wall cells to infection with *Chlamydia pneumoniae*. *J Clin Microbiol* **33**, 2411–2414.

Grayston, J. T. (2000). Background and current knowledge of *Chlamydia pneumoniae* and atherosclerosis. *J Infect Dis* **181** (Suppl. 3), S402–S410.

Grayston, J. T. (2003). Antibiotic treatment of atherosclerotic cardiovascular disease. *Circulation* **107**, 1228–1230.

Grayston, J. T. (2005). *Chlamydia pneumoniae* and atherosclerosis. *Clin Infect Dis* **40**, 1131–1132.

Grayston, J. T., Wang, S.-P., Yeh, L.-J. & Kuo, C.-C. (1985). Importance of reinfection in the pathogenesis of trachoma. *Rev Infect Dis* **7**, 717–725.

Grayston, J. T., Kuo, C.-C., Campbell, L. A. & Wang, S.-P. (1988). *Chlamydia pneumoniae* sp. nov. for *Chlamydia* sp. strain TWAR. *Int J Syst Bacteriol* **39**, 88–90.

Grayston, J. T., Aldous, M. B., Easton, A., Wang, S.-P., Kuo, C.-C., Campbell, L. A. & Altman, J. (1993). Evidence that *Chlamydia pneumonia* causes pneumonia and bronchitis. *J Infect Dis* **168**, 1231–1235.

Grayston, J. T., Kronmal, R. A., Jackson, L. A. & 8 other authors (2005). Azithromycin for the secondary prevention of coronary events. The ACES Investigators. *N Engl J Med* **352**, 1637–1645.

Hauer, A. D., de Vos, P., Peterse, N., Ten Cate, H., van Berkel, Th. J. C., Stassen, F. R. M. & Kuiper, J. (2006). Delivery of *Chlamydia pneumoniae* to the vessel wall aggravates atherosclerosis in LDLr$^{-/-}$ mice. *Cardiovasc Res* **69**, 280–288. Published online ahead of print 17 August 2005.

Heinemann, M., Susa, M., Simnacher, U., Marre, R. & Essig, A. (1996). Growth of *Chlamydia pneumoniae* induces cytokine production and expression of CD14 in a human monocytic cell line. *Infect Immun* **64**, 4872–4875.

Herrera, V. L. M., Shen, L., Lopez, L. V., Didishvili, T., Zhang, Y.-X. & Ruiz-Opazo, N. (2003). *Chlamydia pneumoniae* accelerates coronary artery disease progression in transgenic hyperlipidemia-genetic hypertension rat model. *Mol Med* **9**, 135–142.

Hogg, N. & Berlin, C. (1995). Structure and function of adhesion receptors in leukocyte trafficking. *Immunol Today* **16**, 327–330.

Hu, H., Pierce, G. N. & Zhong, G. (1999). The atherogenic effects of chlamydia are dependent on serum cholesterol and specific to *Chlamydia pneumoniae*. *J Clin Invest* **103**, 747–753.

Iiyama, K., Hajra, L., Iiyama, M., Li, H., DiChiara, M., Medoff, B. D. & Cybulsky, M. I. (1999). Patterns of vascular cell adhesion molecule-1 and intercellular adhesion molecule-1 expression in rabbit and mouse atherosclerotic lesions and at sites predisposed to lesion formation. *Circ Res* **85**, 199–207.

Jackson, L. A., Campbell, L. A., Kuo, C.-C., Rodriguez, D. I., Lee, A. & Grayston, J. T. (1997a). Isolation of *Chlamydia pneumoniae* from a carotid endarterectomy specimen. *J Infect Dis* **176**, 292–295.

Jackson, L. A., Campbell, L. A., Schmidt, R. A., Kuo, C. C., Cappuccio, A. L., Lee, M. J. & Grayston, J. T. (1997b). Specificity of detection of *Chlamydia pneumoniae* in cardiovascular atheroma: evaluation of the innocent bystander hypothesis. *Am J Pathol* **150**, 1785–1790.

Kalayoglu, M. V. & Byrne, G. I. (1998). Induction of macrophage foam cell formation by *Chlamydia pneumoniae*. *J Infect Dis* **177**, 725–729.

Kalayoglu, M. V., Indrawati, Morrison, R. P., Morrison, S. G., Yuan, Y. & Byrne, G. I. (2000). Chlamydial virulence determinants in atherogenesis: the role of chlamydial lipopolysaccharide and heat shock protein 60 in macrophage-lipoprotein interactions. *J Infect Dis* **181** (Suppl. 3), S483–S489.

Kaukoranta-Tolvanen, S. S., Ronni, T., Leinonen, M., Saikku, P. & Laitinen, K. (1996). Expression of adhesion molecules on endothelial cells stimulated by *Chlamydia pneumoniae*. *Microb Pathog* **21**, 407–411.

Kol, A., Sukhova, G. K., Lichtman, A. H. & Libby, P. (1998). Chlamydial heat shock protein 60 localizes in human atheroma and regulates macrophage tumor necrosis factor-alpha and matrix metalloproteinase expression. *Circulation* **98**, 300–307.

Kol, A., Bourcier, T., Lichtman, A. H. & Libby, P. (1999). Chlamydial and human heat shock protein 60s activate human vascular endothelium, smooth muscle cells, and macrophages. *J Clin Invest* **103**, 571–577.

Kothe, H., Dalhoff, K., Rupp, J., Muller, A., Kreuzer, J., Maass, M. & Katus, H. A. (2000). Hydroxymethylglutaryl coenzyme A reductase inhibitors modify the inflammatory response of human macrophages and endothelial cells infected with *Chlamydia pneumoniae*. *Circulation* **101**, 1760–1763.

Kuo, C.-C. & Campbell, L. A. (2000). Detection of *Chlamydia pneumoniae* in arterial tissues. *J Infect Dis* **181** (Suppl. 3), S432–S436.

Kuo, C.-C. & Campbell, L. A. (2003). Chlamydial infections of the cardiovascular system. *Frontiers Biosci* **8**, e36–e43.

Kuo, C. C., Shor, A., Campbell, L. A., Fukushi, H., Patton, D. L. & Grayston, J. T. (1993a). Demonstration of *Chlamydia pneumoniae* in atherosclerotic lesions of coronary arteries. *J Infect Dis* **167**, 841–849.

Kuo, C.-C., Gown, A. M., Benditt, E. P. & Grayston, J. T. (1993b). Detection of *Chlamydia pneumoniae* in aortic lesions of atherosclerosis by immunocytochemical stain. *Arterioscler Thromb* **13**, 1501–1504.

Kuo, C.-C., Jackson, L. A., Campbell, L. A. & Grayston, J. T. (1995). *Chlamydia pneumoniae* (TWAR). *Clin Microbiol Rev* **8**, 451–461.

Lin, T.-M., Campbell, L. A., Rosenfeld, M. E. & Kuo, C.-C. (2000). Monocyte-endothelial cell coculture enhances infection of endothelial cells with *Chlamydia pneumoniae*. *J Infect Dis* **181**, 1096–1100.

Liuba, P., Karnani, P., Pesonen, E., Paakkari, I., Forslid, A., Johansson, L., Persson, K., Wadstrom, T. & Laurini, R._(2000). Endothelial dysfunction after repeated *Chlamydia pneumoniae* infection in apolipoprotein E-knockout mice. *Circulation* **102**, 1039–1044.

Liuba, P., Pesonen, E., Paakkari, I., Batra, S., Forslid, A., Kovanen, P., Pentikainen, M., Persson, K. & Sandstrom, S. (2003). Acute *Chlamydia pneumoniae* infection causes coronary endothelial dysfunction in pigs. *Atherosclerosis* **167**, 215–222.

Maass, M., Bartels, C., Engel, P. M., Mamat, U. & Sievers, H. H. (1998). Endovascular presence of viable *Chlamydia pneumoniae* is a common phenomenon in coronary artery disease. *J Am Coll Cardiol* **31**, 827–832.

McCaffrey, T. A., Fu, C., Du, B. & 8 other authors (2000). High-level expression of Egr-1 and Egr-1-inducible genes in mouse and human atherosclerosis. *J Clin Invest* **105**, 653–662.

Moazed, T. C., Kuo, C.-C., Grayston, J. T. & Campbell, L. A. (1997). Murine models of *Chlamydia pneumoniae* infection and atherosclerosis. *J Infect Dis* **175**, 883–890.

Moazed, T. C., Kuo, C.-C., Grayston, J. T. & Campbell, L. A. (1998). Evidence of systemic dissemination of *Chlamydia pneumoniae* via macrophages in the mouse. *J Infect Dis* **177**, 1322–1325.

Moazed, T. C., Campbell, L. A., Rosenfeld, M. E., Grayston, J. T. & Kuo, C.-C. (1999). *Chlamydia pneumoniae* infection accelerates the progression of atherosclerosis in apolipoprotein E-deficient mice. *J Infect Dis* **180**, 238–241.

Moulder, J. W., Hatch, T. P., Kuo, C.-C., Schachter, J. & Storz, J. (1984). Genus I. *Chlamydia* Jones, Rake and Stearns 1945, 55^{AL}. In *Bergey's Manual of Systematic Bacteriology*, vol. 1, pp. 729–739. Edited by N. R. Krieg & J. G. Holt. Baltimore: Williams & Wilkins.

Muhlestein, J. B., Anderson, J. L., Hammond, E. H., Zhao, L., Trehan, S., Schwobe, E. P. & Carlquist, J. F. (1998). Infection with *Chlamydia pneumoniae* accelerates the development of atherosclerosis and treatment with azithromycin prevents it in a rabbit model. *Circulation* **97**, 633–636.

Nakashima, Y., Plump, A. S., Raines, E. W., Breslow, J. L. & Ross, R. (1994). ApoE-deficient mice develop lesions of all phases of atherosclerosis throughout the arterial tree. *Arterioscler Thromb* **14**, 133–140.

O'Connor, C. M., Dunne, M. W., Pfeffer, M. A., Muhlestein, J. B., Yao, L., Gupta, S., Benner, R. J., Fisher, M. R. & Cook, T. D. (2003). Azithromycin for the secondary prevention of coronary heart disease events: the WIZARD study: a randomized controlled trial. Investigators in the WIZARD Study. *JAMA* **290**, 1459–1466.

Ouchi, K., Fujii, B., Kudo, S., Shirai, M., Yamashita, K., Gondo, T., Ishihara, T., Ito, H. & Nakazawa, T. (2000). *Chlamydia pneumoniae* in atherosclerotic and nonatherosclerotic tissue. *J Infect Dis* **181** (Suppl 3), S441–S443.

Peeling, R. W., Wang, S. P., Grayston, J. T. & 13 other authors (2000). *Chlamydia pneumoniae* serology: interlaboratory variation in microimmunofluorescence assay results. *J Infect Dis* **181** (Suppl. 3), S426–S429.

Pislaru, S. V., Van Ranst, M., Pislaru, C. & 7 other authors (2003). *Chlamydia pneumoniae* induces neointima formation in coronary arteries of normal pigs. *Cardiovasc Res* **57**, 834–842.

Puolakkainen, M., Campbell, L. A., Lin, T. M., Richards, T., Patton, D. L. & Kuo, C. C. (2003). Cell-to-cell contact of human monocytes with infected arterial smooth-muscle cells enhances growth of *Chlamydia pneumoniae. J Infect Dis* **187**, 435–440.

Ramirez, J. A. (1996). Isolation of *Chlamydia pneumoniae* from the coronary artery of a patient with coronary atherosclerosis. The *Chlamydia pneumoniae*/Atherosclerosis Study Group. *Ann Intern Med* **125**, 979–982.

Reddick, R. L., Zhang, S. H. & Maeda, N. (1994). Atherosclerosis in mice lacking *apoE*. Evaluation of lesional development and progression. *Arterioscler Thromb* **14**, 141–147.

Ross, R. (1999). Atherosclerosis – an inflammatory disease. *N Engl J Med* **340**, 115–126.

Rothstein, N. M., Quinn, T. C., Madico, G., Gaydos, C. A. & Lowenstein, C. J. (2001). Effect of azithromycin on murine arteriosclerosis exacerbated by *Chlamydia pneumoniae. J Infect Dis* **183**, 232–238.

Rupp, J., Hellwig-Burgel, T., Wobbe, V., Seitzer, U., Brandt, E. & Maass, M. (2005). *Chlamydia pneumoniae* infection promotes a proliferative phenotype in the vasculature through Egr-1 activation *in vitro* and *in vivo*. *Proc Natl Acad Sci U S A* **102**, 3447–3452.

Saikku, P., Leinonen, M., Mattila, K., Ekman, M. R., Nieminen, M. S., Makela, P. H., Huttunen, J. K. & Valtonen, V. (1988). Serological evidence of an association of a novel *Chlamydia*, TWAR, with chronic coronary heart disease and acute myocardial infarction. *Lancet* ii, 983–986.

Sharma, J., Niu, Y., Ge, J., Pierce, G. N. & Zhong, G. (2004). Heat-inactivated *C. pneumoniae* organisms are not atherogenic. *Mol Cell Biochem* **260**, 147–152.

Shor, A., Kuo, C. C. & Patton, D. L. (1992). Detection of *Chlamydia pneumoniae* in coronary arterial fatty streaks and atheromatous plaques. *S Afr Med J* **82**, 158–161.

Silverman, E. S. & Collins, T. (1999). Pathways of Egr-1-mediated gene transcription in vascular biology. *Am J Pathol* **154**, 665–670.

Smieja, M., Mahony, J., Petrich, A., Boman, J. & Chernesky, M. (2002). Association of circulating *Chlamydia pneumoniae* DNA with cardiovascular disease: a systematic review. *BMC Infect Dis* **2**, 21.

Summersgill, J. T., Molestina, R. E., Miller, R. D. & Ramirez, J. A. (2000). Interactions of

Chlamydia pneumoniae with human endothelial cells. *J Infect Dis* **181** (Suppl 3), S479–S482.

Taylor-Robinson, D. & Boman, J. (2005). The failure of antibiotics to prevent heart attacks. It's not necessarily the end of the road. *BMJ* **331**, 361–362.

Wang, S.-P. & Grayston, J. T. (1986). Microimmunofluorescence serological studies with the TWAR organism. In *Chlamydial Infections: Sixth International Symposium on Human Chlamydial Infections*, pp. 329–332. Edited by D. Oriel, G. Ridgway, J. Schachter, D. Taylor-Robinson & M. Ward. Cambridge: Cambridge University Press.

Wang, S. P. & Grayston, J. T. (1998). *Chlamydia pneumoniae* (TWAR) micro-immunofluorescence antibody studies – 1998 update. In *Chlamydial Infections: Proceedings of the Ninth International Symposium on Human Chlamydial Infection*, pp. 155–159. Edited by R. S. Stephens, G. I. Byrne, G. Christiansen & 7 other editors. San Francisco: International Chlamydia Symposium.

Yaraei, K., Campbell, L. A., Zhu, X., Liles, W. C., Kuo, C.-C. & Rosenfeld, M. E. (2005). *Chlamydia pneumoniae* augments the oxidized low-density lipoprotein-induced death of mouse macrophages by a caspase-independent pathway. *Infect Immun* **73**, 4315–4322.

Unculturable oral bacteria

William G. Wade

King's College London Dental Institute at Guy's, King's College and St Thomas' Hospitals, Infection Research Group, London SE1 9RT, UK

INTRODUCTION

The human mouth is heavily colonized with bacteria. There are around 100 million bacteria in every millilitre of saliva, and it is estimated that around 800 bacterial species are present as part of the commensal microflora. Unlike the normal microflora at other body sites that live in harmony with the host, the oral microflora, at this stage of man's evolution, has to be controlled. If plaque accumulations are not removed by brushing or by other mechanical or chemical means, the gums become inflamed, a condition known as gingivitis. Around some teeth in some individuals, a more serious condition arises, known as periodontitis, which is an inflammatory condition leading to loss of attachment between the gums and teeth and destruction of the supporting structures of the teeth, which can eventually lead to their loss. Individuals with high sucrose intake in their diets are also at risk of dental caries where certain species, notably *Streptococcus mutans* and lactobacilli, ferment the sucrose to produce acids which demineralize the enamel layer. If left untreated, the bacteria invade first the dentine and then the pulp, rendering it non-vital, which in turn can lead to the formation of an abscess around the apex of the tooth. Such abscesses can spread via the tissue planes or the bloodstream to cause serious infections elsewhere in the body, such as abscesses of the liver or brain, which can be fatal. The mouth is also an important reservoir of bacteria capable of causing opportunistic infections, when they infect normally sterile sites such as damaged heart valves, or following bite wounds or in immunocompromised individuals where the bacteria gain access to the bloodstream and can cause septicaemia. The treatment and prevention of oral infections and their complications places a substantial financial burden on society. The key to improved prevention and treatment strategies is

SGM symposium 66: Prokaryotic diversity: mechanisms and significance.
Editors N. A. Logan, H. M. Lappin-Scott & P. C. F. Oyston. Cambridge University Press. ISBN 0 521 86935 8 ©SGM 2006

Table 1. Phyla of the domain *Bacteria* found in the human mouth

Major constituent phyla	Phyla detected rarely
Actinobacteria	*Acidobacteria*
Bacteroidetes	*Chlamydiae*
Firmicutes	*Chloroflexi*
Fusobacteria	'*Deinococcus–Thermus*'
Proteobacteria	OP11
Spirochaetes	
'*Synergistes*'	
TM7	

a better understanding of the composition of the oral microflora and the interactions between the constituent micro-organisms and between the micro-organisms and the host, in health and disease.

THE ORAL HABITAT

The human mouth includes a wide range of habitats. The hardest tissue in the body, the teeth, is found in the mouth alongside the soft shedding surfaces of the mucosa inside the cheeks. The amount of oxygen available varies markedly at different sites in the mouth; the anterior tongue is exposed to oxygen at almost atmospheric levels while, in the gingival crevice, anaerobic conditions prevail (Socransky & Manganiello, 1971). Each different habitat in the mouth is colonized by a diverse, but characteristic, microflora.

THE DIVERSITY OF THE ORAL MICROFLORA

The main focus of this chapter will be the bacterial community of the mouth, but a range of other micro-organisms is also present. Around half of the population harmlessly carry fungi, in the form of *Candida* species, although these yeasts can cause thrush and other infections (Cannon & Chaffin, 1999). The protozoa *Entamoeba gingivalis* and *Trichomonas tenax* are also frequent oral residents, feeding on bacteria and food debris (Wantland *et al.*, 1958). Herpes simplex virus, the cause of cold sores, is the virus most commonly found in the mouth (Youssef *et al.*, 2002). Other viruses can be found in the mouth during systemic viral infections such as mumps, hepatitis and AIDS.

The majority of bacteria found in the mouth belong to the domain *Bacteria*, but two phylotypes of the genus *Methanobrevibacter*, a member of the *Archaea*, have been detected in patients with periodontitis and have been associated with disease activity (Lepp *et al.*, 2004). To date, representatives of 13 phyla of the domain *Bacteria* have been detected in the mouth (Table 1). Members of five phyla, *Acidobacteria*,

Chlamydiae, *Chloroflexi*, *Deinococcus–Thermus* and OP11, have been detected only rarely in the mouth and may not be true residents. The mouth is of course an open system and continually exposed to exogenous bacteria. The oral ecosystem in healthy individuals, however, is extremely resistant to colonization by exogenous organisms (Vollaard & Clasener, 1994) although, if the immune system is compromised by systemic illness or if the oral microflora itself is disrupted by antimicrobial treatment, colonization with opportunistic pathogens such as *Candida* species, coliforms and *Pseudomonas* can occur and infection can result (Sullivan *et al.*, 2001).

CULTURE-INDEPENDENT ANALYSIS OF THE ORAL MICROFLORA

It has long been recognized that not all of the bacteria present in the mouth can be cultured. Comparisons of microscopic and viable counts on agar media suggest that only about half of the oral microflora can be cultured using conventional culture media (Wilson *et al.*, 1997). The methods used to study the unculturable proportion of the oral microflora have been those developed for the study of bacteria in the environment where the proportion of culturable organisms is very much lower. Indeed, it is estimated that less than 1 % of bacteria on Earth can be cultured (Hugenholtz *et al.*, 1998). The methods are based on the amplification by PCR of housekeeping genes, typically that encoding the small-subunit rRNA, the construction of *Escherichia coli* libraries containing cloned genes and the sequencing and phylogenetic identification of the genes. Most studies have used so-called 'universal' PCR primers to amplify the 16S rRNA gene which were designed to amplify targets from either or both of the prokaryotic domains.

Bacterial communities from a wide range of oral habitats and disease states have now been studied by these techniques. For example, Dymock *et al.* (1996) examined the microflora associated with dentoalveolar abscesses and identified five unculturable phylotypes that did not correspond to any previously described bacterial species. Munson *et al.* (2002) performed combined culture and molecular analysis on samples collected from five patients with endodontic infections. The teeth were opened aseptically and samples were collected from the infected root canals by needle and syringe. The two techniques combined demonstrated an average of around 20 bacterial species per sample, considerably more than had been previously reported by culture alone. Only one species, *Dialister invisus*, was found in all five samples, and 27 of the 65 taxa detected among 261 isolates and 624 clones were novel phylotypes. The microflora was dominated by members of the phyla *Firmicutes*, *Bacteroidetes* and *Actinobacteria*. A higher proportion of the *Firmicutes* was detected by the molecular method than by culture, while the converse was true for the *Actinobacteria*. In a study of similar design, the microflora of dental caries penetrating into dentine was investigated (Munson *et al.*, 2004). A microflora dominated by the same three phyla was revealed, albeit in different proportions, with members of the *Actinobacteria* making up a substantially larger proportion of the microflora than in the endodontic lesions. Ninety-five taxa were detected among the 496 isolates and 1577 clones sequenced, of which 44 were only detected by the molecular method and 31 were previously undescribed.

Deposits of plaque around the teeth have also been studied by culture-independent methods. Kroes *et al.* (1999) identified 77 phylotypes in two samples of subgingival plaque collected from one individual with mild gingivitis, 48 % of which were novel. The same habitat was studied by Paster *et al.* (2001) in healthy subjects and patients with a variety of periodontal conditions. Three hundred and forty-seven phylotypes were identified, 215 of which were novel. In a study comparing the microfloras of 15 periodontally diseased and 15 healthy subjects, Kumar *et al.* (2005) identified 274 phylotypes among 4500 clones sequenced, 60 % of which were as yet uncultivated. The genera *Peptostreptococcus*, *Selenomonas* and *Veillonella* were represented primarily by uncultivated taxa and the genera *Desulfobulbus*, *Lachnospira*, *Megasphaera* (*Anaeroglobus*) and *Synergistes* were represented entirely by uncultivated phylotypes. There were marked differences between the microflora found in health and disease and, interestingly, the majority of the taxa associated with disease belonged to Gram-positive genera, a finding at variance with the traditional association of Gram-negative organisms and periodontal disease.

The adoption of molecular methods for the characterization of the oral microflora has benefits in addition to being able to detect the unculturable component. The use of 16S rRNA gene sequence analysis is affording far greater precision to the identification of oral bacteria than has been possible previously using phenotypic tests. In turn, precise identification is revealing that the bacterial communities associated with different oral infections are highly specific to those conditions. A meta analysis (Wade *et al.*, 2005) was performed on studies investigating endodontic lesions, dentinal caries and subgingival plaque in periodontitis (de Lillo *et al.*, 2006; Munson *et al.*, 2002, 2004). Sixty-two taxa were found in the endodontic samples and 98 in the carious lesions, while the periodontal samples were the most species-rich with 161 taxa present. Fifteen taxa were found in both the caries and endodontic samples, 17 in the caries and periodontal samples and 21 in the endodontic and periodontal samples. Only four taxa were found in all three sample sets.

PHYLOGENETIC DISTRIBUTION OF UNCULTURABLE BACTERIA

Unculturable oral bacteria fall into two main categories: those that are related to known, culturable species and those that belong to lineages with no culturable

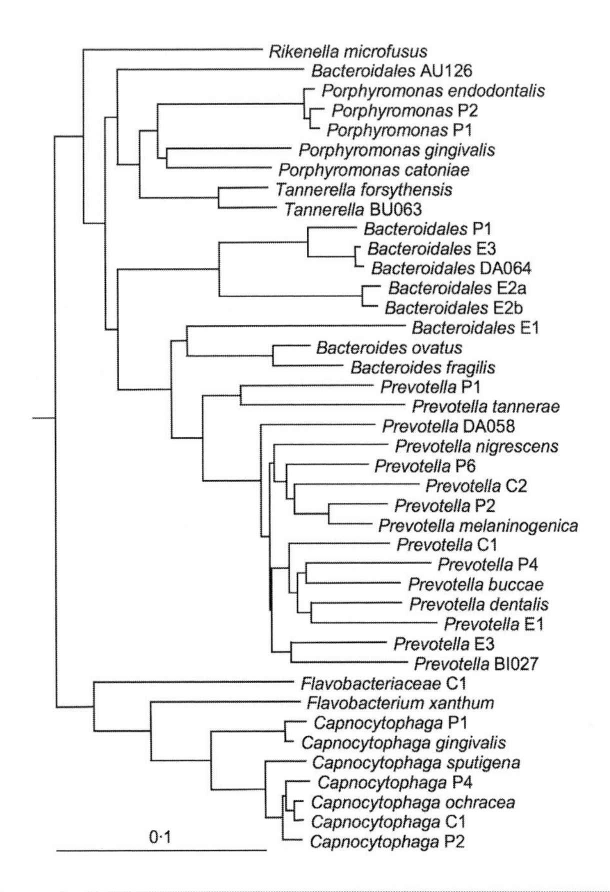

Fig. 1. Phylogenetic tree based on 16S rRNA gene sequence comparisons over 1317 aligned bases showing relationships between novel taxa identified in the phylum *Bacteroidetes* and related species. The tree was constructed using the neighbour-joining method following distance analysis of aligned sequences. Bar, 0·1 nucleotide substitutions per site.

representatives. The former group may, in fact, not be unculturable but may simply represent taxa identified from molecular analyses that are culturable but for which culturable representatives have not yet been identified. For example, there is a large number of un-named species with the genus *Prevotella* (Fig. 1). However, the phylotype PUS3.42, originally described as 'unculturable' (Dymock *et al.*, 1996), has subsequently been shown to correspond to the cultivable species *Prevotella tannerae*. This example and others demonstrate the continuing need for culture-based studies allied to bacterial systematics in order to provide a solid foundation for culture-independent molecular investigations.

A number of examples of the second group have been described. These include oral representatives of entire divisions from which no examples have yet been cultured, such

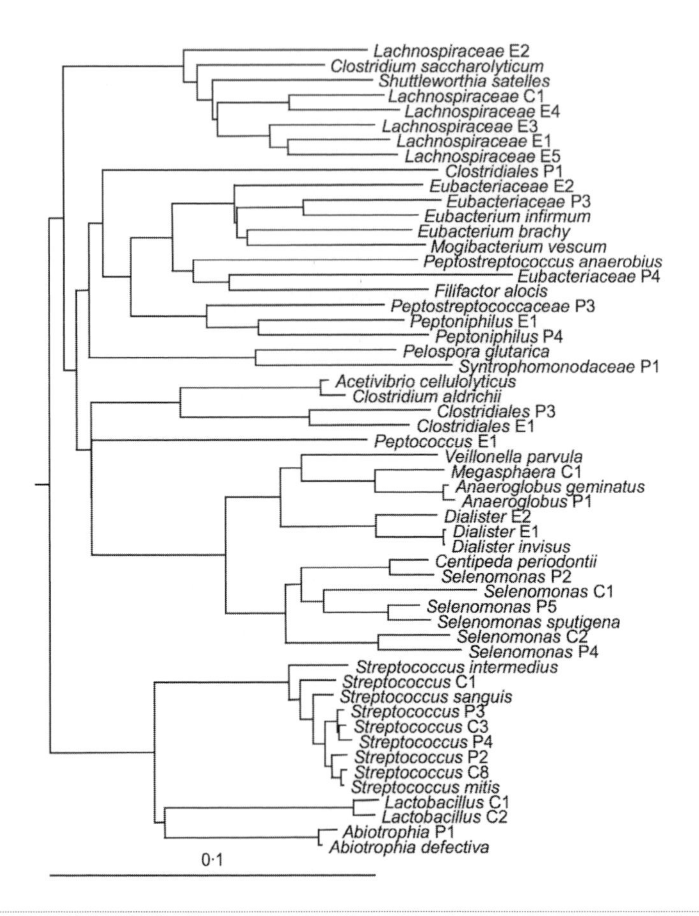

Fig. 2. Phylogenetic tree based on 16S rRNA gene sequence comparisons over 1227 aligned bases showing relationships between novel taxa identified in the phylum *Firmicutes* and related species. The tree was constructed using the neighbour-joining method following distance analysis of aligned sequences. Bar, 0·1 nucleotide substitutions per site.

as candidate division TM7, detected in subgingival plaque (de Lillo *et al.*, 2006; Paster *et al.*, 2001). Numbers of the members of the TM7 division as a whole were found to be largest in mild periodontitis, compared with health and severe periodontitis, and phylotype IO25 appeared to be specifically associated with disease (Brinig *et al.*, 2003). Oligonucleotides specific for this division have been used to detect members of the group in bioreactor sludge by fluorescent *in situ* hybridization, and revealed these organisms to be sheathed filaments (Hugenholtz *et al.*, 2001). Transmission electron microscopy of the filaments revealed them to have Gram-positive cell walls and thus only the third phylum, with *Actinobacteria* and *Firmicutes*, to include Gram-positive organisms (Hugenholtz *et al.*, 2001).

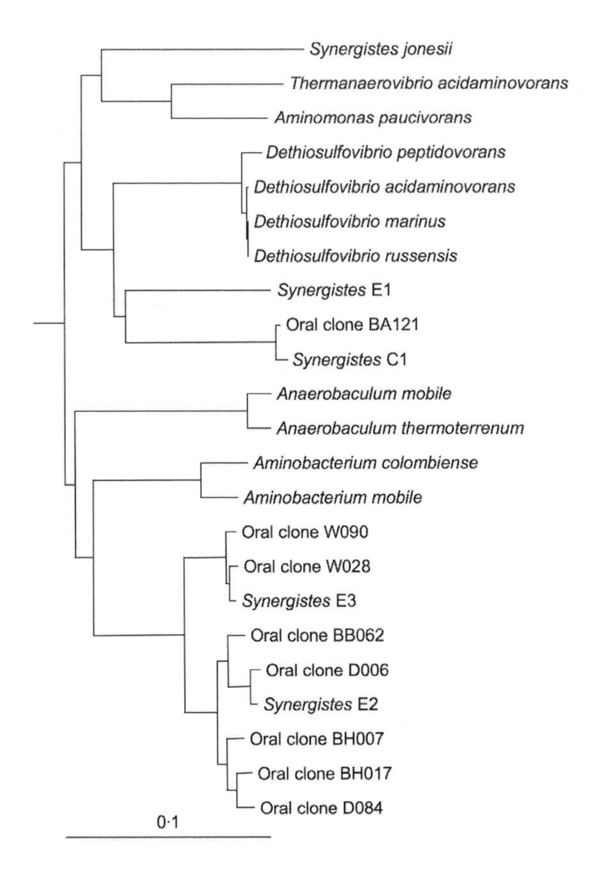

Fig. 3. Phylogenetic tree based on 16S rRNA gene sequence comparisons over 1138 aligned bases showing relationships between members of the proposed phylum '*Synergistes*' and related species. The tree was constructed using the neighbour-joining method following distance analysis of aligned sequences. Bar, 0·1 nucleotide substitutions per site.

A number of lineages within the phylum *Firmicutes* have no culturable representatives (Fig. 2). The largest of these belongs to a single deep branch of the family *Lachnospiraceae*, but others are found within the families *Eubacteriaceae* and *Peptostreptococcaceae* and within the genus *Selenomonas*. Similarly, the phylum *Bacteroidetes* (Fig. 1) includes a deep unculturable branch within the order *Bacteroidales*.

Unculturable lineages are also found within the interesting proposed phylum '*Synergistes*' (Fig. 3). Originally thought to be monophyletic (Hugenholtz *et al*., 1998), it has now been shown to include organisms from a wide variety of habitats, including the human oral cavity (Godon *et al*., 2005). The proposed phylum '*Synergistes*' includes sequences previously assigned to the phyla *Deferribacteres* (Paster *et al*., 2001) and

Firmicutes (Munson *et al.*, 2004), in addition to representatives of the genera *Aminomonas*, *Aminobacterium*, *Anaerobaculum*, *Dethiosulfovibrio* and *Thermanaerovibrio*. Human oral taxa make up two distinct branches within '*Synergistes*'. The first of these, including *Synergistes* C1 and *Synergistes* E1, includes cultivable strains which grow readily on blood agar incubated anaerobically at 37 °C and resemble *Prevotella* species in their cellular and colonial morphology. As yet, no cultivable representatives have been identified among the second branch, which includes *Synergistes* E2 and E3.

The majority of spirochaetes found in the mouth remain unculturable. Choi *et al.* (1994) screened a library constructed with universal 16S rRNA PCR primers and DNA from a single patient with periodontitis, using a treponeme-specific probe. Twenty-three 'species' were found among 81 spirochaetal clones sequenced. Further evidence for the high diversity of oral members of the genus *Treponema* was provided by Dewhirst *et al.* (2000), who identified 47 unculturable taxa of this genus, in addition to the ten cultivable species among around 500 cloned 16S rRNA genes from patients with a variety of periodontal conditions.

The reasons for the unculturability of these lineages remain unknown. Artificial media may not contain nutrients essential for growth, or it may be that members of oral microflora, having evolved in mammals over millions of years as a multispecies biofilm, have become dependent on other members of the community for growth. Such interactions could be nutritional or mediated by bacterial cytokines (reviewed by Wade, 2004). Such interactions have been modelled *in vitro*. Kaeberlein *et al.* (2002) markedly increased the cultivability of the bacterial community in a marine sediment by growing them in a simulated natural environment in a diffusion chamber which allowed chemical interaction between members of the community. Individual taxa that were cultured using this technique could be subsequently grown on solid agar media in co-culture with other organisms.

BIASES IN MOLECULAR CHARACTERIZATION OF COMPLEX BACTERIAL COMMUNITIES

Although the molecular techniques described here are able to characterize those organisms not able to grow on artificial media *in vitro*, they are not without their own biases. Each stage of the process can introduce bias and the important considerations have been discussed extensively elsewhere (von Wintzingerode *et al.*, 1997). Care has to be taken during DNA extraction from the samples to ensure that those with robust cell walls, typically Gram-positive species, are adequately lysed. The PCR is an extremely important step, as the specificity of the primers used will obviously influence which templates are amplified. The number of cycles used in the PCR is also important (Bonnet *et al.*, 2002). A standard PCR of 30–35 cycles will distort the composition of the amplified genes in favour of the dominant members of the community. Between nine and 25 cycles have been recommended for use in molecular microbiology studies.

The sequencing of the cloned genes can also present problems. The sequencing primers may not be specific for templates derived from uncommon phyla, and G+C-rich templates are inherently difficult to sequence. Care must be taken not simply to discard genes that are difficult to sequence, since this will obviously bias the results against these taxa. The G+C content of the DNA of the source organisms appears to be a critical factor in these studies, since it has been observed in a number of studies comparing culture with molecular analysis that proportions of taxa belonging to the *Actinobacteria*, the high-G+C Gram-positives, are underestimated in molecular analyses compared with culture (Munson *et al.*, 2002, 2004). One hypothesis is that, during the PCR, *Taq* polymerase struggles to amplify G+C-rich sections of the template. This phenomenon is well known and such regions cause the polymerase to pause or terminate prematurely (Henke *et al.*, 1997). Because the 16S rRNA gene is highly conserved and all genes in an organism try to match the overall G+C content of their genome, some of the variable regions in the 16S rRNA gene are extremely G+C-rich. Various additives for the PCR have been suggested to destabilize the secondary structures formed by G+C-rich regions which interfere with the polymerase (Baskaran *et al.*, 1996), and studies are in progress to evaluate their usefulness in molecular ecology studies.

UNCULTURABLE BACTERIA ASSOCIATED WITH ORAL DISEASE

PCR/cloning/sequencing studies such as those described above, while useful for the detection of previously undescribed taxa, are time-consuming; such studies then typically only include a small number of subjects and samples. However, when unculturable bacteria have been identified by the PCR/cloning/sequencing methods described above, it is straightforward to use the sequence data obtained to design oligonucleotide probes or PCR primers specific for each phylotype which can then be used to detect them in subsequent studies. For example, primers were designed to three unculturable phylotypes found in acute dental infections and used in PCRs with subgingival plaque collected from patients with periodontitis and healthy controls (Harper-Owen *et al.*, 1999). One phylotype, PUS9.170, was found to be significantly associated with deep pockets in periodontitis. A similar approach was taken by Kumar *et al.* (2003), who used PCR primers specific for 39 taxa, including species associated with health and disease and 10 unculturable phylotypes. Six phylotypes were found to be disease-associated: *Synergistes* D084 and BH017, *Bacteroidales* AU126, *Anaeroglobus* BB166, OP11 division X112 and TM7 division IO25. '*Synergistes*' phylotypes appear to be important in endodontic infections and abscesses, where they are among the most prevalent taxa (Rocas & Siqueira, 2005; Siqueira *et al.*, 2005).

CONCLUSIONS

Culture-independent molecular analyses have shown the oral microflora to be far more species-rich than was previously demonstrable by cultural studies alone. Numerous novel bacterial taxa have been identified and entire lineages without cultivable representatives have been revealed, including some phyla. Some currently unculturable taxa have been shown to be associated with oral disease. Future studies should be directed at elucidating the reasons why so many bacterial species cannot be cultured and improving culture media so that they can be grown *in vitro*. At the same time, metagenomic studies of complex communities should make it possible to sequence and reconstruct the genomes of currently unculturable organisms. This will provide information regarding their pathogenic potential and may provide information that will help efforts to culture them. The mouth provides an excellent model system for complex diseases resulting from a breakdown in the normal homeostasis between the commensal microflora and its human host and deserves further study.

ACKNOWLEDGEMENTS

The contributions of Andrew Weightman, David Dymock, David Spratt, Mark Munson and Ana de Lillo to the work described here are gratefully acknowledged. The studies were supported by grants from the Wellcome Trust (refs 058950 and 061118), Guy's and St Thomas' Charitable Foundation and the South West Regional Health Authority.

REFERENCES

Baskaran, N., Kandpal, R. P., Bhargava, A. K., Glynn, M. W., Bale, A. & Weissman, S. M. (1996). Uniform amplification of a mixture of deoxyribonucleic acids with varying GC content. *Genome Res* **6**, 633–638.

Bonnet, R., Suau, A., Dore, J., Gibson, G. R. & Collins, M. D. (2002). Differences in rDNA libraries of faecal bacteria derived from 10- and 25-cycle PCRs. *Int J Syst Evol Microbiol* **52**, 757–763.

Brinig, M. M., Lepp, P. W., Ouverney, C. C., Armitage, G. C. & Relman, D. A. (2003). Prevalence of bacteria of division TM7 in human subgingival plaque and their association with disease. *Appl Environ Microbiol* **69**, 1687–1694.

Cannon, R. D. & Chaffin, W. L. (1999). Oral colonization by *Candida albicans*. *Crit Rev Oral Biol Med* **10**, 359–383.

Choi, B. K., Paster, B. J., Dewhirst, F. E. & Gobel, U. B. (1994). Diversity of cultivable and uncultivable oral spirochetes from a patient with severe destructive periodontitis. *Infect Immun* **62**, 1889–1895.

de Lillo, A., Ashley, F. P., Palmer, R. M., Munson, M. A., Kyriacou, L., Weightman, A. J. & Wade, W. G. (2006). Novel subgingival bacterial phylotypes detected using multiple universal polymerase chain reaction primer sets. *Oral Microbiol Immunol* **21**, 61–68.

Dewhirst, F. E., Tamer, M. A., Ericson, R. E., Lau, C. N., Levanos, V. A., Boches, S. K., Galvin, J. L. & Paster, B. J. (2000). The diversity of periodontal spirochetes by 16S rRNA analysis. *Oral Microbiol Immunol* **15**, 196–202.

Dymock, D., Weightman, A. J., Scully, C. & Wade, W. G. (1996). Molecular analysis of microflora associated with dentoalveolar abscesses. *J Clin Microbiol* **34**, 537–542.

Godon, J. J., Moriniere, J., Moletta, M., Gaillac, M., Bru, V. & Delgenes, J. P. (2005). Rarity associated with specific ecological niches in the bacterial world: the 'Synergistes' example. *Environ Microbiol* **7**, 213–224.

Harper-Owen, R., Dymock, D., Booth, V., Weightman, A. J. & Wade, W. G. (1999). Detection of unculturable bacteria in periodontal health and disease by PCR. *J Clin Microbiol* **37**, 1469–1473.

Henke, W., Herdel, K., Jung, K., Schnorr, D. & Loening, S. A. (1997). Betaine improves the PCR amplification of GC-rich DNA sequences. *Nucleic Acids Res* **25**, 3957–3958.

Hugenholtz, P., Goebel, B. M. & Pace, N. R. (1998). Impact of culture-independent studies on the emerging phylogenetic view of bacterial diversity. *J Bacteriol* **180**, 4765–4774.

Hugenholtz, P., Tyson, G. W., Webb, R. I., Wagner, A. M. & Blackall, L. L. (2001). Investigation of candidate division TM7, a recently recognized major lineage of the domain Bacteria with no known pure-culture representatives. *Appl Environ Microbiol* **67**, 411–419.

Kaeberlein, T., Lewis, K. & Epstein, S. S. (2002). Isolating "uncultivable" microorganisms in pure culture in a simulated natural environment. *Science* **296**, 1127–1129.

Kroes, I., Lepp, P. W. & Relman, D. A. (1999). Bacterial diversity within the human subgingival crevice. *Proc Natl Acad Sci U S A* **96**, 14547–14552.

Kumar, P. S., Griffen, A. L., Barton, J. A., Paster, B. J., Moeschberger, M. L. & Leys, E. J. (2003). New bacterial species associated with chronic periodontitis. *J Dent Res* **82**, 338–344.

Kumar, P. S., Griffen, A. L., Moeschberger, M. L. & Leys, E. J. (2005). Identification of candidate periodontal pathogens and beneficial species by quantitative 16S clonal analysis. *J Clin Microbiol* **43**, 3944–3955.

Lepp, P. W., Brinig, M. M., Ouverney, C. C., Palm, K., Armitage, G. C. & Relman, D. A. (2004). Methanogenic archaea and human periodontal disease. *Proc Natl Acad Sci U S A* **101**, 6176–6181.

Munson, M. A., Pitt-Ford, T., Chong, B., Weightman, A. & Wade, W. G. (2002). Molecular and cultural analysis of the microflora associated with endodontic infections. *J Dent Res* **81**, 761–766.

Munson, M. A., Banerjee, A., Watson, T. F. & Wade, W. G. (2004). Molecular analysis of the microflora associated with dental caries. *J Clin Microbiol* **42**, 3023–3029.

Paster, B. J., Boches, S. K., Galvin, J. L., Ericson, R. E., Lau, C. N., Levanos, V. A., Sahasrabudhe, A. & Dewhirst, F. E. (2001). Bacterial diversity in human subgingival plaque. *J Bacteriol* **183**, 3770–3783.

Rocas, I. N. & Siqueira, J. F., Jr (2005). Detection of novel oral species and phylotypes in symptomatic endodontic infections including abscesses. *FEMS Microbiol Lett* **250**, 279–285.

Siqueira, J. F., Jr, Rocas, I. N., Cunha, C. D. & Rosado, A. S. (2005). Novel bacterial phylotypes in endodontic infections. *J Dent Res* **84**, 565–569.

Socransky, S. S. & Manganiello, S. D. (1971). The oral microbiota of man from birth to senility. *J Periodontol* **42**, 485–496.

Sullivan, A., Edlund, C. & Nord, C. E. (2001). Effect of antimicrobial agents on the ecological balance of human microflora. *Lancet Infect Dis* **1**, 101–114.

Vollaard, E. J. & Clasener, H. A. (1994). Colonization resistance. *Antimicrob Agents Chemother* **38**, 409–414.

von Wintzingerode, F., Gobel, U. B. & Stackebrandt, E. (1997). Determination of microbial diversity in environmental samples: pitfalls of PCR-based rRNA analysis. *FEMS Microbiol Rev* **21**, 213–229.

Wade, W. G. (2004). Non-culturable bacteria in complex commensal populations. In *Advances in Applied Microbiology*, pp. 93–103. Edited by A. I. Laskin, J. W. Bennett & G. M. Gadd. San Diego: Elsevier Academic Press.

Wade, W. G., Munson, M. A., de Lillo, A. & Weightman, A. J. (2005). Specificity of the oral microflora in dentinal caries, endodontic infections and periodontitis. In *Interface Oral Health Science*, pp. 150–157. International Congress Series. Edited by M. Watanabe, N. Takahashi & H. Takada. Amsterdam: Elsevier.

Wantland, W. W., Wantland, E. M., Remo, J. W. & Winquist, D. L. (1958). Studies on human protozoa. *J Dent Res* **37**, 949–950.

Wilson, M. J., Weightman, A. J. & Wade, W. G. (1997). Applications of molecular ecology in the characterisation of uncultured microorganisms associated with human disease. *Rev Med Microbiol* **8**, 91–101.

Youssef, R., Shaker, O., Sobeih, S., Mashaly, H. & Mostafa, W. Z. (2002). Detection of herpes simplex virus DNA in serum and oral secretions during acute recurrent herpes labialis. *J Dermatol* **29**, 404–410.

Comparative genomics – what do such studies tell us about the emergence and spread of key pathogens?

Richard W. Titball and Melanie Duffield

Defence Science and Technology Laboratory, Porton Down, Salisbury, Wiltshire SP4 0JQ, UK

THE EXAMPLE OF *YERSINIA PESTIS*

Plague is a disease that has shaped the social, genetic and industrial makeup of many populations and especially those in Europe. The pandemic of disease which had the greatest impact on Europe occurred during the 14th to 16th centuries and is usually referred to as the 'Black Death' (Perry & Fetherston, 1997). During this pandemic it is estimated that 30 % of the population of Europe died. The Justinian plague occurred in AD 541 to 544, originally in Egypt but then spreading through the Mediterranean and Middle East eventually to involve most of the known world (Perry & Fetherston, 1997). The third pandemic of plague originated in China in the 1850s but the impact of this pandemic was minimal in Europe. However, during the peak of the outbreak in India at the end of the 19th century, the disease killed a million individuals a year (Perry & Fetherston, 1997). Nowadays, the 2000–3000 cases of plague which are reported to the World Health Organization each year (Anonymous, 2000) are considered to be the vestigial remnants of the third plague pandemic.

The aetiological agent of plague is *Yersinia pestis*, one of three human-pathogenic *Yersinia* species. The human pathogenic *Yersinia* are closely related but cause very different diseases (Brubaker, 1991). Both *Yersinia enterocolitica* and *Yersinia pseudotuberculosis* cause relatively mild, self-limiting infections of the gastrointestinal tract, whereas *Y. pestis* causes an acute systemic infection which is often fatal. *Y. enterocolitica* and *Y. pseudotuberculosis* are normally transmitted to humans via infected foodstuffs, whereas *Y. pestis* is an obligate pathogen which is usually dependent on fleas for transmission between animal hosts (Brubaker, 1991). Strains of *Y. pestis* have

SGM symposium 66: Prokaryotic diversity: mechanisms and significance.
Editors N. A. Logan, H. M. Lappin-Scott & P. C. F. Oyston. Cambridge University Press. ISBN 0 521 86935 8

Table 1. Comparison of the genomes of *Y. pestis* and *Y. pseudotuberculosis*

Feature	*Y. pestis* CO92	*Y. pseudotuberculosis* IP32953
Sizes (bp):		
Chromosome	4 653 728	4 744 671
Plasmid pYV1	70 305	68 526
Plasmid pMT1	96 210	Not present
Plasmid pPCP1	9 612	Not present
Numbers of:		
Coding sequences (chromosome)	4 012	3 974
Pseudogenes	298	62
IS elements	138	20
Total unique coding sequences	304	124
Phage-related coding sequences	183	71

conventionally been assigned to one of three biovars: orientalis, medievalis or antiqua. Additionally, there are reports of rhamnose-positive strains of *Y. pestis* endemic to some parts of China and the former Soviet Union. These strains reportedly cause mild disease in rodents or occasionally in humans. Some have suggested that these strains constitute new biovars named pestoides or microtus (Anisimov, 2002; Zhou *et al.*, 2004).

Although *Y. pestis* and *Y. pseudotuberculosis* have very different lifestyles and pathogenic potentials, DNA hybridization studies reveal that these species are closely related (Brubaker, 1991; Perry & Fetherston, 1997). This relationship has also been confirmed by comparing the sequences of selected regions of five housekeeping genes in a range of *Y. pestis* and *Y. pseudotuberculosis* strains (Achtman *et al.*, 1999). Significant sequence diversity was found in *Y. pseudotuberculosis* strains, but in *Y. pestis* the gene fragments were identical. Surprisingly, the evolutionary distance between strains in the *Y. pseudotuberculosis* group was greater than the evolutionary distance between *Y. pestis* and *Y. pseudotuberculosis* (Achtman *et al.*, 1999). These findings indicate that *Y. pestis* is essentially a clonal population and that, strictly, it should be classified as a variant of *Y. pseudotuberculosis*.

Comparative genomics of *Y. pestis* and *Y. pseudotuberculosis*

A comparison of the genome sequences of *Y. pestis* strain CO92 (Parkhill *et al.*, 2001) and *Y. pseudotuberculosis* strain IP32953 reveals both expected similarities and unexpected differences (Table 1). The chromosomes are similar in size, and 75 % of the encoded genes have 97 % sequence identity (Chain *et al.*, 2004). However, the chromosomes are not precisely collinear (Fig. 1). This is consistent with the observation that

Fig. 1. Comparison of the chromosomes of *Y. pestis* strain CO92 (upper) and *Y. pseudotuberculosis* IP32953 (lower) using the Artemis comparison tool. Numbers indicated are base pairs from the origin of replication.

the *Y. pestis* genome is in a fluid state (Radnedge *et al.*, 2002), with the inversion of at least three regions during growth of the bacterium (Parkhill *et al.*, 2001). The fluidity of the genome is also confirmed by comparison of the genomes of *Y. pestis* strains CO92 (biovar orientalis), KIM (biovar medievalis) and 91001 (biovar medievalis). There are significant intrachromosomal rearrangements and, surprisingly, there is more sequence similarity between strains KIM and CO92 than between the strains belonging to biovar medievalis, indicating significant genome fluidity within the biovars.

Some of the chromosomal genes in both *Y. pestis* and *Y. pseudotuberculosis* appear to have been horizontally acquired. These genes are clustered into chromosomal islands and are frequently phage-associated (Hinchliffe *et al.*, 2003; Radnedge *et al.*, 2002). All of the human-pathogenic *Yersinia* species possess the pYV plasmid which encodes the type III secretion system critical for virulence (Brubaker, 1991; Cornelis *et al.*, 1998). *Y. pestis* possesses two additional plasmids, pMT1 and pPCP1. These plasmids encode a range of required proteins which are critical for the new lifestyle of *Y. pestis*, and the significance and origins of these plasmids are discussed in more detail below.

A striking difference between the genomes is the relative abundance of pseudogenes in *Y. pestis* with a functional counterpart in *Y. pseudotuberculosis*. These pseudogenes

Table 2. Functions in *Y. pseudotuberculosis* of pseudogenes common to *Y. pestis* strains KIM, CO92 and 91001

Gene function	No. of pseudogenes
Surface (inner and outer membrane, surface structures)	42
Conserved hypothetical	28
Adaptation/pathogenicity/chaperones	22
Central/intermediary/miscellaneous metabolism	18
Hypothetical	17
Regulators	10
Small molecule degradation	7
Information transfer (transcription/translation, DNA/RNA modification)	3
Phage/IS elements	2
Large molecule degradation	1
Energy metabolism	1

Table 3. Examples of genes which have been acquired, inactivated or deleted during the evolution of *Y. pestis*

Gene; gene product	Status in *Y. pestis*	Possible function(s) of gene product
inv; invasin	Inactivated; IS element	Invasion of M cells in the gut
yadA; *Yersinia* adherence factor	Inactivated; frameshift mutation	Attachment to host cells; resistance to serum killing
ureD; urease	Inactivated; frameshift mutation	Survival in the gut
LPS O-antigen	Inactivated; multiple mutations	Survival in the gut; resistance to serum killing
Haemin storage	Acquired; chromosomal island	Survival in the flea
Plasminogen activator	Acquired; plasmid pPCP1	Dissemination from peripheral sites (e.g. flea bite)
Murine toxin	Acquired; plasmid pMT1	Survival in the flea
F1 capsular antigen	Acquired; plasmid pMT1	Avoidance of phagocytosis

appear to have arisen via a range of mechanisms. Some are inactivated because they are interrupted by IS elements, and there appears to have been a massive proliferation of IS elements during the evolution of *Y. pestis*. Other genes are inactivated by frameshift mutations or deletions, some of which are likely to be mediated by IS elements.

By comparing the genomes of *Y. pseudotuberculosis* IP32953 and *Y. pestis* strains CO92, KIM and 91001, we have identified a total of 380 pseudogenes. Of these, 62 pseudogenes were found in *Y. pseudotuberculosis*, but 34 of these were also found in all three *Y. pestis* strains. The 28 pseudogenes genes which are unique to *Y. pseudotuberculosis* have presumably occurred after the divergence of the species.

In total, 151 pseudogenes (Table 2) are conserved in all three *Y. pestis* strains but are not found in *Y. pseudotuberculosis*. This finding suggests that they occurred soon after the divergence of *Y. pseudotuberculosis* and *Y. pestis*. Most of these genes are involved in either membrane transport, metabolism, pathogenicity or unknown functions (i.e. hypothetical proteins) and their inactivation may be related to the adaptation of *Y. pestis* to a more restricted niche as a mammalian blood-borne pathogen.

Ten regulatory genes were inactivated in all three *Y. pestis* strains. This is consistent with the reduced number of niches occupied by *Y. pestis* in comparison with *Y. pseudotuberculosis*. A further 42 pseudogenes were predicted to encode cell-surface proteins. Many of these are likely to play a function in transport and uptake. Loss of these genes is consistent with the reduced need for a range of solutes in restricted niche environments.

Twenty-two pseudogenes which could play a role in virulence in *Y. pseudotuberculosis* were found in all three *Y. pestis* strains. These include five with similarity to genes encoding an insecticidal toxin that are thought to be implicated in the adaptation to the flea life cycle. A further 10 pseudogenes were associated with flagella formation and four were involved in chemotaxis. Other potential virulence pseudogenes were involved in pili formation, adhesion, haemolysin activity, invasion and O-antigen biosynthesis.

The presence of many of the pseudogenes in *Y. pestis* can be explained in terms of the new lifestyle of the pathogen (Table 3). For example, the gene cluster which would encode the O-antigen is disrupted by frameshift mutations in the *ddhB*, *gmd*, *fcl* and *ushA* genes (Prior *et al.*, 2001). This is consistent with the observation that O-antigen mutants of *Y. enterocolitica* are attenuated when given orally (al Hendy *et al.*, 1992; Reeves, 1995). The restoration of an O-antigen gene cluster into *Y. pestis* did not affect virulence when the bacterium was given by the intravenous route (Oyston *et al.*, 2003), supporting the suggestion that this gene cluster was inactivated because it was no longer required for the lifestyle of the pathogen. Indeed, the lack of an O-antigen has been shown to be essential for plasminogen activation, without which the bacterium cannot disseminate from the site of infection (Kukkonen *et al.*, 2004). Similarly, *Y. pestis* contains a mutation in the *ureD* gene and is normally urease-negative (Sebbane *et al.*, 2001). Urease is essential for the virulence of orally administered *Y. enterocolitica* (Gripenberg-Lerche *et al.*, 2000), and restoring the urease-positive phenotype in *Y. pestis* did not affect virulence (Sebbane *et al.*, 2001). Several other pseudogenes in *Y. pestis* can be explained by the altered lifestyle of the pathogen. For example, genes which play a role in adherence of bacteria to intestinal cells, such as *yadA* and *inv*, are inactivated in *Y. pestis*. Similarly, the loss of motility by *Y. pestis* might be consistent with the inability of this bacterium to survive in the environment outside of the mammalian or flea host.

Origins of the *Y. pestis* plasmids

The plasmids that are unique to *Y. pestis* encode a range of functions which are required for the lifestyle of this pathogen (Table 3). For example, plasmid pPCP1 encodes a plasminogen-activator protease which is essential for the dissemination of *Y. pestis* from peripheral sites of infection (Lahteenmaki *et al.*, 2001, 2003). The pMT1 plasmid encodes the phospholipase D (murine toxin), which is essential for survival of the bacterium in the flea (Hinnebusch *et al.*, 2002), and strains of *Y. pestis* lacking the pMT1 plasmid are not able to colonize fleas (Hinnebusch *et al.*, 1998). This plasmid also encodes the polypeptide capsule (F1 antigen) which plays a role in resistance to phagocytosis (Du *et al.*, 2002). Clearly, the lifestyle of *Y. pestis* is dependent on the possession of these plasmids. However, it is not clear how these plasmids were acquired. One possibility is that they were acquired in the gut of a mammal infected with *Y. pseudotuberculosis* from another pathogen or a member of the gut flora. Alternatively, it is possible that the plasmids were acquired in the environment or even in an insect.

So far, there are no indications of the origin of plasmid pPCP1. However, some insight into the origin of plasmid pMT1 comes from the finding that many strains of *Salmonella enterica* serovar Typhi, and especially those isolated in South-East Asia, possess a cryptic plasmid (pHCM2) which is 52 % similar (Prentice *et al.*, 2001). Although these plasmids show striking similarity, it is notable that the region of pMT1 which encodes F1 antigen and murine toxin is absent from pHCM2. Therefore, although there is some evidence pointing towards acquisition of the pMT1 plasmid in the gut, the precise origins of the plasmid are not clear.

Y. pestis evolved from *Y. pseudotuberculosis*

All of the evidence points towards the evolution of *Y. pestis* from *Y. pseudotuberculosis* as a consequence of the acquisition of plasmids pMT1 and pPCP1 and the large-scale inactivation of genes. Several studies have provided significant insight into the detail of this evolutionary step. It is known that *Y. pestis* possesses a non-functional gene cluster which would otherwise encode the lipopolysaccharide O-antigen. Conversely, it is known that different strains of *Y. pseudotuberculosis* produce serologically different O-antigens and therefore possess very different encoding gene clusters. The ancestral strain of *Y. pseudotuberculosis* should therefore be revealed as having the functional form of the O-antigen cluster in *Y. pestis*. By characterizing the O-antigen cluster in 21

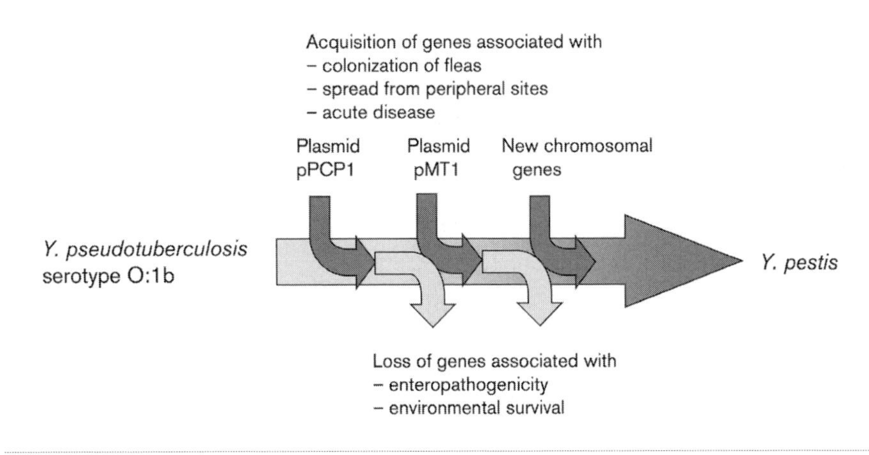

Fig. 2. Schematic representation of the evolution of *Y. pestis* from *Y. pseudotuberculosis* serotype O : 1b.

serotypes of *Y. pseudotuberculosis*, Skurnik *et al.* (2000) were able to show that *Y. pestis* evolved from *Y. pseudotuberculosis* serotype O : 1b.

Molecular methods have also been used to estimate the likely time of evolution of *Y. pestis*. Initially, it was proposed that the three biovars of *Y. pestis* were associated with the three pandemics of plague, and a preliminary analysis using molecular methods supported this suggestion (Achtman *et al.*, 1999). However, the pattern of evolution of *Y. pestis* was subsequently shown to be more complex. Using three different multilocus typing methods, Achtman *et al.* (2004) presented evidence that *Y. pestis* evolved from *Y. pseudotuberculosis* approximately 10 000 years ago. The initial evolutionary steps appear to have given rise to intermediate forms of *Y. pestis*, which might be closely related to the rhamnose-positive isolates of *Y. pestis* which are proposed for inclusion in biovars pestoides or microtus (Achtman *et al.*, 2004). Equally importantly, Achtman *et al.* (2004) have shown that typing of strains using a range of multilocus molecular methods does not result in the clustering of strains into biovars along discrete evolutionary branches (Fig. 2). Overall, it seems that the assignment of strains into one of three biovars is no longer appropriate, and the evidence that the three biovars arose during the three pandemics of plague (Achtman *et al.*, 1999) now seems less convincing (Achtman *et al.*, 2004).

Compiling all of the information outlined above, one can derive a summary of the events which occurred during the evolution of *Y. pestis* from *Y. pseudotuberculosis*. It is clear that *Y. pestis* evolved from *Y. pseudotuberculosis* serotype O : 1b relatively recently and, during passage through an evolutionary bottleneck, there was both genome reduction and the acquisition of new DNA (Fig. 3). However, the genome of *Y. pestis* is

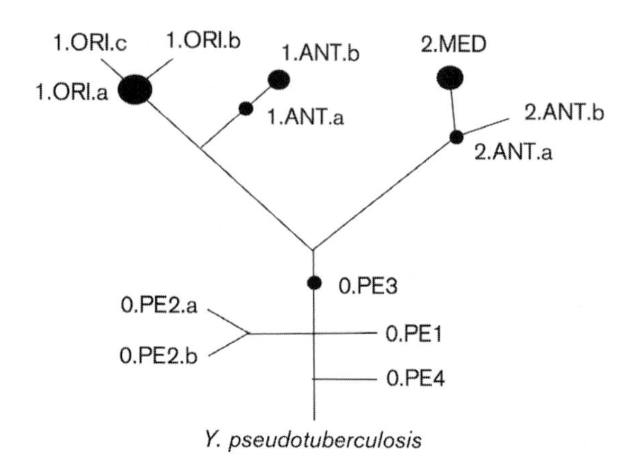

Fig. 3. Consensus evolutionary branch order in *Y. pestis*, derived from analysis of synonymous single nucleotide polymorphisms (SNPs), variations in the number of tandem repeats and insertion of IS*100* elements. PE, *Y. pestis* biovar pestoides; ORI, *Y. pestis* biovar orientalis; MED, *Y. pestis* biovar medievalis; ANT, *Y. pestis* biovar antiqua. The sizes of circles at nodes represent the number of isolates. Reproduced from Achtman *et al.* (2004) with permission (© 2004, National Academy of Sciences, USA).

still fluid, undergoing additional gene inactivation events, gene acquisition events and genome rearrangements. It has been suggested that the periodic appearance of hyper-virulent strains might explain the periodic pandemics of disease, with the pathogen evolving towards a less virulent form during the pandemic. Clearly, at the moment, such suggestions are not proven.

IS THIS PATTERN OF EVOLUTION APPARENT WITH OTHER KEY PATHOGENS?

Prior to the availability of genome sequence information, it had generally been assumed that pathogens causing severe diseases would have evolved by acquiring additional DNA encoding virulence determinants. For pathogens like *Bacillus anthracis*, this is essentially the case. Compared with *Bacillus cereus*, which is the presumed ancestor, most of the differences are found on the two virulence plasmids (Ivanova *et al.*, 2003; Read *et al.*, 2003). Indeed, the recent finding that a strain of *Bacillus cereus* capable of causing an anthrax-like disease possessed two virulence plasmids supports the conclusion that *Bacillus anthracis* evolved mainly by gene acquisition (Hoffmaster *et al.*, 2004). However, an emerging theme is that the evolution of many pathogens is dependent on gene inactivation. Clearly, the evolution of *Y. pestis* is partially dependent on gene inactivation and the acquisition of new DNA encoding virulence determinants. This pattern of evolution is also likely to be true for *Francisella tularensis* (Larsson *et al.*, 2005), although the progenitor species is not known at this time.

Other pathogens appear to have evolved solely as a result of genome 'downsizing'. One of the first and most profound examples of this type of evolution is found in *Mycobacterium leprae* (Cole *et al.*, 2001). The progenitor of *M. leprae* is not known, but is assumed to be a *Mycobacterium tuberculosis*-like species. Less than half of the *M. leprae* genome contains functional genes, with over 1000 pseudogenes of functional counterparts in *M. tuberculosis* (Cole *et al.*, 2001). The large-scale inactivation of genes has disrupted a wide range of metabolic functions. The disruption of these pathways explains the inability of the bacterium to survive outside an animal host (Cole *et al.*, 2001).

The progenitor of *Burkholderia mallei* is much more clearly defined as *Burkholderia pseudomallei*. The evolution of *Burkholderia mallei* appears to have been accompanied by the loss of 1·4 Mbp of DNA (Holden *et al.*, 2004; Nierman *et al.*, 2004). Downsizing of genomes is especially associated with the evolution of pathogens with a broad host range to host-specific ones. For example, genome downsizing has been accompanied by the inability of *Burkholderia mallei* to survive outside a mammalian host, whereas *Burkholderia pseudomallei* is distributed widely in soil and watercourses in endemic regions of the world. More importantly, whereas *Burkholderia pseudomallei* causes disease in a wide range of animal species, *Burkholderia mallei* only causes acute disease in horses and in humans. Similarly, *Bordetella pertussis* and *Bordetella parapertussis* are thought to be independent derivatives of *Bordetella bronchiseptica*-like ancestors, and there is evidence of large-scale deactivation of genes in these two derivative species. The limited host ranges of *Bordetella pertussis* and *Bordetella parapertussis* can be described as a loss of host-interaction mechanisms (Parkhill *et al.*, 2003). *S. enterica* serovar Typhimurium is able to infect a range of animal species and pseudogenes are found to make up only about 1 % of its coding DNA, whilst *S. enterica* serovars Typhi and Paratyphi have both evolved into human-specific pathogens and pseudogenes account for ~5 % of each of their genomes. This suggests that genome degradation occurs in genes that are no longer required within the host-specific niche adopted by such bacteria (McClelland *et al.*, 2004).

The availability of genome sequences has revolutionized our understanding of the ways in which pathogens have evolved. No single mechanism explains how pathogens evolve but, in general, the evidence is that the most dangerous pathogens evolved from lower-virulence ancestral species. It is clear that the emergence of pathogens merely reflects the myriad of genetic interchanges between bacterial species and genetic changes within species which are occurring all the time. On this basis, we can be certain that other pathogens will emerge in the future.

REFERENCES

Achtman, M., Zurth, K., Morelli, C., Torrea, G., Guiyoule, A. & Carniel, E. (1999). *Yersinia pestis*, the cause of plague, is a recently emerged clone of *Yersinia pseudotuberculosis*. *Proc Natl Acad Sci U S A* **96**, 14043–14048.

Achtman, M., Morelli, G., Zhu, P. & 14 other authors (2004). Microevolution and history of the plague bacillus, *Yersinia pestis*. *Proc Natl Acad Sci U S A* **101**, 17837–17842.

al Hendy, A., Toivanen, P. & Skurnik, M. (1992). Lipopolysaccharide O side chain of *Yersinia enterocolitica* O : 3 is an essential virulence factor in an orally infected murine model. *Infect Immun* **60**, 870–875.

Anisimov, A. P. (2002). *Yersinia pestis* factors ensuring circulation and persistence of the plague pathogen in ecosystems of natural foci. Communication 1. *Mol Genet Microbiol Virol* (English translation of *Mol Genet Mikrobiol Virusol*) **4**, 1–30.

Anonymous (2000). Human plague in 1998 and 1999. *World Health Organ Wkly Epidemiol Rec* **75**, 338–343.

Brubaker, R. R. (1991). Factors promoting acute and chronic diseases caused by yersiniae. *Clin Microbiol Rev* **4**, 309–324.

Chain, P. S. G., Carniel, E., Larimer, F. W. & 20 other authors (2004). Insights into the evolution of *Yersinia pestis* through whole-genome comparison with *Yersinia pseudotuberculosis*. *Proc Natl Acad Sci U S A* **101**, 13826–13831.

Cole, S. T., Eiglmeier, K., Parkhill, J. & 41 other authors (2001). Massive gene decay in the leprosy bacillus. *Nature* **409**, 1007–1011.

Cornelis, G. R., Boland, A., Boyd, A. P., Geuijen, C., Iriarte, M., Neyt, C., Sory, M. P. & Stainier, I. (1998). The virulence plasmid of *Yersinia*, an antihost genome. *Microbiol Mol Biol Rev* **62**, 1315–1352.

Du, Y., Rosqvist, R. & Forsberg, A. (2002). Role of fraction 1 antigen of *Yersinia pestis* in inhibition of phagocytosis. *Infect Immun* **70**, 1453–1460.

Gripenberg-Lerche, C., Zhang, L., Ahtonen, P., Toivanen, P. & Skurnik, M. (2000). Construction of urease-negative mutants of *Yersinia enterocolitica* serotypes O : 3 and O : 8: role of urease in virulence and arthritogenicity. *Infect Immun* **68**, 942–947.

Hinchliffe, S. J., Isherwood, K. E., Stabler, R. A. & 7 other authors (2003). Application of DNA microarrays to study the evolutionary genomics of *Yersinia pestis* and *Yersinia pseudotuberculosis*. *Genome Res* **13**, 2018–2029.

Hinnebusch, B. J., Fischer, E. R. & Schwan, T. G. (1998). Evaluation of the role of the *Yersinia pestis* plasminogen activator and other plasmid-encoded factors in temperature-dependent blockage of the flea. *J Infect Dis* **178**, 1406–1415.

Hinnebusch, B. J., Rudolph, A. E., Cherepanov, P., Dixon, J. E., Schwan, T. G. & Forsberg, A. (2002). Role of *Yersinia* murine toxin in survival of *Yersinia pestis* in the midgut of the flea vector. *Science* **296**, 733–735.

Hoffmaster, A. R., Ravel, J., Rasko, D. A. & 19 other authors (2004). Identification of anthrax toxin genes in a *Bacillus cereus* associated with an illness resembling inhalation anthrax. *Proc Natl Acad Sci U S A* **101**, 8449–8454.

Holden, M. T. G., Titball, R. W., Peacock, S. J. & 45 other authors (2004). Genomic plasticity of the causative agent of melioidosis, *Burkholderia pseudomallei*. *Proc Natl Acad Sci U S A* **101**, 14240–14245.

Ivanova, N., Sorokin, A., Anderson, I. & 20 other authors (2003). Genome sequence of *Bacillus cereus* and comparative analysis with *Bacillus anthracis*. *Nature* **423**, 87–91.

Kukkonen, M., Suomalainen, M., Kyllonen, P., Lahteenmaki, K., Lang, H., Virkola, R., Helander, I. M., Holst, O. & Korhonen, T. K. (2004). Lack of O-antigen is essential for plasminogen activation by *Yersinia pestis* and *Salmonella enterica*. *Mol Microbiol* **51**, 215–225.

Lahteenmaki, K., Kukkonen, M. & Korhonen, T. K. (2001). The Pla surface protease/adhesin of *Yersinia pestis* mediates bacterial invasion into human endothelial cells. *FEBS Lett* **504**, 69–72.

Lahteenmaki, K., Kukkonen, M., Jaatinen, S., Suomalainen, M., Soranummi, H., Virkola, R., Lang, H. & Korhonen, T. K. (2003). *Yersinia pestis* Pla has multiple virulence-associated functions. In *Genus Yersinia: Entering the Functional Genomic Era*, pp. 141–145. Proceedings of the 8th International Symposium on *Yersinia*, 4–8 September 2002, Turku, Finland. Edited by M. Skurnik, J. A. Bengoechea & K. Granfors. New York & London: Kluwer Academic.

Larsson, P., Oyston, P. C. F., Chain, P. & 24 other authors (2005). The complete genome sequence of *Francisella tularensis*, the causative agent of tularemia. *Nat Genet* **37**, 153–159.

McClelland, M., Sanderson, K. E., Clifton, S. W. & 32 other authors (2004). Comparison of genome degradation in Paratyphi A and Typhi, human-restricted serovars of *Salmonella enterica* that cause typhoid. *Nat Genet* **36**, 1268–1274.

Nierman, W. C., DeShazer, D., Kim, H. S. & 30 other authors (2004). Structural flexibility in the *Burkholderia mallei* genome. *Proc Natl Acad Sci U S A* **101**, 14246–14251.

Oyston, P. C. F., Prior, J. L., Kiljunen, S., Skurnik, M., Hill, J. & Titball, R. W. (2003). Expression of heterologous O-antigen in *Yersinia pestis* KIM does not affect virulence by the intravenous route. *J Med Microbiol* **52**, 289–294.

Parkhill, J., Wren, B. W., Thomson, N. R. & 32 other authors (2001). Genome sequence of *Yersinia pestis*, the causative agent of plague. *Nature* **413**, 523–527.

Parkhill, J., Sebaihia, M., Preston, A. & 50 other authors (2003). Comparative analysis of the genome sequences of *Bordetella pertussis*, *Bordetella parapertussis* and *Bordetella bronchiseptica*. *Nat Genet* **35**, 32–40.

Perry, R. D. & Fetherston, J. D. (1997). *Yersinia pestis* – etiologic agent of plague. *Clin Microbiol Rev* **10**, 35–66.

Prentice, M. B., James, K. D., Parkhill, J. & 13 other authors (2001). *Yersinia pestis* pFra shows biovar-specific differences and recent common ancestry with a *Salmonella enterica* serovar Typhi plasmid. *J Bacteriol* **183**, 2586–2594.

Prior, J. L., Parkhill, J., Hitchen, P. G. & 8 other authors (2001). The failure of different strains of *Yersinia pestis* to produce lipopolysaccharide O-antigen under different growth conditions is due to mutations in the O-antigen gene cluster. *FEMS Microbiol Lett* **197**, 229–233.

Radnedge, L., Agron, P. G., Worsham, P. L. & Andersen, G. L. (2002). Genome plasticity in *Yersinia pestis*. *Microbiology* **148**, 1687–1698.

Read, T. D., Peterson, S. N., Tourasse, N. & 49 other authors (2003). The genome sequence of *Bacillus anthracis* Ames and comparison to closely related bacteria. *Nature* **423**, 81–86.

Reeves, P. (1995). Role of O-antigen variation in the immune response. *Trends Microbiol* **3**, 381–386.

Sebbane, F., Devalckenaere, A., Foulon, J., Carniel, E. & Simonet, M. (2001). Silencing and reactivation of urease in *Yersinia pestis* is determined by one G residue at a specific position in the *ureD* gene. *Infect Immun* **69**, 170–176.

Skurnik, M., Peippo, A. & Ervela, E. (2000). Characterization of the O-antigen gene clusters of *Yersinia pseudotuberculosis* and the cryptic O-antigen gene cluster of *Yersinia pestis* shows that the plague bacillus is most closely related to and has evolved from *Y. pseudotuberculosis* serotype O : 1b. *Mol Microbiol* **37**, 316–330.

Zhou, D. S., Han, Y. P., Song, Y. J., Huang, P. T. & Yang, R. F. (2004). Comparative and evolutionary genomics of *Yersinia pestis*. *Microbes Infect* **6**, 1226–1234.

Spread of genomic islands between clinical and environmental strains

Jens Klockgether, Oleg N. Reva and Burkhard Tümmler

Klinische Forschergruppe, OE 6711, Medizinische Hochschule Hannover, D-30625 Hannover, Germany

COMMON FEATURES OF GENOMIC ISLANDS

The genome of a bacterium consists of a core that is common to all strains of a taxon and an accessory part that varies within and among clones of a taxon. The accessory genome represents the flexible gene pool that frequently undergoes acquisition and loss of genetic information and hence plays an important role for the adaptive evolution of bacteria (Dobrindt *et al.*, 2004). The flexible gene pool is made up of accessory elements such as bacteriophages, plasmids, IS elements, transposons, conjugative transposons, integrons and genomic islands (GEIs).

GEIs are chromosomal regions that are typically flanked by direct repeats and inserted at the 3′ end of a tRNA gene. They contain transposase or integrase genes that are required for chromosomal integration and excision and further mobility-related genes. GEIs are clone- or strain-specific and are never found in all clones of a taxon. Most GEIs are easily differentiated from the core genome by their atypical G+C contents and atypical oligonucleotide composition, with steep gradients thereof at their boundaries (Reva & Tümmler, 2005). First identified in pathogenic bacteria ('pathogenicity islands'), GEIs have since been detected in numerous non-pathogenic species. GEIs may confer fitness traits, increase metabolic versatility or adaptability or promote bacterium–host interaction in terms of symbiosis, commensalism or virulence (Dobrindt *et al.*, 2004).

GEIs have been found in the majority of all currently completely sequenced bacterial genomes, but there is a bias towards Gram-negative bacteria and life in microbial

SGM symposium 66: Prokaryotic diversity: mechanisms and significance.
Editors N. A. Logan, H. M. Lappin-Scott & P. C. F. Oyston. Cambridge University Press. ISBN 0 521 86935 8 ©SGM 2006

communities. Colonization of habitats with large and complex bacterial populations such as the rhizosphere or the gastrointestinal tract of animals seems to be associated with a higher frequency of GEIs than life in an isolated and/or sparsely populated niche. In other words, access to the gene pool of other taxa facilitates the emergence of GEIs.

THE PARADIGM: GENOMIC ISLANDS IN THE *ENTEROBACTERIACEAE*

Acquisition of GEIs by horizontal gene transfer could be a key factor for the adaptation of a bacterium to a particular host or niche. This issue has been addressed for the various pathovars of *Escherichia coli* (Dobrindt *et al*., 2003; Welch *et al*., 2002). For uropathogenic strains of *E. coli*, island acquisition has resulted in the capability to infect the urinary tract and bloodstream and evade host defences without compromising the ability for harmless colonization of the intestine. For different intestinal pathogens, acquired genes promote the colonization of specific regions of the intestine and new modes of interaction with the host tissue that produce clinically distinct variations of gastrointestinal disease. Each type of *E. coli* possesses combinations of island genes that confer its characteristic lifestyle or disease-causing traits. Examination of extraintestinal pathogenic, intestinal pathogenic and commensal *E. coli* isolates revealed a weak association between the tropism for a niche and the repertoire of pathogenicity islands (Dobrindt *et al*., 2003). However, even though similar virulence genes come into play, their linkage relationships and chromosomal locations vary considerably from strain to strain within a pathovar (Welch *et al*., 2002). In other words, two GEIs inserted at the same tRNA site do not necessarily contain the same set of genes.

A further example of the role of GEIs for adaptation to a host is the evolution of *Salmonella* serovars (Kingsley *et al*., 2000). The acquisition of an island containing the *shdA* gene by the major evolutionary lineage within the genus *Salmonella*, *Salmonella enterica* subspecies I, was accompanied by an expansion in host range to include the warm-blooded mammals and birds in addition to reptiles.

Most GEIs have only been described for the sequenced index case. An interesting exception is the high-pathogenicity island (HPI) of enterobacteria. This 35–45 kb island carries genes involved in synthesis, regulation and transport of the siderophore yersiniabactin. Initially detected in pathogenic *Yersinia* species (Carniel *et al*., 1996), the HPI was found to be harboured with high frequency by *S. enterica* subspecies III and VI (Oelschlaeger *et al*., 2003) and enteroaggregative and extraintestinal pathogenic *E. coli* (Schubert *et al*., 1998). However, the HPI was also detected in non-pathogenic and commensal members of the family *Enterobacteriaceae* such as *Citrobacter diversus* and *Klebsiella* species (Bach *et al.*, 2000). Iron is essential for almost all bacteria and, consequently, siderophore production has a dual role to increase fitness in iron-restricted environments and to contribute to virulence in infected hosts, which explains the widespread distribution of the HPI in pathogenic and non-pathogenic members of the *Enterobacteriaceae*.

Most HPIs are stably integrated into the chromosome, probably because of deletion or mutations in essential mobility genes. However, an HPI was recently identified in the *E. coli* strain ECOR31 that carries a complete set of conjugative plasmid functions (Schubert *et al.*, 2004). HPI_{ECOR31} was found to be inserted into a tRNA gene and carries an intact *attR* site. The GEI carries an integrase gene, an origin of transfer and DNA-processing region and a complete and functional mating-pair formation system. Induction of the HPI integrase results in precise excision and circularization of HPI_{ECOR31}. Replication as an autonomous plasmid is not possible, however, because HPI_{ECOR31} lacks the essential *repABC* genes. In summary, HPI_{ECOR31} may be considered as the progenitor of the contemporary HPIs, most of which have lost the ability to be excised because of the loss of the *attR* site, truncation of the GEI and/or deletions within the integrase gene (Schubert *et al.*, 2004).

GENOMIC ISLANDS IN *BURKHOLDERIA* AND *PSEUDOMONAS*

GEIs shared by clinical and environmental isolates

The occurrence of a GEI in pathogenic and non-pathogenic isolates has been observed not only for the HPI, but also for genome islands that are prevalent in *Pseudomonas* and *Burkholderia* species. *Burkholderia pseudomallei* is the causative agent of melioidosis. The genome of strain K96243 is composed of two chromosomes (Holden *et al.*, 2004). Sixteen GEIs together make up 6·1 % of the genome, only one of which is also present in the genome of the clonally related organism *Burkholderia mallei*. Further analysis revealed these islands to be variably present in a collection of clinical and soil isolates (Holden *et al.*, 2004). One GEI was present in all tested strains, four GEIs were detected in 10–70 % of strains, three islands were in 10 % or fewer of strains and one island was unique to the sequenced strain.

The ubiquitous and metabolically versatile *Pseudomonas aeruginosa* is an important opportunistic pathogen for humans, plants and animals. Several large GEIs have been detected in strains from human infections and aquatic habitats. The only large genome island known so far that is not associated with a tRNA gene is the 49 kb PAGI-1. PAGI-1 contains genes potentially involved in oxidative stress resistance. This first described GEI in *P. aeruginosa* was found to be present in 85 % of tested North

American clinical isolates from septicaemia, airways and urinary tract infections (Liang *et al.*, 2001). The island was probably assembled from two ancestral components of different G+C content. Thirty-five kb of the higher G+C content portion is also found in the genome of the biosafety strain *Pseudomonas putida* KT2440 (Nelson *et al.*, 2002). *P. putida* is a metabolically versatile, rapidly growing bacterium frequently isolated from moist temperate soils and waters, particularly polluted soils. The sequenced KT2440 strain is a plasmid-cured derivative of the Japanese soil isolate mt-2 (Nakazawa, 2002). In other words, PAGI-1 was identified in pseudomonads of unrelated habitats and geographical origin.

The other known large GEIs of *P. aeruginosa* integrate into tRNA genes. Two different types have been identified, the prototypes of which are the islands PAGI-2/PAGI-3 (Larbig *et al.*, 2002) and pKLC102 (Klockgether *et al.*, 2004)/PAPI-1 (He *et al.*, 2004), respectively. PAGI-2 and PAGI-3 were sequenced in two strains of the major clone C (Römling *et al.*, 1997), an isolate from the lungs of a patient with cystic fibrosis and an isolate from a river. In both strains, the region consists of an individual strain-specific GEI and a shorter stretch of clone-specific sequence. The left boundary of the islands is a cluster of tRNA genes comprising one $tRNA^{Glu}$ gene followed by two identical $tRNA^{Gly}$ genes, one serving as the integration site for the *P. aeruginosa* genome island PAGI-2, the other for PAGI-3. PAGI-2 and PAGI-3 terminate at the right end with the terminal 16 and 24 nucleotides of the 3′ end of the $tRNA^{Gly}$ gene, respectively. In both islands, the first open reading frame (ORF) adjacent to the $tRNA^{Gly}$ gene encodes a bacteriophage P4-related multidomain integrase with an unusual transposase-like C terminus. PAGI-2 and PAGI-3 have a modular bipartite structure. The first part adjacent to the tRNA gene consists of strain-specific ORFs encoding metabolic functions and transporters, the majority of which has homologues of known function in other eubacteria (ORFs C2–C35; cargo region). The second part is made up of a syntenic set of ORFs the majority of which are either classified as conserved hypotheticals or related to DNA replication or mobility genes (ORFs C1, C36–C111; conserved part). Forty-seven of these ORFs are arranged in the same order in the two islands, with an amino acid sequence identity of 35–88 %.

The incidence of islands of PAGI-2 type in the *P. aeruginosa* population has been investigated by Southern analysis (J. Klockgether, unpublished data). An array of PAGI-2 ORFs was probed with genomic DNA from 71 *P. aeruginosa* strains, each representing a separate clone (Morales *et al.*, 2004). The collection consisted of 55 clinical isolates and 16 environmental isolates of diverse geographical origin. PAGI-2-like islands were identified in 31 of the 71 strains. The ORFs of the cargo region were absent in all but five strains. ORFs of the conserved part, however, were present to different extents. Twenty-five ORFs were present in all PAGI-2-type islands, a further

23 ORFs were detected in the majority of strains and 19 ORFs were variably present. The copy number of PAGI-2-type islands in the chromosome was one (12 strains), two (11 strains), three (seven strains) or four (one strain). PAGI-2 was completely conserved in five analysed *P. aeruginosa* strains, an isolate from an ear infection from the USA isolated in 1980, an isolate from 1985 from the airways of a patient with cystic fibrosis living in Lower Saxony, Germany, and three isolates from river and sanitary facilities in households sampled in Northrhine-Westphalia, Germany, in 1992. In other words, PAGI-2 was identified in unrelated strains from diverse habitats and geographical origin.

The *P. aeruginosa* chromosome consists of three large hypervariable regions (Römling *et al.*, 1995, 1997), one of which is the PAGI-2-like GEI. The integration and excision of GEIs into the two copies of a $tRNA^{Lys}$ gene make up the other two hypervariable regions. The sequenced prototypes are the pathogenicity island PAPI-1 (He *et al.*, 2004) and the mobile genetic element pKLC102 (Klockgether *et al.*, 2004). PAPI-1 and pKLC102 share numerous features: approximate size (108 and 102 kb), a $tRNA^{Asp}$, $tRNA^{Pro}$ and $tRNA^{Lys}$ gene cluster at their leftward PAO1 junction and a direct repeat of the 3′ half of the $tRNA^{Lys}$ gene at their right border and the presence of the integrase and chromosome-partitioning genes at the ends of the island, similar to PAGI-2 and PAGI-3. PAPI-1 encodes at least 19 virulence factors that occur on GEIs found in a wide spectrum of other pathogenic bacteria (He *et al.*, 2004). PAPI-1 was first identified in wound and cystic fibrosis isolates from the USA (He *et al.*, 2004), but meanwhile has also been detected in clinical and wastewater isolates from Europe (L. Wiehlmann, personal communication). pKLC102 is more abundant than PAPI-1 in the *P. aeruginosa* population. More than 85 % of strains in our reference collection harbour pKLC102-type GEIs. Clone C strains from the environment were found to harbour chromosomal and episomal copies of pKLC102. pKLC102 contains the 8·5 kb *chvB* gene, homologues of which are known to confer host tropism and virulence and to be essential for the interaction of the bacterium with its eukaryotic host. PAPI-1 and pKLC102 encode type IV group B pili and type IV thin sex pili, respectively, and share a set of homologues found as island-specific genes in PAGI-2 and PAGI-3. The mobile pKLC102 shares with PAPI-1 the phage module that confers integrase, the *att* element and the syntenic set of conserved genes that was first detected in PAGI-2 and PAGI-3, but it differs from PAPI-1 in carrying a plasmid module that confers *oriV* and genes for replication, partitioning and conjugation (Klockgether *et al.*, 2004). In summary, PAPI-1 and pKLC102 can be found in *P. aeruginosa* isolates from both clinical and environmental habitats. PAPI-1 is associated with one abundant epidemic clone represented by the sequenced strain PA14, whereas the mobile pKLC102 is widespread in the *P. aeruginosa* population. Moreover, pKLC102-like GEIs were detected in two sequenced *Pseudomonas syringae* strains (Feil *et al.*, 2005).

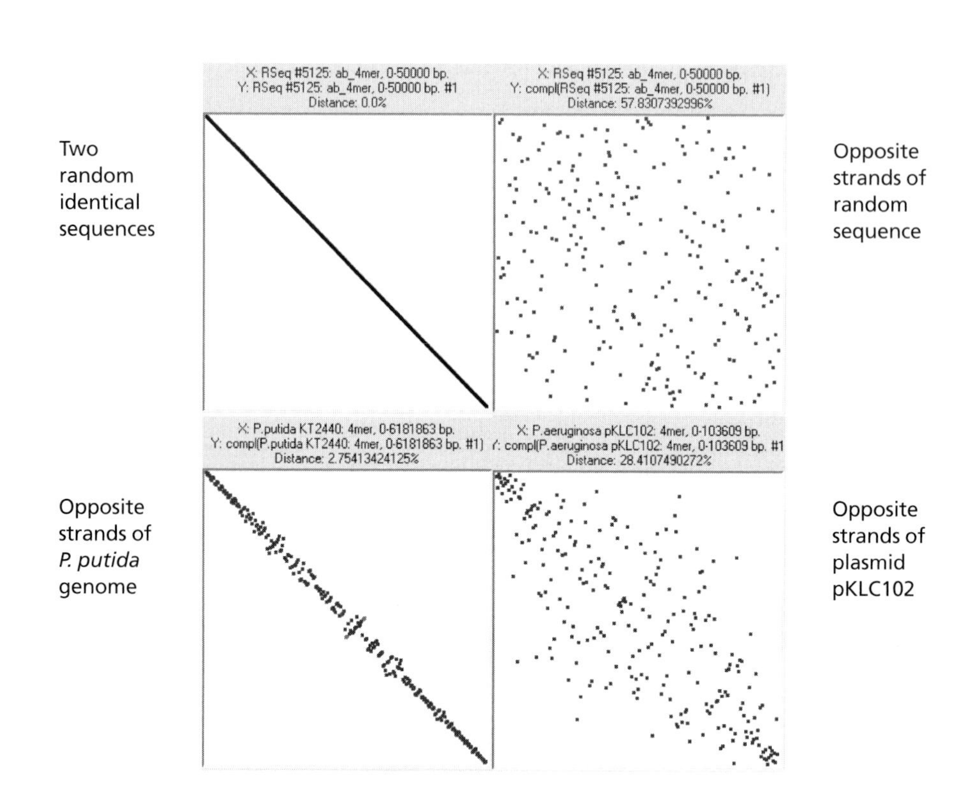

Fig. 1. Frequency of tetranucleotides in two identical sequences (50 kb, upper left) and the two complementary strands of a random sequence (50 kb, upper right), the *P. putida* KT2440 chromosome (6182 kbp, lower left) and the conjugative GEI pKLC102 (104 kbp, lower right). The frequency of each of the 256 tetranucleotides in each strand was counted and then the tetranucleotides were sorted by increasing frequency. The global difference between the oligonucleotide patterns of two strands is described by the distance *D*, i.e. the ratio of the observed to the maximal possible difference between the oligonucleotide patterns. *D* was calculated to be 0 for the identical sequence, 57·8 % for the random sequence, 2·8 % for the KT2440 chromosome and 28·4 % for pKLC102. Definition of *D* and algorithms are described in detail in the original publication by Reva & Tümmler (2004).

The promiscuous pKLC102 is a conjugative GEI of plasmid and phage origin. Compared with the typical GEI, it is endowed with exceptionally high mobility. Inspection of its oligonucleotide composition provides a hint of why pKLC102 has not been stably captured by the host chromosome. Bacterial chromosomes are characterized by strand symmetry and intrastrand parity of complementary oligonucleotides (Reva & Tümmler, 2004). In other words, oligonucleotides and their reverse complements occur with similar frequency in bacterial chromosomes. In contrast, no such correlation is observed for a random sequence (Fig. 1). Conjugative GEIs such as pKLC102 are intermediate between chromosome and random sequence. Oligonucleotide frequency on the two strands is only weakly correlated in pKLC102, whereas each oligonucleotide occurs with almost identical frequency on the two strands of a bacterial chromosome, as shown in Fig. 1 for *P. putida* KT2440. Stably integrated GEIs have an atypical oligonucleotide composition compared with the core genome, but strand symmetry is locally maintained (Reva & Tümmler, 2005). Conjugative GEIs, like most phages (Reva & Tümmler, 2004), do not adhere to this rule, suggesting that they behave as selfish parasitic DNA that is prone to horizontal spread within and among taxa.

PAGI-2 and pKLC102 belong to a family of ancient GEIs

PAGI-2, PAGI-3, PAPI-1 and pKLC102 share a syntenic set of homologous genes the majority of which are conserved hypotheticals of unknown function. A database search revealed sequence similarity to 30 further contiguous sequences in beta- and gamma-proteobacteria (Table 1; updated 10 October 2005). All GEIs were integrated into tRNA genes. Fifteen of 33 core genes were congruent, with the phylogenetic relationships of each of the individual genes indicating that all five large genome islands known so far in *P. aeruginosa* belong to one family of related syntenic GEIs with a deep evolutionary origin (Mohd-Zain *et al.*, 2004). As an example of a conserved core gene, Fig. 2 shows the phylogenetic relationships of all homologues of *P. aeruginosa* C ORF47. The core genes are likely to account for the conjugative transfer of the GEIs and may even encode autonomous replication.

Accessory gene clusters are nestled among the core genes and encode the following diverse major attributes: antibiotic, metal and antiseptic resistance, degradation of chemicals, type IV secretion systems, two-component signalling systems, Vi antigen capsule synthesis, toxin production and a wide range of metabolic functions. Examples are the large conjugative *Haemophilus influenzae* antibiotic-resistance element, ICE*Hin1056* (Mohd-Zain *et al.*, 2004), the pathogenicity island SPI-7 found in *S. enterica* serovar Typhi C18 and Ty2 (Pickard *et al.*, 2003) and the *clc* element that was originally discovered in *Pseudomonas* sp. B13 and encodes the enzymes for 3- and 4-chlorocatechol degradation (Frantz & Chakrabarty, 1987). The $tRNA^{Gly}$-integrated *clc* element is an example of a xenobiotic-degradation island that allows bacteria to handle compounds that have been recently released into the environment by humans. This type of element has been detected in numerous chloroaromatic-degrading bacteria (van der Meer & Sentchilo, 2003) including the completely sequenced *Burkholderia xenovorans* strain LB400, a micro-organism known for its capabilities to degrade polychlorobiphenyls (Bopp, 1986).

Spread of GEIs among species

Spread of GEIs among taxa has been detected for HPI (see above), the *clc* element (Ravatn *et al.*, 1998; Springael *et al.*, 2002), pKLC102 (Klockgether *et al.*, 2004; Feil *et al.*, 2005) and PAGI-2 (J. Klockgether, unpublished data). PAGI-2 and the *clc* element of *Pseudomonas* sp. B13 share 85–100 % nucleotide sequence identity in the conserved

Table 1. pKLC102-type GEIs in proteobacteria

Strain	GEI	Source/habitat	GenBank accession no.
Azoarcus sp. EbN1*		Anoxic freshwater mud	NC_006513
Azotobacter vinelandii AvOP		Soil	NZ_AAAU00000000
Burkholderia pseudomallei S13		Reference isolate	NZ_AAHW00000000
Erwinia carotovora subsp. *atroseptica* SCRI1043		Potato stem	NC_004547
Haemophilus ducreyi 35000HP		Human	NC_002940
Haemophilus influenzae 1056†	ICE*Hin1056*	Human blood	AJ627386
Histophilus somni 129PT		Animal mucous membranes	NZ_AABO00000000
Histophilus somni 2336		Pneumoniac calf	NZ_AACJ00000000
Legionella pneumophila Paris		Endemic isolate	NC_006368
Legionella pneumophila Lens		Epidemic isolate	NC_006369
Legionella pneumophila subsp. *pneumophila* Philadelphia 1		Lung	NC_002942
Neisseria gonorrhoeae MS11	GGI	Human	AY803022
Nitrosomonas eutropha C71		Soil	NZ_AAJE00000000
Photorhabdus luminescens subsp. *laumondii* TT01		Nematode symbiont	NC_005126
Pseudomonas aeruginosa C‡	PAGI-2	Cystic fibrosis patient	AF440523
Pseudomonas aeruginosa C	pKLC102	Cystic fibrosis patient	AY257538
Pseudomonas aeruginosa C	PAGI-5	Cystic fibrosis patient	Unpublished
Pseudomonas aeruginosa PA14	PAPI-1	Clinical isolate	AY273869
Pseudomonas aeruginosa SG17M	PAGI-3	River	AF440524
Pseudomonas aeruginosa TB	PAGI-6	Cystic fibrosis patient	Unpublished
Pseudomonas fluorescens Pf-5		Plant commensal	NC_004129
Pseudomonas fluorescens PfO-1		Agricultural soil	NZ_AAAT00000000
Pseudomonas fluorescens SBW25		Rhizosphere	Unpublished
Pseudomonas sp. B13§	*clc* transposon	Soil	AJ617740
Pseudomonas syringae pv. phaseolicola 1302A	PPHGI-1	Bean plant	AJ870974
Pseudomonas syringae pv. syringae B728a		Snap bean leaflet	NC_007005
Pseudomonas syringae pv. tomato DC3000		Tomato plant	NC_004578
Rubrivivax gelatinosus PM1 (*Methylobium petroleophilum* PM1)		Water pollution treatment plant	NZ_AAEM00000000
Salmonella enterica subsp. *enterica* serovar Typhi CT18	SPI-7	Typhoid fever patient	NC_003198

Table 1. *cont.*

Strain	GEI	Source/habitat	GenBank accession no.
Xanthomonas axonopodis pv. citri 306		Citrus plant	NC_003919
Xanthomonas campestris pv. campestris 8004		Cauliflower	NC_007086
Xylella fastidiosa 9a5c		Orange plant	NC_002488
Yersinia enterocolitica 8081		Gastroenteritis	Unpublished
Yersinia pseudotuberculosis 32777	YAPI	Infection	AJ627388

*Two related islands (a and b) are present in the genome (see Fig. 2).

†Nearly identical islands are present in *H. influenzae* strains R2866 and 86-028NP.

‡An identical island is present in *Cupriavidus metallidurans* CH34.

§100 % sequence identity with *clc* plus further genes in *Ralstonia* sp. JS705 and *Burkholderia xenovorans* LB400.

region (Fig. 3a, b). The *clc* element was found to be capable to self-transfer to other beta- and gammaproteobacteria (Ravatn *et al.*, 1998), whereby it excised from the donor chromosome at a low frequency and self-transferred to the new host in which it reintegrated. *clc*-like elements have a worldwide distribution; for example, an element was detected in a *Ralstonia* sp. isolate from contaminated groundwater in Texas (Müller *et al.*, 2003). PAGI-2 also seems to be widely distributed in diverse species, geographical regions and habitats. PAGI-2 was originally detected in the lungs of a German cystic fibrosis patient (Larbig *et al.*, 2002) who has been chronically carrying the genome island stably integrated into the chromosome for more than 20 years. Despite its apparently stable chromosomal integration, PAGI-2 was found to be widespread in the *P. aeruginosa* population (see above), indicating that it can be mobilized and transferred to other strains. This hypothesis is supported by the fact that a copy of PAGI-2 was identified with 100 % nucleotide sequence identity in the recently sequenced *Cupriavidus* (formerly *Ralstonia*) *metallidurans* CH34 chromosome. This strain was isolated in 1976 from the sludge of a zinc decantation tank in Belgium that was polluted with high concentrations of several heavy metals. A small collection of *C. metallidurans* and *Cupriavidus campiniensis* environmental isolates was screened for the presence of PAGI-2 by hybridization onto the PAGI-2 macroarray. Three of five *C. metallidurans* strains (Fig. 3c) and both of two *C. campiniensis* strains harboured close homologues to all PAGI-2 ORFs in their chromosomes. One *C. metallidurans* strain was carrying a truncated version of PAGI-2 (J. Klockgether, unpublished data). All *Cupriavidus* strains were isolated from polluted environmental habitats. In summary, closely related *clc*-like and PAGI-2-like genome islands (Fig. 2) were identified in clinical and environmental beta- and gammaproteobacteria of diverse geographical

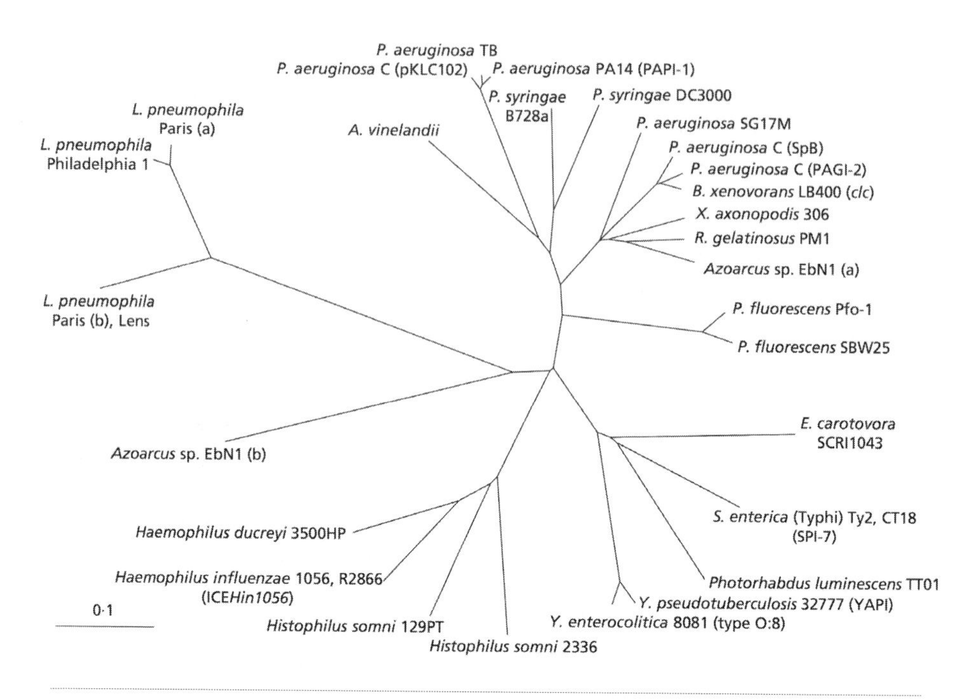

Fig. 2. Unrooted tree based on CLUSTAL X 1.83 analysis of 32 GEIs. The amino acid sequences of all homologues of the *P. aeruginosa* C ORF47 (Larbig *et al.*, 2002) were aligned. Each strain containing a GEI was identified and, where available, the designation of the GEI (e.g. PAPI-1) is given in parentheses. See Table 1 for strain and sequence details.

origin. These GEIs are made up of a module that encodes strain-specific features and a module of conserved syntenic homologues, the majority of which are conserved hypotheticals of yet unknown function. The syntenic gene contig is probably involved in the excision, transfer, integration and stabilization of the genome island (Sentchilo *et al.*, 2003a, b).

CONCLUSIONS

The gene repertoire of a bacterial cell consists of genes that have been transmitted vertically over long periods of time and genes that were acquired or generated at various points of the lineage, including some very recently (Lerat *et al.*, 2005). Horizontal gene transfer provides most of the diversity in the genomic repertoire, but the majority of the horizontally acquired genes that persist in genomes are transmitted strictly vertically (Lerat *et al.*, 2005). Hence, despite substantial horizontal gene transfer, the phylogenetic relationships between taxa are robust, as indicated by the congruence of gene trees based on rRNA gene sequences, gene contents or average amino acid identity of shared genes (Konstantinidis & Tiedje, 2005). This chapter reports on an exception to this rule. GEIs are part of the variable bacterial gene pool. They are only present in

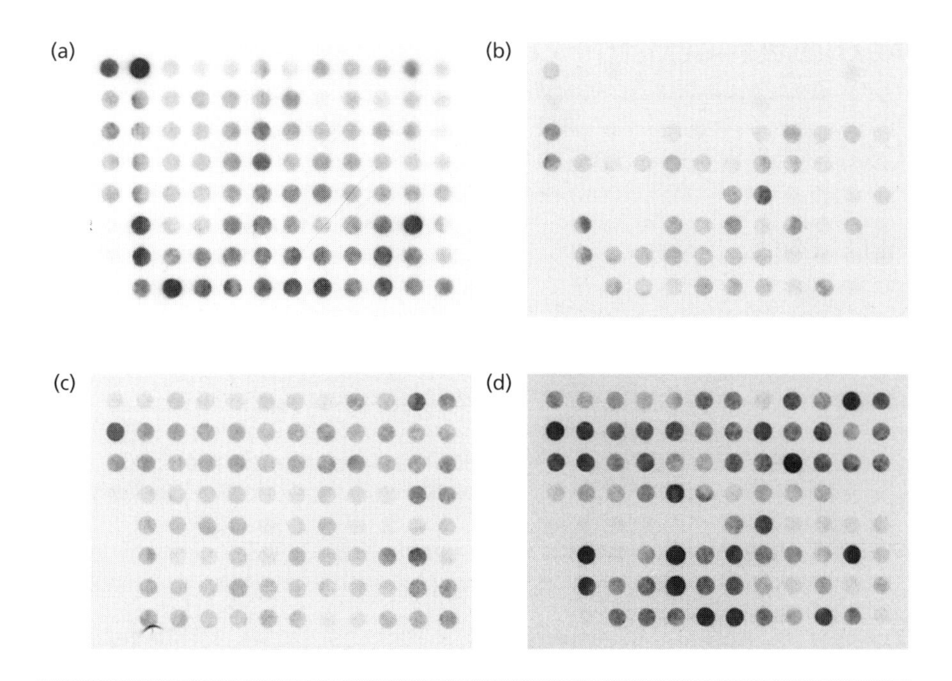

Fig. 3. Hybridization of the PAGI-2 macroarray with genomic DNA from clinical and environmental isolates. (a) *P. aeruginosa* strain C, the initial source of PAGI-2, isolated from the airways of a patient with cystic fibrosis in Hannover, Germany (Larbig *et al.*, 2002). (b) *Ralstonia* sp. strain JS705, isolated from groundwater at Kelly Air Force Base, near San Antonio, TX, USA, which uses chlorobenzene as sole carbon and energy source (Müller *et al.*, 2003). (c) *Cupriavidus metallidurans* KT21, a wastewater isolate from Germany. (d) *C. metallidurans* CH79, recovered from polluted sediment at a zinc factory in Liège, Belgium. Strain JS705 was obtained from Jan Roelof van der Meer (University of Lausanne, Switzerland); the two *C. metallidurans* strains were provided by Max Mergeay (Belgian Nuclear Research Centre, Mol, Belgium).

some strains or clones of a species and hence contribute to intra- and interclonal diversity. Moreover, some island types have been identified in numerous taxa. The most prominent example known to date is an evolutionarily ancient GEI that is widespread among beta- and gammaproteobacteria and has been identified in isolates from clinical and environmental habitats. This GEI consists of two modules: one module endows strain-specific features and the other consists of a set of conserved syntenic genes. Sequence identity of the syntenic set is less than that of vertically transmitted genes, indicating that these genes and probably also the encoded functions are more diverse than those encoded by orthologues of core genomes. Most genes of the conserved module are classified as conserved hypotheticals of yet unknown function, although at least a subset should be involved in the excision, transfer, integration or stabilization of the island (Sentchilo *et al.*, 2003a, b). Future research will unravel whether the syntenic gene set is not only essential for the maintenance of the GEI, but also influences the expression, regulation and function of gene products of the core genome.

ACKNOWLEDGEMENTS

This work was supported by grants from the Deutsche Forschungsgemeinschaft (Europäisches Graduiertenkolleg 'Pseudomonas: Pathogenicity and Biotechnology' and the Schwerpunktprogramm 'Ökologie bakterieller Krankheitserreger - molekulare und evolutionäre Aspekte').

REFERENCES

Bach, S., de Almeida, A. & Carniel, E. (2000). The *Yersinia* high-pathogenicity island is present in different members of the family *Enterobacteriaceae*. *FEMS Microbiol Lett* **183**, 289–294.

Bopp, L. H. (1986). Degradation of highly chlorinated PCBs by *Pseudomonas* strain LB400. *J Ind Microbiol* **1**, 23–29.

Carniel, E., Guilvout, I. & Prentice, M. (1996). Characterization of a large chromosomal "high-pathogenicity island" in biotype 1B *Yersinia enterocolitica*. *J Bacteriol* **178**, 6743–6751.

Dobrindt, U., Agerer, F., Michaelis, K. & 7 other authors (2003). Analysis of genome plasticity in pathogenic and commensal *Escherichia coli* isolates by use of DNA arrays. *J Bacteriol* **185**, 1831–1840.

Dobrindt, U., Hochhut, B., Hentschel, U. & Hacker, J. (2004). Genomic islands in pathogenic and environmental microorganisms. *Nat Rev Microbiol* **2**, 414–424.

Feil, H., Feil, W. S., Chain, P. & 17 other authors (2005). Comparison of the complete genome sequences of *Pseudomonas syringae* pv. syringae B728a and pv. tomato DC3000. *Proc Natl Acad Sci U S A* **102**, 11064–11069.

Frantz, B. & Chakrabarty, A. M. (1987). Organization and nucleotide sequence determination of a gene cluster involved in 3-chlorocatechol degradation. *Proc Natl Acad Sci U S A* **84**, 4460–4464.

He, J., Baldini, R. L., Deziel, E. & 7 other authors (2004). The broad host range pathogen *Pseudomonas aeruginosa* strain PA14 carries two pathogenicity islands harboring plant and animal virulence genes. *Proc Natl Acad Sci U S A* **101**, 2530–2535.

Holden, M. T., Titball, R. W., Peacock, S. J. & 45 other authors (2004). Genomic plasticity of the causative agent of melioidosis, *Burkholderia pseudomallei*. *Proc Natl Acad Sci U S A* **101**, 14240–14245.

Kingsley, R. A., van Amsterdam, K., Kramer, N. & Bäumler, A. J. (2000). The *shdA* gene is restricted to serotypes of *Salmonella enterica* subspecies I and contributes to efficient and prolonged fecal shedding. *Infect Immun* **68**, 2720–2727.

Klockgether, J., Reva, O., Larbig, K. & Tümmler, B. (2004). Sequence analysis of the mobile genome island pKLC102 of *Pseudomonas aeruginosa* C. *J Bacteriol* **186**, 518–534.

Konstantinidis, K. T. & Tiedje, J. M. (2005). Towards a genome-based taxonomy for prokaryotes. *J Bacteriol* **187**, 6258–6264.

Larbig, K. D., Christmann, A., Johann, A., Klockgether, J., Hartsch, T., Merkl, R., Wiehlmann, L., Fritz, H. J. & Tümmler, B. (2002). Gene islands integrated into $tRNA^{Gly}$ genes confer genome diversity on a *Pseudomonas aeruginosa* clone. *J Bacteriol* **184**, 6665–6680.

Lerat, E., Daubin, V., Ochman, H. & Moran, N. A. (2005). Evolutionary origins of genomic repertoires in bacteria. *PLoS Biol* **3**, e130.

Liang, X., Pham, X. Q., Olson, M. V. & Lory, S. (2001). Identification of a genomic island present in the majority of pathogenic isolates of *Pseudomonas aeruginosa*. *J Bacteriol* **183**, 843–853.

Mohd-Zain, Z., Turner, S. L., Cerdeno-Tarraga, A. M. & 7 other authors (2004). Transferable antibiotic resistance elements in *Haemophilus influenzae* share a common evolutionary origin with a diverse family of syntenic genomic islands. *J Bacteriol* **186**, 8114–8122.

Morales, G., Wiehlmann, L., Gudowius, P., van Delden, C., Tümmler, B., Martinez, J. L. & Rojo, F. (2004). Structure of *Pseudomonas aeruginosa* populations analyzed by single nucleotide polymorphism and pulsed-field gel electrophoresis genotyping. *J Bacteriol* **186**, 4228–4237.

Müller, T. A., Werlen, C., Spain, J. & van der Meer, J. R. (2003). Evolution of a chlorobenzene degradative pathway among bacteria in a contaminated groundwater mediated by a genomic island in *Ralstonia*. *Environ Microbiol* **5**, 163–173.

Nakazawa, T. (2002). Travels of a *Pseudomonas*, from Japan around the world. *Environ Microbiol* **4**, 782–786.

Nelson, K. E., Weinel, C., Paulsen, I. T. & 40 other authors (2002). Complete genome sequence and comparative analysis of the metabolically versatile *Pseudomonas putida* KT2440. *Environ Microbiol* **4**, 799–808.

Oelschlaeger, T. A., Zhang, D., Schubert, S., Carniel, E., Rabsch, W., Karch, H. & Hacker, J. (2003). The high-pathogenicity island is absent in human pathogens of *Salmonella enterica* subspecies I but present in isolates of subspecies III and VI. *J Bacteriol* **185**, 1107–1111.

Pickard, D., Wain, J., Baker, S. & 12 other authors (2003). Composition, acquisition, and distribution of the Vi exopolysaccharide-encoding *Salmonella enterica* pathogenicity island SPI-7. *J Bacteriol* **185**, 5055–5065.

Ravatn, R., Zehnder, A. J. & van der Meer, J. R. (1998). Low-frequency horizontal transfer of an element containing the chlorocatechol degradation genes from *Pseudomonas* sp. strain B13 to *Pseudomonas putida* F1 and to indigenous bacteria in laboratory-scale activated-sludge microcosms. *Appl Environ Microbiol* **64**, 2126–2132.

Reva, O. N. & Tümmler, B. (2004). Global features of sequences of bacterial chromosomes, plasmids and phages revealed by analysis of oligonucleotide usage patterns. *BMC Bioinformatics* **5**, 90.

Reva, O. N. & Tümmler, B. (2005). Differentiation of regions with atypical oligonucleotide composition in bacterial genomes. *BMC Bioinformatics* **6**, 251.

Römling, U., Greipel, J. & Tümmler, B. (1995). Gradient of genomic diversity in the *Pseudomonas aeruginosa* chromosome. *Mol Microbiol* **17**, 323–332.

Römling, U., Schmidt, K. D. & Tümmler, B. (1997). Large genome rearrangements discovered by the detailed analysis of 21 *Pseudomonas aeruginosa* clone C isolates found in environment and disease habitats. *J Mol Biol* **271**, 386–404.

Schubert, S., Rakin, A., Karch, H., Carniel, E. & Heesemann, J. (1998). Prevalence of the "high-pathogenicity island" of *Yersinia* species among *Escherichia coli* strains that are pathogenic to humans. *Infect Immun* **66**, 480–485.

Schubert, S., Dufke, S., Sorsa, J. & Heesemann, J. (2004). A novel integrative and conjugative element (ICE) of *Escherichia coli*: the putative progenitor of the *Yersinia* high-pathogenicity island. *Mol Microbiol* **51**, 837–848.

Sentchilo, V., Ravatn, R., Werlen, C., Zehnder, A. J. & van der Meer, J. R. (2003a). Unusual integrase gene expression on the *clc* genomic island in *Pseudomonas* sp. strain B13. *J Bacteriol* **185**, 4530–4538.

Sentchilo, V., Zehnder, A. J. & van der Meer, J. R. (2003b). Characterization of two alternative promoters for integrase expression in the *clc* genomic island of *Pseudomonas* sp. strain B13. *Mol Microbiol* **49**, 93–104.

Springael, D., Peys, K., Ryngaert, A. & 8 other authors (2002). Community shifts in a seeded 3-chlorobenzoate degrading membrane biofilm reactor: indications for involvement of in situ horizontal transfer of the *clc*-element from inoculum to contaminant bacteria. *Environ Microbiol* **4**, 70–80.

van der Meer, J. R. & Sentchilo, V. (2003). Genomic islands and the evolution of catabolic pathways in bacteria. *Curr Opin Biotechnol* **14**, 248–254.

Welch, R. A., Burland, V., Plunkett, G., III & 16 other authors (2002). Extensive mosaic structure revealed by the complete genome sequence of uropathogenic *Escherichia coli*. *Proc Natl Acad Sci U S A* **99**, 17020–17024.

Evolving gene clusters in soil bacteria

Alice Morningstar, William H. Gaze, Sahar Tolba and Elizabeth M. H. Wellington

Department of Biological Sciences, University of Warwick, Coventry CV4 7AL, UK

INTRODUCTION

Soil is heterogeneous in nearly all respects and contains a huge diversity of micro-organisms. The availability of carbon and other energy sources, mineral nutrients and water varies considerably over space and time, as does temperature. Adaptations to nutrient poverty including oligotrophy and zymogeny (upsurge in growth when nutrients are available) are common. The water films essential for microbial life in soil are discontinuous, and only clay particles have the necessary charges to hold water against the pull of gravity. Clay-coated soil particles cluster together to form aggregates, and these aggregates or clusters of aggregates with their adjacent water form the microhabitats in which bacteria function (Stotzky, 1997). The result of the discrete microhabitats in soil is that microbial population dynamics and interactions are very different from those in well-mixed substrates such as some aquatic environments. Soil is also a reservoir for pesticides and other chemical and microbiological inputs from slurry application, all of which will have a selective impact on the indigenous bacteria.

Bacterial evolutionary histories are difficult to untangle. Different scales of evolution occur simultaneously, from events possible over a few generations (chromosomal rearrangement, gene deletion and acquisition of genes via horizontal transfer) to the eon-scale generative evolution which creates diversity from which the novel functional genes of the future will be selected. In the age of genomics we are developing the tools to study the ecology of microbes in soil. The few metagenomic projects undertaken so far illustrate the diversity of bacteria in soil (>3000 ribotypes in a Minnesota farm soil

SGM symposium 66: Prokaryotic diversity: mechanisms and significance.
Editors N. A. Logan, H. M. Lappin-Scott & P. C. F. Oyston. Cambridge University Press. ISBN 0 521 86935 8 ©SGM 2006

sample) and the technical difficulties in producing overlapping sequence; five billion base pairs of sequence would be necessary to obtain the eightfold coverage traditionally targeted for draft genome assemblies, even for the single most predominant genome (Tringe *et al.*, 2005).

This chapter focuses on the evolutionary processes that are revealed by our knowledge of gene clusters studied in soil-dwelling bacteria. In soil, large communities of diverse bacteria exist where the horizontal gene pool can provide access to adaptive traits such as xenobiotic degradation, antibiotic production and resistance. The heterogeneity of the soil environment also requires flexibility, allowing bacteria to adapt to constantly changing environmental conditions and respond to challenges from anthropogenic inputs. Emphasis is placed on antibiotic production and resistance as examples of clustered genes which undoubtedly play a key role in the competitiveness of bacteria in soil, while being non-essential for growth under laboratory conditions. Soil-dwelling bacterial groups such as actinobacteria, pseudomonads and bacilli produce an impressive range of antibiotics and other secondary metabolites. Streptomycetes in particular show the ability to produce multiple secondary metabolites from diverse chemical classes (Hopwood, 2004; Weber *et al.*, 2003). It has been argued (Challis & Hopwood, 2003) that synergy and contingency are important in driving evolution of multiple pathways for production of secondary metabolites and give a competitive advantage to the producer. Studying the evolution of these gene clusters will improve our understanding of bacterial growth and survival in soil.

Defining gene clusters

Like many other terms in biology, 'cluster' is frequently used but seldom precisely defined. A useful definition includes co-ordinated regulation of a number of adjacent transcription units which may be found in both senses and strands and any frame. Collections of genes with related function can undergo reassortment to form new pathways and are likely to have undergone horizontal transfer at some stage of their evolution. Operons could be distinguished from clusters, as bacterial operons have been described as one polycistronic transcriptional unit initiated at one promoter, whereas gene clusters might have more than one promoter and several transcriptional units in various senses. Singleton & Sainsbury (2001) describe prokaryotic operons as two or more genes whose expression is co-ordinated from (i) a single promoter, from which a polycistronic mRNA is produced, or (ii) a common regulatory region, mRNAs 'being formed by divergent transcription from different promoters'. The definition of gene cluster used here is closer to the second definition.

A method has recently been proposed to search genomes and apply an analytical statistical test for clustering (Hoberman *et al.*, 2005) where searches are made for contiguous regions containing genes with homology, separated by no more than a certain number of non-homologous genes. The important contribution of this method is in the definition of clustering used. Some degree of homology is assumed to indicate functionally related genes, which would have an origin in duplication followed by divergence. Non-homologous genes situated within clusters of related genes may function as part of the cluster. The max-gap cluster analysis (Hoberman *et al.*, 2005) allows a certain number of unrelated genes to be passed over in order to identify the higher-level functional organization which constitutes a cluster.

Origin of gene clusters – tandem duplication and divergence

Gene clusters are formed by selection, and linkage would be expected whenever this produces a selective advantage. Variation released in recombination must be sufficiently conservative to ensure that at least some viable cells are produced. Hence we expect a linkage effect preventing recombination between factors which interact intensely, since disruption of the linkage would have a viability effect (Stahl & Murray, 1966).

Duplication of gene clusters has been observed to create massive gene dosages in *Escherichia coli* and *Salmonella enterica* serovar Typhimurium, where short clusters have been observed to be reversibly duplicated up to a hundred times. This suggests that another mechanism of recombination other than *recA*-mediated sister chromosome exchanges is taking place (Hughes, 1998). In streptomycetes, genomic instability seems very common, with extensive chromosomal deletions and intense amplification taking place in the apparent absence of selection pressure (Birch *et al.*, 1990). It has been proposed that gene duplication acts as a 'dynamic and reversible regulatory mechanism that facilitates adaptation' (Reams & Neidle, 2004), so the selective advantage of such tandem duplications as a response to variable conditions needs to be included in consideration of gene clustering in environments such as soil.

Gene duplication followed by divergence is likely to be a feature of the histories of genes within functional clusters. Bioinformatic comparison methods can test whether pairs of genes chosen are more or less likely to be found side by side in genomes. Paralogous genes (thought to have originated from a duplication event within an organism) are found adjacent to each other across the genomes of *E. coli* and *Bacillus subtilis*. Unrelated genes located side by side across the boundary between cistrons are used as controls in the comparison as they are assumed not to be under selection for adjacency (Janga & Moreno-Hagelsieb, 2004). The conservation of adjacency in related genes compared to those considered to be unrelated genes across the boundary of a polycistronic transcription unit appears to extend across a large number of genomes (Moreno-Hagelsieb *et al.*, 2001).

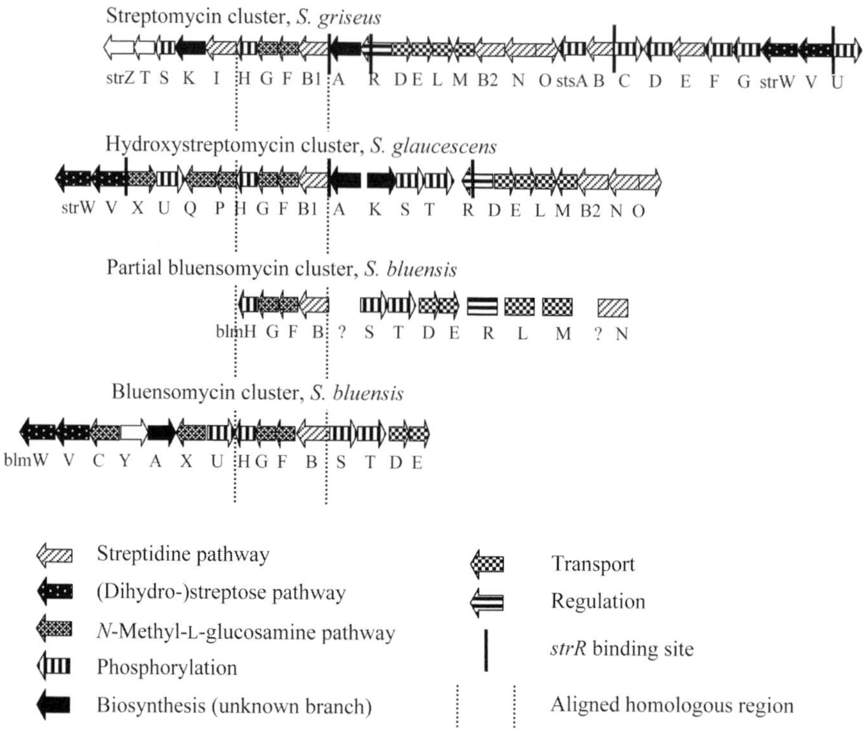

Fig. 1. Biosynthetic gene clusters for streptomycin and bluensomycin. The streptomycin gene cluster from *Streptomyces griseus*, the hydroxystreptomycin cluster from *Streptomyces glaucescens*, the partial cluster of bluensomycin from *Streptomyces bluensis* and the bluensomycin gene cluster from *S. bluensis* as reported by Jung *et al.* (2003) are illustrated. The clusters are aligned according to the homologous region around *strB1*. Adapted from Tolba (2004), after Piepersberg (1997) and Jung *et al.* (2003).

Genes in the streptomycin cluster (Fig. 1) are found in all reading frames on both strands and in both senses. The retention of clusters through evolutionary history is intriguing, given that partial loss would not necessarily result in death of the organism. It is clear that some partial clusters exist (Egan *et al.*, 2001), and this will be discussed further.

Other researchers have used the tendency of related genes to be found adjacent to each other in reverse to predict gene pairs likely to be in an operon from their adjacent occurrence across many genomes (Dandekar *et al.*, 1998; Ermolaeva *et al.*, 2001; Overbeek *et al.*, 1999, 2000, 2003). Although it is clear that overall genomic order is not preserved to a great extent across bacteria (Bentley *et al.*, 2002; Mushegian & Koonin, 1996), the identification of related pairs from duplication and divergence is a starting point for tracing the evolution of a gene cluster.

The 'selfish operon'

It has been proposed that gene clusters form through repeated horizontal gene transfer (HGT) events, between and within taxa, the so-called 'selfish operon' model (Lawrence & Roth, 1996). Formation through HGT would account for retention in clusters of genes that are expected to be selectively neutral or under only weak selection. It is not likely that this applies to all gene clusters, because many of the xenobiotic-degrading gene clusters that have evolved in soil bacteria result directly from selection in response to exposure and are thought to be valuable for survival and hence under selection. Genes 'not essential' under laboratory conditions might still be essential for survival in soil, and hence selectable and able to cluster via linkage effects.

A sustained attack on the selfish operon model has resulted in some interesting work testing hypotheses regarding the expected distribution of clustering. It has been found in whole genome analysis of *E. coli* that essential genes with related functions have a stronger tendency to cluster than non-essential genes (Pal & Hurst, 2004). This is of course the expected result, as linkage is expected between genes with related function. Another test examined whether horizontally transferred genes were more or less likely to be found in operons (Price *et al.*, 2005), and such investigations into clustering mechanisms have obvious relevance to clustering processes in soil bacteria.

It seems that organization in bacterial genomes arises at the operon level and above despite frequent gene rearrangements (Reams & Neidle, 2004). Co-transcription appears to be insufficient to account for the structure of gene clusters in soil bacteria, with genes organized together on the genome but in both senses and different frames. Gene amplification can work alongside HGT as an evolutionary mechanism involved in gene clustering. Gene clusters are often found on mobile genetic elements (MGEs) such as plasmids, phages and genomic islands (GEIs) and can confer a new phenotypic trait allowing improved adaptation to a specific niche. However, it has been argued that HGT is not a cause of operon formation but instead promotes the prevalence of pre-existing operons (Price *et al.*, 2005).

SECONDARY METABOLISM AND ANTIBIOTIC GENE CLUSTERS

Antibiotics are synthesized by bacteria in soil (Anukool *et al.*, 2004) and the biosynthetic gene clusters responsible for such metabolite production are co-ordinately regulated (Chater & Bibb, 1997), have been subject to HGT (Egan *et al.*, 2001) and are discontinuous in their distribution within closely related groups of species. The gene clusters usually have a chromosomal location, but several have been found on plasmids, particularly linear plasmids in the case of streptomycetes (Kinashi *et al.*, 1994; Mochizuki *et al.*, 2003). Within the *Streptomycetaceae* there is an extensive diversity of secondary metabolites: some, like streptomycin, have a defined biological activity, while others such as geosmin, a volatile isoprenoid compound (responsible for the soil's 'earthy' odour), have unknown activity. Production of secondary metabolites provides a selective advantage in soil where the roles of the metabolites may be as molecular weapons, repellents, attractants, 'nutriophores' (e.g. siderophores), components of detoxification/replacement metabolism, co-catalysts or enzyme or transport inhibitors, shaped by a long evolutionary history for highly specific interaction with their molecular targets (Piepersberg, 1993, 2002).

Streptomycetes are ubiquitous soil bacteria and can be readily isolated from the majority of soils. They are filamentous, having a complex life cycle: spores germinate and form vegetative and then aerial hyphae which mature into spore chains that are subsequently released. Sequencing of the *Streptomyces coelicolor* chromosome has revealed an 'unprecedented proportion of regulatory genes' thought to be involved in response to external stimuli and stresses (Bentley *et al.*, 2002). More than 20 gene clusters have been found in the *S. coelicolor* genome. Sets of duplicated genes are suggested to provide 'tissue-specific' responses depending on the developmental phase of the cell surrounding the nucleoid.

More than a million base pairs of genetic material at either end of the *S. coelicolor* linear chromosome can be deleted without affecting viability under lab conditions (Bentley *et al.*, 2002). These 'arm' regions of the chromosome are found to be variable and seem to contain an increasing proportion of functions related to secondary metabolism and resistance towards the ends. The core region close to the origin of replication by contrast contains most of the essential housekeeping genes involved in processes such as cell cycling, energy, protein and DNA metabolism. The core has been found to be more stable, with gene order conserved across different strains and even amongst distantly related organisms such as *Mycobacterium tuberculosis* and *Corynebacterium diphtheriae* (Bentley *et al.*, 2002).

When different strains encounter each other, formation of partial diploids as the chromosomes align followed by crossovers between chromosomes and homologous recombination can result in recombination of up to a fifth of the genome, which is a large amount of genetic material in an organism with an 8 Mb chromosome. The linear chromosome, with bound telomeres, means that a single crossover can cause exchange of genetic material. It might be assumed that this is a rare event but the formation of one crossover instead of two for an exchange is uniquely created by the linear structure of the chromosome. The more frequently occurring case where both telomeres are inherited from the same parent causes a high degree of linkage to be observed between the telomeres; hence the circular linkage map (Wang *et al.*, 1999). The trigger for crossovers and subsequent rearrangements is still obscure, but the elevated frequency of such events under the stress of mutagenic treatment and the involvement of illegitimate recombination indicate that they may be the result of an SOS-like response (Birch *et al.*, 1990), with little evidence to support a connection between mobile DNA elements and the phenomenon of genetic instability.

Recent advances in the study of recombination in *Acinetobacter* (de Vries & Wackernagel, 2002; de Vries *et al.*, 2004) reveal one possible basis for such extraordinary feats of recombination. It is proposed that homologous recombination within an anchor region couples the donor and recipient DNA, followed by an illegitimate recombination event integrating the section. The DNA mismatch repair system in *Pseudomonas stutzeri*, another ubiquitous soil bacterium, has been implicated in the acquisition of foreign DNA from plants as well as other microbes, and seems to affect this homology-facilitated illegitimate recombination process (Meier & Wackernagel, 2005). Another group report homology-independent illegitimate recombination, also in *Acinetobacter* (Reams & Neidle, 2004). Although the detail has yet to be elucidated, the evidence of such transfer processes supports the conclusions from phylogenetic studies of HGT as a major factor in the evolution and adaptation of microbes (Vining, 1992).

In some gene clusters (e.g. streptomycin biosynthesis; Fig. 1), genes are found in opposing senses and on different strands within the cluster. It would appear that genes are transcribed in opposite directions. When this occurs amongst closely spaced promoters, it has been suggested that the supercoiling caused by multiple polymerases affects expression of genes or operons (Opel *et al.*, 2001). So the arrangement of genes in the streptomycin biosynthesis cluster (Fig. 1) could be associated with divergent transcription effects altering the expression of genes.

Transcriptional coupling of this kind is likely to serve a physiological purpose: in *E. coli*, global superhelical density of the chromosome is controlled by the energy charge of the cell, which is affected by environmental stresses and transition between growth states (Opel *et al.*, 2001). The twin-domain model (Liu & Wang, 1987) suggests that closely spaced, divergent, superhelically sensitive promoters can affect the transcriptional activity of one another by transcriptionally induced negative DNA supercoiling generated in the divergent promoter region (Rhee *et al.*, 1999).

Transcriptional interference is thought to be very widespread, where the activity of one transcription complex affects the activity of others. Studies in *E. coli* (Callen *et al.*, 2004) show effects on transcription due to the formation of other transcriptional complexes, even at distant promoters. Although the dynamics of transcriptional interference do not seem to conform to a simple model (Shearwin *et al.*, 2005; Sneppen *et al.*, 2005), perhaps consideration of multiple polymerase effects may throw light on the arrangement of genes in gene clusters.

In the streptomycin biosynthesis cluster in *Streptomyces griseus*, several pairs of genes lie in opposing senses. This would prevent polycistronic transcription, but divergent transcription might account for another level of regulation from the arrangement of the genes. However, the specific arrangement of genes in an antibiotic cluster might not affect antibiotic synthesis. Biosynthetic clusters producing chromomycin in *S. griseus* (Menendez *et al.*, 2004) and mithramycin in *Streptomyces argillaceus* (Rohr *et al.*, 1999), despite making very similar chemical structures and having nearly identical genes, are ordered quite differently (O'Connor, 2004).

Selective pressures favouring gene clustering

Firn & Jones (2000) put forward the view that not every product of secondary metabolism has to be bioactive in order for the ability to create these substances to be retained by selection. The selective advantage of the rare bioactive molecule is clear, and it seems reasonable to suppose that natural selection favours organisms able to generate and maintain chemical novelties at the lowest cost. By analogy with the animal immune system, possession of the mechanism for generating metabolites is crucial, and the waste resulting from the many non-active molecules generated, like the many non-useful antibodies, can be tolerated.

Alteration of substrate specificity is likely to be a favourable mutation in an enzyme involved in metabolism. In primary metabolism, enzymes seem to have very precise substrate specificity, probably due to the high cost of inefficient processing in the cell's most essential functions. Retention in secondary metabolism of enzymes with broad substrate specificity would be expected, given the different selection pressures on such products (Firn & Jones, 2000). Enzymes with broad specificity will have a greater chance to generate novel products: a single mutation in spearmint produced novel peppermint flavours (Croteau, 1991). Enzymes further down the monoterpene production pathway were apparently able to act on the modified substrates to produce the novel products, and perhaps this broad specificity would not be expected in primary metabolism.

Vining (1992) focuses on the survival value of secondary metabolites for the organism. He suggests that secondary metabolism is found when the organism exists in a competitive environment. The soil environment is likely to fit these criteria. Different kinds of organisms, plants, fungi and arthropods, as well as many kinds of microbes are competing for the nutrient resources available in soil. In harsher environments where primarily anaerobes or chemolithotrophs are found, resources are so scarce that few types of organism are found, competition is lower and metabolism is more streamlined with fewer secondary metabolic products. Vining (1992) points out that many of the bioactive metabolites produced have specificity for a particular kind of competitor, whether a grazer of mycelium (e.g. arthropods targeted by avermectin produced by *Streptomyces avermitilis*; Hotson, 1982) or a more direct competitor for nutrition. For example, siderophore pseudobactin 358, produced by *Pseudomonas putida* in the rhizosphere, seems to limit growth of other bacteria and fungi by sequestering available iron (de Weger *et al*., 1988).

Genome comparison between *S. coelicolor* and *S. avermitilis* has been used to provide a model for the evolution of these large linear chromosomes. The central 'core' region is derived from a common actinobacterial ancestor and contains mostly essential and housekeeping genes, while the chromosome 'arms' comprise laterally acquired contingency genes, which have been added and retained for their circumstantially advantageous qualities (Karoonuthaisiri *et al*., 2005). The linear chromosome could be seen as the heritable mechanism which allows this division of the chromosome into two gene pools.

The genome of *S. avermitilis* clearly supports this model (Ikeda *et al*., 2003). Comparison with *S. coelicolor* shows that the core regions are very similar while the arms are almost completely different and are likely to have developed independently. Ikeda *et al*. (2003) also note that the origin of replication for the *S. avermitilis* chromosome is 776 kb away from the centre. However, it does appear to lie central to the core region, so it might be the uneven size of the arms that gives the impression of asymmetry. Intriguingly, Ikeda *et al*. (2003) also note that, although most of the secondary metabolism clusters are located in the arms, the few that lie in the core tend to be common to several streptomycete species, suggesting that they were fixed in the chromosome at an evolutionarily early stage.

The streptomycete linear chromosome could provide a distinct advantage for horizontal transfer and clustering via duplication due to the unstable arm regions stabilized by protein-bound telomeres. Rare recombinations near one end of the chromosome could result in exchange of the unstable arm region between strains, as depicted in Fig. 2. The molecular-genetic system of linear chromosomes could facilitate the assembly of secondary metabolite gene clusters including those of antibiotics. It could provide both increased rate of variation through HGT for selection to act upon and the necessary conservation of function by stable telomeres and a core region distant enough from the arms for essential metabolic function to stay intact in the event of telomere exchange.

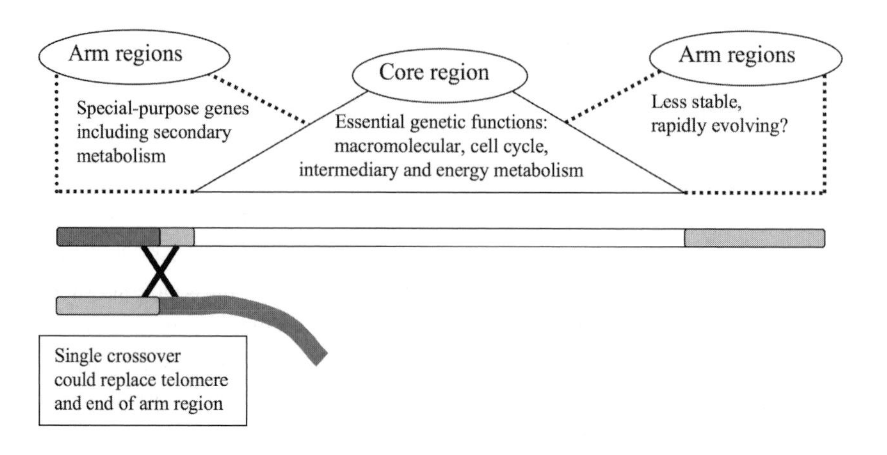

Fig. 2. Linear chromosome of *Streptomyces*. Arm and core regions seem to contain genes for different metabolic functions. The linear chromosome structure might occasionally recombine from only one crossover after partial diploid formation to exchange a telomere-protein-bound end with another chromosome fragment. However, the telomere protein ends might be bound together, holding the chromosome in circular form. Adapted from Piepersberg (2002).

Origin of gene clusters – HGT

Vining (1992) has compared nucleotide sequences from polyketide and phenazine antibiotic synthesis pathways with related genes in primary pathways. Where sequence similarity exists to primary pathway genes in distantly related organisms, it seems likely that the secondary metabolite gene originated in that organism. Vining (1992) suggests from his study of fatty acid synthase and polyketide synthase (PKS) pathways that certain core secondary metabolite pathways have 'arisen early in evolution and been maintained in reservoir organisms' from which they have been donated via gene transfer. If this is the case, he suggests we would expect to see an ancient core component in widely distributed secondary metabolic pathways, with more recently evolved terminal and specificity-conferring reactions 'grafted on' to the ancient core of the pathway. The early steps in which intermediates of little selective value are generated might have arisen only once, but horizontal transfer allows their acquisition once they are linked in a pathway which provides a selective advantage.

The analogy between polyketide and long-chain fatty acid biosynthesis has been extended by studies that have demonstrated similarity between the products of several PKS genes (Hopwood & Sherman, 1990; Katz & Donadio, 1993) and their fatty acid synthase congeners in *E. coli* (Magnuson *et al.*, 1993; Vanden Boom & Cronan, 1989), yeast (Schweizer *et al.*, 1986) and mammals (Witkowski *et al.*, 1991). Examples have been found of PKSs resembling each of the classical classes of fatty acid synthases (FASs). The PKSs are classified in to three types of which types I and II are found mainly in bacteria and fungi. It is believed that type I and II PKSs share the same evolutionary origin, but their sequences are too far diverged for significant DNA hybridization (McCarthy *et al.*, 1983).

PKSs are multifunctional enzymes which carry out repeated rounds of carbon chain building by condensation of carboxylic acids to form a polyketide chain. Metsa-Ketela *et al.* (1999) studied the molecular diversity of a portion of the ketosynthase gene, KSα, responsible for the building of the carbon chain (in the aromatic polyketide pathway) in a wide range of soil isolates of *Streptomyces* species. Comparison of KSα gene sequence phylogeny with that obtained from analysis of the γ-variable 16S rRNA gene indicated extensive HGT. There was no evidence of correlation between the two gene phylogenies and very similar KSα sequences were recovered in distantly related species while high KSα sequence diversity was evident within species. Gene clusters for glycopeptide antibiotic biosynthesis showed strong evidence for mosaicism; Donadio *et al.* (1991) reported on the comparison of five gene clusters involved in biosynthesis of structurally related antibiotics which shared a similar mechanism of action on the bacterial cell wall. Conserved synteny was observed between clusters and preceding intergenic regions were often conserved. The clusters consisted of distinct gene cassettes encoding enzymes for subpathways which expanded by gene acquisition adding specific tailoring steps, regulatory and resistance genes. Extensive conserved synteny has also been observed in comparative analysis of gene clusters (PKSs, post-PK modifications and regulatory genes) involved in polyene antibiotic biosynthesis (Aparicio *et al.*, 2003). The clusters for nystatin, primaricin, amphotericin and candicidin were compared and evidence for gene duplications was noted and thought to have played a recent role in the evolution of the gene clusters. Many polyenes show potent antifungal activity, and Aparicio *et al.* (2003) hypothesized that the large polyene-producing PKS systems could have evolved from the smaller PKSs that assemble macrolides that inhibit prokaryotic ribosomes.

The streptomycin cluster – ecological context

The streptomycin biosynthetic gene cluster is most studied in its ecological context. Streptomycin use in horticulture has never been widespread in Europe and has been banned, although it was used to treat fireblight (*Erwinia*) infection in orchards and was also occasionally used in veterinary medicine. Streptomycin is one of the best studied antibiotics; several studies of the ecology of producers have elucidated the evolutionary context in which the genes are evolving (Egan *et al.*, 2001; Tolba *et al.*, 2002). Density and spatial clustering of the producing strain appears to have an impact on antibiotic production (Wiener, 2000), which also seems to be highly variable (Davelos *et al.*, 2004a, b). In genetic analysis of *Streptomyces* strains implicated in potato scab disease, pathogenicity and phenotype do not seem to correlate well with phylogeny inferred from 16S rRNA gene sequence (Bramwell *et al.*, 1998). This complicates the task of inferring past, present and future evolutionary processes. Comparison of strains from sites where streptomycin has been applied (as plantomycin) and those from untreated sites or sites where sewage sludge has been applied (Tolba *et al.*, 2002) showed that the level of resistance to streptomycin in recovered strains was similar regardless of treatment. Screening of these soil isolates from various European locations (Tolba *et al.*, 2002) for *strA* (thought to have a role in streptomycin resistance) and *strB1* (the neighbouring gene in *S. griseus* which is involved in biosynthesis) revealed increased prevalence of both genes in the streptomycin-treated soil.

The presence of streptomycin biosynthesis genes in non-producing strains is interesting. Antibiotic production by streptomycetes has been shown to have a role in preventing the invasion of sensitive competitors under lab conditions, compared with either mediation between co-inoculated *S. griseus* and *B. subtilis* or invasion of *S. griseus* into an established population of *B. subtilis* (Wiener, 1996). When production as well as resistance is conferred by the gene cluster, it is reasonable to suggest that production has a role in defending localized nutrients from sensitive strains. When streptomycin is present in the environment without direct metabolic cost (i.e. from anthropogenic application), the cost of producing streptomycin is less worthwhile, as fewer sensitive strains can be assumed to be present. Acquisition of partial clusters may allow resistance to streptomycin (e.g. via modification of the streptomycin molecule); hence we could expect those genes to be retained, unless a ribosome mutation also conferring resistance (target modification) was found in the organism.

Although prevalence of *strA* and *strB1* genes in DNA extracted from soil at the plantomycin-treated site was high, few producers were isolated. This could indicate that some strains had acquired the genes but not the regulatory apparatus necessary for production. Since the usefulness of production is reduced by anthropogenic application, it might be expected that some strains would lose production capability, since the selective advantage would have disappeared. However, if we expect that evolution favours organisms able to produce non-essential metabolites at lower cost (Firn & Jones, 2000, 2003), we might expect to recover producers, non-producers with some production genes and strains with ribosome mutations conferring resistance without the presence of production genes alongside each other.

The fitness cost of maintaining biosynthetic genes would not be sufficient to confer an advantage to strains which had lost them. The possibilities for selection of production and resistance are summarized in Fig. 3 where, in case 1, streptomycin is produced and, in case 2, streptomycin is produced but anthropogenic streptomycin is also encountered. Case 3 indicates that the loss of production capacity reduces metabolic

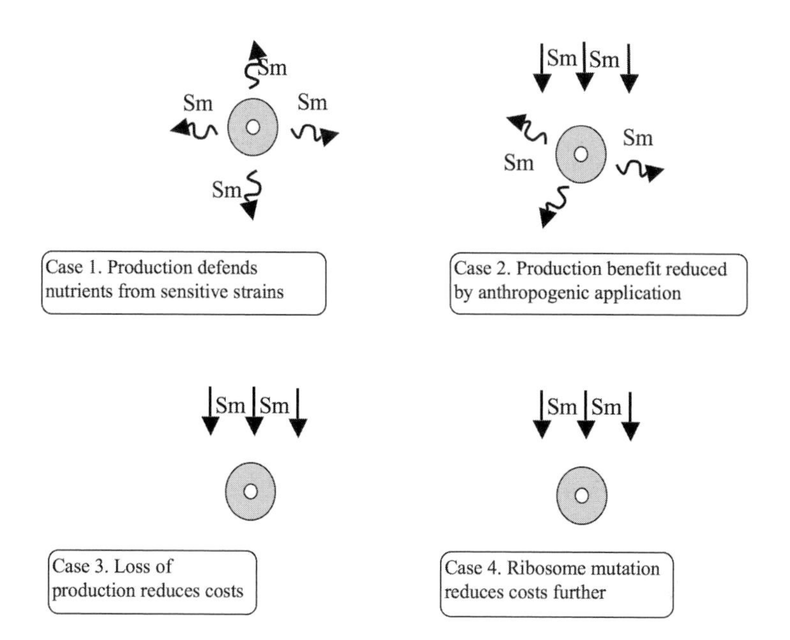

Fig. 3. Evolutionary pressures on streptomycin (Sm) producers during anthropogenic streptomycin application. All three cases (producers with full biosynthetic cluster, resistant organisms with some biosynthesis genes conferring resistance and ribosome mutants) might be expected in isolates if, as Firn & Jones (2000, 2003) suggest, natural selection favours organisms able to produce and innovate secondary metabolites at little cost.

costs and case 4 illustrates that the acquisition of a 'target mutation' in ribosome sequence confers resistance without the cost of maintaining any biosynthetic genes. The study (Tolba *et al.*, 2002) also showed that streptomycetes genetically similar to *S. coelicolor* had *strA* and *strB1* genes with high homology to those found in *S. griseus* type strains, indicative of horizontal transfer. Other incidences have been found of transfer between streptomycetes, e.g. from Brazilian soils (Huddleston *et al.*, 1997); this would be similar to case 3 in Fig. 3 above. MGEs have not yet been implicated in the mobility of the *str* cluster, although linear plasmids and GEIs are thought to play a role in mobilizing functional antibiotic gene clusters.

Future work might investigate the suggestion (Vining, 1992) that the biosynthetic cluster had assembled in one strain followed by transfer to others. Comparison of sugar-handling genes in the pathway might quantify mutation rates of genes involved in polyketide sugar-unit synthesis and nucleotide sugar metabolism compared with those in streptomycin biosynthesis clusters. This kind of study might identify the sources of the genes which have diverged to perform streptomycin biosynthesis by which strains have the highest homology between those and sugar-handling genes.

EVOLUTION OF XENOBIOTIC-DEGRADING GENE CLUSTERS

The ability to degrade xenobiotic compounds added to soil has evolved in a wide range of bacteria and has resulted in the development of gene clusters. Various mechanisms of gene acquisition, involving both MGEs and HGT, have been implicated to allow catabolism of anthropogenic compounds. It is also evident that pathways for degradation of man-made pesticides and other xenobiotics have evolved from pre-existing clusters of genes required for degradation of naturally occurring, chemically similar metabolic products.

The possible evolutionary events involved in development of catabolic pathways for combating potential toxicity and novelty of xenobiotics have been extensively reviewed (Copley, 2000; Johnson & Spain, 2003; Reams & Neidle, 2004; Springael & Top, 2004). Studies of these catabolic pathways have revealed much about how bacteria evolve and adapt to new substrates; a striking example of this is the gene cluster for catabolism of 2,4-dinitrotoluene (2,4-DNT) in *Burkholderia cepacia* (Johnson & Spain, 2003). The 2,4-DNT gene cluster is composed of three modules, each with specific dioxygenases; their phylogeny relates to aromatic acid, benzenoid and naphthalene dioxygenases, respectively. Gene recruitment resulting in a cluster of genes from different, existing pathways was evident in the evolution of the pathway for pentachlorophenol (PCP) degradation (Copley, 2000). Soil bacteria have evolved the ability to degrade this pesticide since its introduction in the early part of the 20th century. The *pcp* genes have been found clustered on two chromosomal fragments in *Sphingobium chlorophenolicum* (Dai & Copley, 2004) and there is evidence that these genes have been subject to HGT within bacterial communities (Tiirola *et al.*, 2002).

Dejonghe *et al.* (2000) proved that bioaugmentation of soil could be used to enhance the rate of 2,4-dichlorophenoxyacetic acid (2,4-D) degradation by addition of donor strains carrying the *tfd* genes on plasmids. The catabolic genes were transferred to the indigenous community via conjugation, while the donor strain did not persist in the soil. Other MGEs have also been described in the dissemination of catabolic activities in soil and include GEIs (Dobrindt *et al.*, 2004), defined as large chromosomal regions flanked by repeat sequences; they contain integrases or transposases and may achieve integration via a tRNA gene as in the case of the *clc* element, encoding the genes for degradation of 3-chlorobenzoate (Muller *et al.*, 2003). GEIs clearly play a role in facilitating further evolution of gene clusters and subsequent dissemination.

INTEGRONS: CO-ORDINATELY REGULATED MOBILE GENE CLUSTERS

An interesting example of gene clustering is represented by a grouping of genes into structures known as integrons; these are recombination and expression systems that capture genes as part of a genetic element known as a gene cassette (Recchia & Hall, 1995). There are different types of integrons: one type sometimes referred to as 'super integrons', which are host-specific, chromosomally encoded and are more stable in their gene order and number of genes, and highly mobile 'resistance integrons' situated on plasmids, transposons and the bacterial chromosome such as class 1 integrons which carry a diverse range and number of genes (Fluit & Schmitz, 2004). These definitions of 'super integrons' and 'resistance integrons' are subject to debate (Hall & Stokes, 2004), but are still widely used. Most cassettes of known function found within non-species-specific resistance integrons confer antibiotic or quaternary ammonium compound resistance. Multidrug resistance among members of the *Enterobacteriaceae* is strongly linked with the presence of integrons (Leverstein-van Hall *et al.*, 2003). In this discussion the term integron will be used to refer to resistance integrons unless otherwise stated.

Integrons are examples of genetic elements that carry clusters or groupings of genes due to their ability to mobilize and regulate expression of cassette genes. A promoter in the 5′ conserved segment of class 1 integrons directs transcription of nearly all the cassette-encoded genes (Hall & Collis, 1995), with a few exceptions such as *qacE*, which confers biocide resistance and possesses its own promoter (Ploy *et al.*, 1998). Class 1 integrons (which are the most widely studied class) are found with varying numbers of gene cassettes or none at all (Rosser & Young, 1999). The maintenance of empty integrons may allow integration of cassette genes if bacteria encounter conditions where cassette-encoded genes confer an advantage, as cassette genes are known to be excised and reintegrated in response to selective pressure. Class 1 integrons are often mobile, either being active transposons or being derivatives that are defective in self-transposition (Partridge *et al.*, 2002). Loss of transposase and/or resolvase genes from the 3′ conserved segment prevents self-transposition, although experimental evidence of transposition in non-active transposon integrons suggests that movement is still possible if requisite gene products are supplied *in trans* (from a gene not on the integron). Integrons are also often situated on plasmids, which increases their already considerable potential for HGT.

Recent studies on integrons from soil total community DNA (TCDNA) have revealed that there are numerous different integron classes, with 14 new undescribed integrase genes discovered in heavy-metal-contaminated mine tailings (Nemergut *et al.*, 2004). A gene cassette showing high identity to a gene encoding a step in a pathway for nitroaromatic catabolism was also discovered in DNA extracted from mine tailings, suggesting that integrons may be important in the HGT of genes other than antibiotic and biocide resistance.

The vast majority of class 1 integrons have been reported from members of the *Enterobacteriaceae*, although they have been occasionally reported from Gram-positive bacteria such as *Corynebacterium glutamicum* (Nesvera *et al.*, 1998). However, a recent study of class 1 integron prevalence in chicken litter revealed that *Corynebacterium*, *Staphylococcus*, *Aerococcus* and *Brevibacterium* spp. harboured the genetic element. These species comprised >85 % of the litter community compared with <2 % made up of members of the *Enterobacteriaceae* (Nandi *et al.*, 2004). Remarkably, 'the concentration of *intI1* genes ranged from 50- to 500-fold greater than the concentration of cultivatable aerobic Gram-negative bacteria, the assumed major hosts for class 1 integrons' (Nandi *et al.*, 2004). The chickens from which the litter was sampled were treated with a range of antibiotics, antibacterials and coccidiostats; it is thought that the use of antimicrobials in medicine and the farm environment may have resulted in an increased prevalence of integrons and the diversity of gene cassettes they carry. A study presently under way in the authors' laboratory has highlighted a possible correlation between anthropogenic activity and class 1 integron prevalence in TCDNA extracted from soils (N. Abdouslam, personal communication). A previous study suggested a correlation between exposure to industrial biocides and the prevalence of class 1 integrons in culturable aerobes (Gaze *et al.*, 2005).

It is becoming clear that integrons are distributed far more widely in bacterial taxa and that the functional diversity of cassette genes is greater than previously thought. Coupled with the ability of cassette genes, integrons and MGEs bearing integrons to undergo HGT, the potential for these gene clusters to proliferate in soil bacteria exposed to a wide variety of selective pressures is considerable and is contributing to the evolution of antibiotic resistance in clinically important taxa.

CONCLUSIONS

Bacteria in soil must respond to constant changes in growth conditions and survive competition from a diverse microbial community. The ability to acquire a useful phenotypic trait, or multiple traits, encoded by a gene cluster provides the opportunity to acquire a large number of genes in one step. The importance of the horizontal gene pool has featured constantly in explanations for the evolution of gene clusters with the development of patchwork catabolic pathways combining enzymes from two or more pathways. The streptomycete linear chromosome may be part of the mechanism by which gene clusters are assembled amongst soil bacteria, as it allows high variation rates, acquisition and reassortment of genes in unstable arm regions as well as the stabilizing mechanisms and conservation of essential function by protein-bound telomeres and the conserved core region. In addition, the formation of gene cluster mosaics has resulted from the expansion of biosynthetic diversity achieved by recruitment of a cassette of genes encoding a specific part of a subpathway. Gene duplication and mutation play important roles in allowing changes in enzyme specificity and the evolution of new traits.

Evolution of antibiotic biosynthetic pathways may be achieved by continuous mixing of subsets of gene clusters and the formation of new ones. Certain core secondary metabolic pathways which had arisen early in evolution could have been maintained in reservoir organisms that served as donors for gene transfer. Widely distributed pathways such as polyketide formation with a common biochemistry may prove to have a relatively ancient core component on to which the more recent processing reactions responsible for the species specificity of the product have been recruited. Clearly, the organization of antibiotic biosynthesis genes in clusters has greatly facilitated the transfer and expansion of diversity within antibiotic-producing bacteria. Specialized MGEs have evolved to facilitate acquisition of a series of genes often encoding multiple defence mechanisms such as antibiotic resistance and to enable high levels of gene expression. In soil, anthropogenic inputs are dramatically changing the horizontal gene pool and this may have serious impacts on dissemination of certain traits such as antibiotic resistance.

ACKNOWLEDGEMENTS

We gratefully acknowledge financial support from the Natural Environment Research Council (grant NER/A/S/2000/01253) and the Biotechnology and Biological Sciences Research Council (grant 88/GM114200) plus a NERC studentship (A. M.).

REFERENCES

Anukool, U., Gaze, W. H. & Wellington, E. M. (2004). In situ monitoring of streptothricin production by *Streptomyces rochei* F20 in soil and rhizosphere. *Appl Environ Microbiol* **70**, 5222–5228.

Aparicio, J. F., Caffrey, P., Gil, J. A. & Zotchev, S. B. (2003). Polyene antibiotic biosynthesis gene clusters. *Appl Microbiol Biotechnol* **61**, 179–188.

Bentley, S. D., Chater, K. F., Cerdeno-Tarraga, A. M. & 40 other authors (2002). Complete genome sequence of the model actinomycete *Streptomyces coelicolor* A3(2). *Nature* **417**, 141–147.

Birch, A., Hausler, A. & Hutter, R. (1990). Genome rearrangement and genetic instability in *Streptomyces* spp. *J Bacteriol* **172**, 4138–4142.

Bramwell, P. A., Wiener, P., Akkermans, A. D. & Wellington, E. M. (1998). Phenotypic, genotypic and pathogenic variation among streptomycetes implicated in common scab disease. *Lett Appl Microbiol* **27**, 255–260.

Callen, B. P., Shearwin, K. E. & Egan, J. B. (2004). Transcriptional interference between convergent promoters caused by elongation over the promoter. *Mol Cell* **14**, 647–656.

Challis, G. L. & Hopwood, D. A. (2003). Synergy and contingency as driving forces for the evolution of multiple secondary metabolite production by *Streptomyces* species. *Proc Natl Acad Sci U S A* **100** (Suppl. 2), 14555–14561.

Chater, K. F. & Bibb, M. J. (1997). Regulation of bacterial antibiotic production. In *Products of Secondary Metabolism*, pp. 57–105. Edited by H. Kleinkauf & H. von Dohren. Weinheim: VCH.

Copley, S. D. (2000). Evolution of a metabolic pathway for degradation of a toxic xenobiotic: the patchwork approach. *Trends Biochem Sci* **25**, 261–265.

Croteau, R. (1991). Metabolism of monoterpenes in mint (*Mentha*) species. *Planta Med* **57** (Suppl.), 10–14.

Dai, M. & Copley, S. D. (2004). Genome shuffling improves degradation of the anthropogenic pesticide pentachlorophenol by *Sphingobium chlorophenolicum* ATCC 39723. *Appl Environ Microbiol* **70**, 2391–2397.

Dandekar, T., Snel, B., Huynen, M. & Bork, P. (1998). Conservation of gene order: a fingerprint of proteins that physically interact. *Trends Biochem Sci* **23**, 324–328.

Davelos, A. L., Kinkel, L. L. & Samac, D. A. (2004a). Spatial variation in frequency and intensity of antibiotic interactions among streptomycetes from prairie soil. *Appl Environ Microbiol* **70**, 1051–1058.

Davelos, A. L., Xiao, K., Flor, J. M. & Kinkel, L. L. (2004b). Genetic and phenotypic traits of streptomycetes used to characterize antibiotic activities of field-collected microbes. *Can J Microbiol* **50**, 79–89.

Dejonghe, W., Goris, J., El Fantroussi, S., Hofte, M., De Vos, P., Verstraete, W. & Top, E. M. (2000). Effect of dissemination of 2,4-dichlorophenoxyacetic acid (2,4-D) degradation plasmids on 2,4-D degradation and on bacterial community structure in two different soil horizons. *Appl Environ Microbiol* **66**, 3297–3304.

de Vries, J. & Wackernagel, W. (2002). Integration of foreign DNA during natural transformation of *Acinetobacter* sp. by homology-facilitated illegitimate recombination. *Proc Natl Acad Sci U S A* **99**, 2094–2099.

de Vries, J., Herzfeld, T. & Wackernagel, W. (2004). Transfer of plastid DNA from tobacco to the soil bacterium *Acinetobacter* sp. by natural transformation. *Mol Microbiol* **53**, 323–334.

de Weger, L. A., van Arendonk, J. J., Recourt, K., van der Hofstad, G. A., Weisbeek, P. J. & Lugtenberg, B. (1988). Siderophore-mediated uptake of Fe^{3+} by the plant growth-stimulating *Pseudomonas putida* strain WCS358 and by other rhizosphere microorganisms. *J Bacteriol* **170**, 4693–4698.

Dobrindt, U., Hochhut, B., Hentschel, U. & Hacker, J. (2004). Genomic islands in pathogenic and environmental microorganisms. *Nat Rev Microbiol* **2**, 414–424.

Donadio, S., Staver, M. J., McAlpine, J. B., Swanson, S. J. & Katz, L. (1991). Modular organization of genes required for complex polyketide biosynthesis. *Science* **252**, 675–679.

Egan, S., Wiener, P., Kallifidas, D. & Wellington, E. M. (2001). Phylogeny of *Streptomyces* species and evidence for horizontal transfer of entire and partial antibiotic gene clusters. *Antonie van Leeuwenhoek* **79**, 127–133.

Ermolaeva, M. D., White, O. & Salzberg, S. L. (2001). Prediction of operons in microbial genomes. *Nucleic Acids Res* **29**, 1216–1221.

Firn, R. D. & Jones, C. G. (2000). The evolution of secondary metabolism – a unifying model. *Mol Microbiol* **37**, 989–994.

Firn, R. D. & Jones, C. G. (2003). Natural products – a simple model to explain chemical diversity. *Nat Prod Rep* **20**, 382–391.

Fluit, A. C. & Schmitz, F. J. (2004). Resistance integrons and super-integrons. *Clin Microbiol Infect* **10**, 272–288.

Gaze, W. H., Abdouslam, N., Hawkey, P. M. & Wellington, E. M. (2005). Incidence of class 1 integrons in a quaternary ammonium compound-polluted environment. *Antimicrob Agents Chemother* **49**, 1802–1807.

Hall, R. M. & Collis, C. M. (1995). Mobile gene cassettes and integrons: capture and spread of genes by site-specific recombination. *Mol Microbiol* **15**, 593–600.

Hall, R. M. & Stokes, H. W. (2004). Integrons or super integrons? *Microbiology* **150**, 3–4.

Hoberman, R., Sankoff, D. & Durand, D. (2005). The statistical significance of max-gap clusters. In *Proceedings of RECOMB 2004 Workshop on Comparative Genomics*, LNBI 3388, pp. 55–71. Edited by J. Lagergren. Berlin & Heidelberg: Springer.

Hopwood, D. A. (2004). Cracking the polyketide code. *PLoS Biol* **2**, E35.

Hopwood, D. A. & Sherman, D. H. (1990). Molecular genetics of polyketides and its comparison to fatty acid biosynthesis. *Annu Rev Genet* **24**, 37–66.

Hotson, I. K. (1982). The avermectins: a new family of antiparasitic agents. *J S Afr Vet Assoc* **53**, 87–90.

Huddleston, A. S., Cresswell, N., Neves, M. C., Beringer, J. E., Baumberg, S., Thomas, D. I. & Wellington, E. M. (1997). Molecular detection of streptomycin-producing streptomycetes in Brazilian soils. *Appl Environ Microbiol* **63**, 1288–1297.

Hughes, D. (1998). Impact of homologous recombination on genome organization and stability. In *Organization of the Prokaryotic Genome*, pp. 109–128. Edited by R. L. Charlebois. Washington, DC: American Society for Microbiology.

Ikeda, H., Ishikawa, J., Hanamoto, A., Shinose, M., Kikuchi, H., Shiba, T., Sakaki, Y., Hattori, M. & Omura, S. (2003). Complete genome sequence and comparative analysis of the industrial microorganism *Streptomyces avermitilis*. *Nat Biotechnol* **21**, 526–531.

Janga, S. C. & Moreno-Hagelsieb, G. (2004). Conservation of adjacency as evidence of paralogous operons. *Nucleic Acids Res* **32**, 5392–5397.

Johnson, G. R. & Spain, J. C. (2003). Evolution of catabolic pathways for synthetic compounds: bacterial pathways for degradation of 2,4-dinitrotoluene and nitrobenzene. *Appl Microbiol Biotechnol* **62**, 110–123.

Jung, Y. G., Kang, S. H., Hyun, C. G., Yang, Y. Y., Kang, C. M. & Suh, J. W. (2003). Isolation and characterization of bluensomycin biosynthetic genes from *Streptomyces bluensis*. *FEMS Microbiol Lett* **219**, 285–289.

Karoonuthaisiri, N., Weaver, D., Huang, J., Cohen, S. N. & Kao, C. M. (2005). Regional organization of gene expression in *Streptomyces coelicolor*. *Gene* **353**, 53–66.

Katz, L. & Donadio, S. (1993). Polyketide synthesis: prospects for hybrid antibiotics. *Annu Rev Microbiol* **47**, 875–912.

Kinashi, H., Mori, E., Hatani, A. & Nimi, O. (1994). Isolation and characterization of linear plasmids from lankacidin-producing *Streptomyces* species. *J Antibiot* **47**, 1447–1455.

Lawrence, J. G. & Roth, J. R. (1996). Selfish operons: horizontal transfer may drive the evolution of gene clusters. *Genetics* **143**, 1843–1860.

Leverstein-van Hall, M. A., Blok, H. E. M., Donders, A. R. T., Paauw, A., Fluit, A. C. & Verhoef, J. (2003). Multidrug resistance among *Enterobacteriaceae* is strongly associated with the presence of integrons and is independent of species or isolate origin. *J Infect Dis* **187**, 251–259.

Liu, L. F. & Wang, J. C. (1987). Supercoiling of the DNA template during transcription. *Proc Natl Acad Sci U S A* **84**, 7024–7027.

Magnuson, K., Jackowski, S., Rock, C. O. & Cronan, J. E., Jr (1993). Regulation of fatty acid biosynthesis in *Escherichia coli*. *Microbiol Rev* **57**, 522–542.

McCarthy, A. D., Goldring, J. P. & Hardie, D. G. (1983). Evidence that the multifunctional polypeptides of vertebrate and fungal fatty acid synthases have arisen by independent gene fusion events. *FEBS Lett* **162**, 300–304.

Meier, P. & Wackernagel, W. (2005). Impact of *mutS* inactivation on foreign DNA acquisition by natural transformation in *Pseudomonas stutzeri*. *J Bacteriol* **187**, 143–154.

Menendez, N., Nur-e-Alam, M., Brana, A. F., Rohr, J., Salas, J. A. & Mendez, C. (2004). Biosynthesis of the antitumor chromomycin A3 in *Streptomyces griseus*: analysis of the gene cluster and rational design of novel chromomycin analogs. *Chem Biol* **11**, 21–32.

Metsa-Ketela, M., Salo, V., Halo, L., Hautala, A., Hakala, J., Mantsala, P. & Ylihonko, K. (1999). An efficient approach for screening minimal PKS genes from *Streptomyces*. *FEMS Microbiol Lett* **180**, 1–6.

Mochizuki, S., Hiratsu, K., Suwa, M., Ishii, T., Sugino, F., Yamada, K. & Kinashi, H. (2003). The large linear plasmid pSLA2-L of *Streptomyces rochei* has an unusually condensed gene organization for secondary metabolism. *Mol Microbiol* **48**, 1501–1510.

Moreno-Hagelsieb, G., Trevino, V., Perez-Rueda, E., Smith, T. F. & Collado-Vides, J. (2001). Transcription unit conservation in the three domains of life: a perspective from *Escherichia coli*. *Trends Genet* **17**, 175–177.

Muller, T. A., Werlen, C., Spain, J. & Van Der Meer, J. R. (2003). Evolution of a chlorobenzene degradative pathway among bacteria in a contaminated groundwater mediated by a genomic island in *Ralstonia*. *Environ Microbiol* **5**, 163–173.

Mushegian, A. R. & Koonin, E. V. (1996). Gene order is not conserved in bacterial evolution. *Trends Genet* **12**, 289–290.

Nandi, S., Maurer, J. J., Hofacre, C. & Summers, A. O. (2004). Gram-positive bacteria are a major reservoir of class 1 antibiotic resistance integrons in poultry litter. *Proc Natl Acad Sci U S A* **101**, 7118–7122.

Nemergut, D. R., Martin, A. P. & Schmidt, S. K. (2004). Integron diversity in heavy-metal-contaminated mine tailings and inferences about integron evolution. *Appl Environ Microbiol* **70**, 1160–1168.

Nesvera, J., Hochmannova, J. & Patek, M. (1998). An integron of class 1 is present on the plasmid pCG4 from gram-positive bacterium *Corynebacterium glutamicum*. *FEMS Microbiol Lett* **169**, 391–395.

O'Connor, S. (2004). Aureolic acids: similar antibiotics with different biosynthetic gene clusters. *Chem Biol* **11**, 8–10.

Opel, M. L., Arfin, S. M. & Hatfield, G. W. (2001). The effects of DNA supercoiling on the expression of operons of the *ilv* regulon of *Escherichia coli* suggest a physiological rationale for divergently transcribed operons. *Mol Microbiol* **39**, 1109–1115.

Overbeek, R., Fonstein, M., D'Souza, M., Pusch, G. D. & Maltsev, N. (1999). The use of gene clusters to infer functional coupling. *Proc Natl Acad Sci U S A* **96**, 2896–2901.

Overbeek, R., Larsen, N., Pusch, G. D., D'Souza, M., Selkov, E., Jr, Kyrpides, N., Fonstein, M., Maltsev, N. & Selkov, E. (2000). WIT: integrated system for high-throughput genome sequence analysis and metabolic reconstruction. *Nucleic Acids Res* **28**, 123–125.

Overbeek, R., Larsen, N., Walunas, T. & 19 other authors (2003). The ERGO genome analysis and discovery system. *Nucleic Acids Res* **31**, 164–171.

Pal, C. & Hurst, L. D. (2004). Evidence against the selfish operon theory. *Trends Genet* **20**, 232–234.

Partridge, S. R., Brown, H. J. & Hall, R. M. (2002). Characterization and movement of the class 1 integron known as Tn*2521* and Tn*1405*. *Antimicrob Agents Chemother* **46**, 1288–1294.

Piepersberg, W. (1993). Streptomycetes and corynebacteria. In *Biological Fundamentals*, pp. 434–468. Edited by H. Sahm. Weinheim: Verlag Chemie.

Piepersberg, W. (1997). Molecular biology, biochemistry, and fermentation of aminoglycoside antibiotics. In *Biotechnology of Antibiotics*, 2nd edn, pp. 81–163. Edited by W. R. Strohl. New York: Marcel Dekker.

Piepersberg, W. (2002). Endogenous antimicrobial molecules: an ecological perspective. In *Molecular Medical Microbiology*, pp. 561–584. Edited by M. Sussman. San Diego: Academic Press.

Ploy, M. C., Courvalin, P. & Lambert, T. (1998). Characterization of In40 of *Enterobacter aerogenes* BM2688, a class 1 integron with two new gene cassettes, *cmlA2* and *qacF*. *Antimicrob Agents Chemother* **42**, 2557–2563.

Price, M. N., Huang, K. H., Arkin, A. P. & Alm, E. J. (2005). Operon formation is driven by co-regulation and not by horizontal gene transfer. *Genome Res* **15**, 809–819.

Reams, A. B. & Neidle, E. L. (2004). Selection for gene clustering by tandem duplication. *Annu Rev Microbiol* **58**, 119–142.

Recchia, G. D. & Hall, R. M. (1995). Gene cassettes: a new class of mobile element. *Microbiology* **141**, 3015–3027.

Rhee, K. Y., Opel, M., Ito, E., Hung, S. P., Arfin, S. M. & Hatfield, G. W. (1999). Transcriptional coupling between the divergent promoters of a prototypic LysR-type regulatory system, the *ilvYC* operon of *Escherichia coli*. *Proc Natl Acad Sci U S A* **96**, 14294–14299.

Rohr, J., Méndez, C. & Salas, J. A. (1999). The biosynthesis of aureolic acid group antibiotics. *Bioorg Chem* **27**, 41–54.

Rosser, S. J. & Young, H. K. (1999). Identification and characterization of class 1 integrons in bacteria from an aquatic environment. *J Antimicrob Chemother* **44**, 11–18.

Schweizer, M., Roberts, L. M., Holtke, H. J., Takabayashi, K., Hollerer, E., Hoffmann, B., Muller, G., Kottig, H. & Schweizer, E. (1986). The pentafunctional FAS1 gene of yeast: its nucleotide sequence and order of the catalytic domains. *Mol Gen Genet* **203**, 479–486.

Shearwin, K. E., Callen, B. P. & Egan, J. B. (2005). Transcriptional interference – a crash course. *Trends Genet* **21**, 339–345.

Singleton, P. & Sainsbury, D. (2001). *Dictionary of Microbiology and Molecular Biology*, 3rd edn. Chichester: Wiley.

Sneppen, K., Dodd, I. B., Shearwin, K. E., Palmer, A. C., Schubert, R. A., Callen, B. P. & Egan, J. B. (2005). A mathematical model for transcriptional interference by RNA polymerase traffic in *Escherichia coli*. *J Mol Biol* **346**, 399–409.

Springael, D. & Top, E. M. (2004). Horizontal gene transfer and microbial adaptation to xenobiotics: new types of mobile genetic elements and lessons from ecological studies. *Trends Microbiol* **12**, 53–58.

Stahl, F. W. & Murray, N. E. (1966). The evolution of gene clusters and genetic circularity in microorganisms. *Genetics* **53**, 569–576.

Stotzky, G. (1997). Soil as an environment for microbial life. In *Modern Soil Microbiology*, pp. 1–20. Edited by J. D. van Elsas, J. T. Trevors & E. Wellington. New York: Marcel Dekker.

Tiirola, M. A., Wang, H., Paulin, L. & Kulomaa, M. S. (2002). Evidence for natural horizontal transfer of the *pcpB* gene in the evolution of polychlorophenol-degrading sphingomonads. *Appl Environ Microbiol* **68**, 4495–4501.

Tolba, S. T. M. (2004). *Distribution of streptomycin resistance and biosynthesis genes in streptomycetes recovered from different soil sites and the role of horizontal gene transfer in their dissemination*. PhD thesis, University of Warwick, UK.

Tolba, S., Egan, S., Kallifidas, D. & Wellington, E. M. H. (2002). Distribution of streptomycin resistance and biosynthesis genes in streptomycetes recovered from different soil sites. *FEMS Microbiol Ecol* **42**, 269–276.

Tringe, S. G., von Mering, C., Kobayashi, A. & 10 other authors (2005). Comparative metagenomics of microbial communities. *Science* **308**, 554–557.

Vanden Boom, T. & Cronan, J. E., Jr (1989). Genetics and regulation of bacterial lipid metabolism. *Annu Rev Microbiol* **43**, 317–343.

Vining, L. C. (1992). Roles of secondary metabolites from microbes. In *Secondary Metabolites: their Function and Evolution*, pp. 184–198. Edited by D. J. Chadwick & J. Whelan. Chichester: Wiley.

Wang, S.-J., Chang, H.-M., Lin, Y.-S., Huang, C.-H. & Chen, C. W. (1999). *Streptomyces* genomes: circular genetic maps from the linear chromosomes. *Microbiology* **145**, 2209–2220.

Weber, T., Welzel, K., Pelzer, S., Vente, A. & Wohlleben, W. (2003). Exploiting the genetic potential of polyketide producing streptomycetes. *J Biotechnol* **106**, 221–232.

Wiener, P. (1996). Experimental studies on the ecological role of antibiotic production in bacteria. *Evol Ecol* **10**, 405–421.

Wiener, P. (2000). Antibiotic production in a spatially structured environment. *Ecol Lett* **3**, 122–130.

Witkowski, A., Rangan, V. S., Randhawa, Z. I., Amy, C. M. & Smith, S. (1991). Structural organization of the multifunctional animal fatty-acid synthase. *Eur J Biochem* **198**, 571–579.

Unusual micro-organisms from unusual habitats: hypersaline environments

Antonio Ventosa

Department of Microbiology and Parasitology, Faculty of Pharmacy, University of Seville, 41012 Seville, Spain

INTRODUCTION

Thomas D. Brock defined extreme environments, considering that there are environments with high species diversity and others with low species diversity. Those environments with low species diversity, in which whole taxonomic groups are missing, are called 'extreme' (Brock, 1979). It is not easy to find a definition that is completely acceptable for all environments that are considered as extreme, but we observe that in some habitats environmental conditions such as pH, temperature, pressure, nutrients or saline concentrations are extremely high or low and that only limited numbers of species (that may grow at high cell densities) are well adapted to those conditions.

Hypersaline environments are typical extreme habitats, in which the high salt concentration is not the only environmental factor that may limit their biodiversity; they have low oxygen concentrations, depending on the geographical area, high or low temperatures, and are sometimes very alkaline. Other factors that may influence their biodiversity are the pressure, low nutrient availability, solar radiation or the presence of heavy metals and other toxic compounds (Rodriguez-Valera, 1988). With a few exceptions, most inhabitants of these environments are micro-organisms that are called 'halophiles'. However, different groups can be distinguished on the basis of their physiological responses to salt. Several classifications have been proposed; one that is very well accepted considers the optimum growth of the micro-organisms at different salt concentrations. Thus, Kushner & Kamekura (1988) defined several categories of micro-organisms on the basis of their optimal growth: non-halophiles are those that grow best in media containing less than 0·2 M NaCl (some of which, the halotolerant,

SGM symposium 66: Prokaryotic diversity: mechanisms and significance.
Editors N. A. Logan, H. M. Lappin-Scott & P. C. F. Oyston. Cambridge University Press. ISBN 0 521 86935 8 ©SGM 2006

can tolerate high salt concentrations), slight halophiles (marine bacteria) grow best in media with 0·2 to 0·5 M NaCl, moderate halophiles grow best with 0·5 to 2·5 M NaCl and extreme halophiles show optimal growth in media containing 2·5 to 5·2 M (saturated) NaCl.

In hypersaline habitats, especially in those in which salinities exceed 1·5 M (about 10 %), the two main groups of micro-organisms that predominate are the moderately halophilic bacteria and the extremely halophilic micro-organisms (archaea and bacteria). Archaea have been associated with extreme environments and, although today they are also recognized as normal inhabitants of other non-extreme environments, they constitute a large proportion of the microbial biota of hypersaline environments. Most are haloarchaea, but some methanogenic species have also been described from these environments.

Halophiles are found in many saline environments; the most important are hypersaline waters and soils, but the latter are much less studied. They can also be isolated from salt or salt deposits and from a variety of salted products, from salted fish or meats to fermented foods, as well as other materials such as salted animal hides. Hypersaline waters, those with higher concentrations of salt than sea water, can be divided into thalassohaline, which have a marine origin, if they have a composition similar to that of sea water, or athalassohaline, if their composition reflects the composition of the surrounding geology, topography and climatic conditions, often particularly influenced by the dissolution of mineral deposits; thus the composition of such waters varies widely (Rodriguez-Valera, 1988; Grant, 1990). Typical examples of thalassohaline water systems are solar salterns, used for the natural evaporation of sea water for the production of salt. They are excellent models for the study of halophiles, providing a series of ponds with different salinities, from sea water to salt saturation (Rodriguez-Valera, 1988; Grant, 1990). Typical examples of athalassohaline waters that have been studied in more detail are the Dead Sea, Great Salt Lake, some cold hypersaline lakes in Antarctica or alkaline lakes, particularly East African lakes, like Lake Magadi or the lakes of Wadi Natrun (Rodriguez-Valera, 1988; Javor, 1989; Grant, 1990).

In this chapter, I will review the micro-organisms that inhabit hypersaline environments and the features that make them unique. Some have very special characteristics that will be emphasized in this review. I will devote special attention to some recent studies on the square archaea and the extremely halophilic bacterium *Salinibacter*.

EXTREMELY HALOPHILIC ARCHAEA

The haloarchaea (also designated halobacteria) constitute a large group of extremely halophilic, aerobic archaea that are placed in the order *Halobacteriales*, family *Halo-*

bacteriaceae (Grant *et al.*, 2001). Classically, and for a long period of time, they were easily differentiated microscopically as rods or cocci that were respectively included within the genus *Halobacterium* or *Halococcus*, with a very limited number of species (Gibbons, 1974), probably due to the homogeneity of the isolation techniques and culture media in use as well as the limited number of hypersaline environments studied. When different approaches were used and many new hypersaline habitats were studied, the diversity found within this microbial group increased considerably. In fact, they are currently represented by 20 different genera and a large number of species, summarized in Table 1. Several other genera have recently been reported: *Halovivax asiaticus* (Castillo *et al.*, 2006) and '*Halostagnum larsenii*' (A. M. Castillo, M. C. Gutierrez, M. Kamekura, Y. Xue, Y. Ma, D. A. Cowan, B. E. Jones, W. D. Grant and A. Ventosa, manuscript in preparation) were isolated from a saline lake in Inner Mongolia, China. In addition to morphological and phenotypic features and comparison of 16S rRNA gene sequences, as is current practice in the systematics of other prokaryotic micro-organisms, the taxonomy of haloarchaea is also largely based on chemotaxonomic features: the polar lipid composition has proven to be an important marker for differentiation at the genus level (Grant *et al.*, 2001).

Haloarchaea require at least 1·5 M NaCl for growth and most species grow optimally in media with 3·5–4·5 M NaCl; many are able to grow in saturated NaCl (5·2 M) (Grant *et al.*, 2001). However, haloarchaea isolated from coastal salt-marsh sediments able to grow at lower salinities (around that of sea water) have been reported (Purdy *et al.*, 2004). They produce red- to pink-pigmented colonies due to the presence of bacterioruberins, C_{50} carotenoids that are partially responsible for the typical coloration of many natural environments in which they may develop in large numbers. However, there are a few exceptions that are not pink- to red-pigmented, including species of the genus *Natrialba*. Another interesting aspect of the haloarchaea is the presence in some of them of retinal-based pigments, such as bacteriorhodopsin, that act as a proton pump driven by light energy (Lanyi, 1995). They are found in many hypersaline environments, such as salt lakes, soda lakes, salterns, solar salt and subterranean salt deposits and salted foods (Grant *et al.*, 2001).

Haloarchaea have typical archaeal characteristics such as the presence of ether-linked phosphoglycerides that can be easily detected by thin-layer chromatography. All haloarchaea contain phytanyl ether analogues of phosphatidylglycerol and phosphatidylglycerol phosphate methyl ester. Many species also have phosphatidylglycerol sulfate and one or more glycolipids and sulfated glycolipids (Grant *et al.*, 2001; Kates, 1993). All haloarchaea have diphytanyl ($C_{20}C_{20}$) glycerol ether core lipids and some may have additional phytanyl-sesterterpanyl ($C_{20}C_{25}$) glycerol core lipids; furthermore, some haloalkaliphiles have disesterterpanyl ($C_{25}C_{25}$) glycerol ether lipids (Grant *et al.*, 2001).

Table 1. Validly published genus and species names within the family *Halobacteriaceae*

The first species name listed corresponds to the type species of each genus (updated 31 December 2005). The three-letter genus abbreviations recommended by the ICSP Subcommittee on the taxonomy of the family *Halobacteriaceae* have been used. Basonyms/synonyms of organisms that have been transferred to other genera are not included.

Genus	Reference(s)	Genus	Reference(s)
Halobacterium	Elazari-Volcani (1957); Grant (2001a)	*Halorubrum* (cont.)	
Hbt. salinarum	Ventosa & Oren (1996); Grant (2001a)	*Hrr. distributum*	Zvyagintseva & Tarasov (1987); Oren & Ventosa (1996)
Hbt. noricense	Gruber *et al.* (2004)	*Hrr. lacusprofundi*	Franzmann *et al.* (1988); McGenity & Grant (1995)
Halalkalicoccus	Xue *et al.* (2005)	*Hrr. sodomense*	Oren (1983); McGenity & Grant (1995)
Hac. tibetensis	Xue *et al.* (2005)	*Hrr. tebenquichense*	Lizama *et al.* (2002)
Haloarcula	Torreblanca *et al.* (1986)	*Hrr. terrestre*	Ventosa *et al.* (2004)
Har. vallismortis	Gonzalez *et al.* (1979); Torreblanca *et al.* (1986)	*Hrr. tibetense*	Fan *et al.* (2004)
Har. argentinensis	Ihara *et al.* (1997)	*Hrr. trapanicum*	Petter (1931); McGenity & Grant (1995)
Har. hispanica	Juez *et al.* (1986)	*Hrr. vacuolatum*	Mwatha & Grant (1993); Kamekura *et al.* (1997)
Har. japonica	Takashina *et al.* (1990)	*Hrr. xinjiangense*	Feng *et al.* (2004)
Har. marismortui	Oren *et al.* (1990)	*Halosimplex*	Vreeland *et al.* (2002)
Har. quadrata	Oren *et al.* (1999)	*Hsx. carlsbadense*	Vreeland *et al.* (2002)
Halobaculum	Oren *et al.* (1995)	*Haloterrigena*	Ventosa *et al.* (1999)
Hbl. gomorrense	Oren *et al.* (1995)	*Htg. turkmenica*	Zvyagintseva & Tarasov (1987); Ventosa *et al.* (1999)
Halobiforma	Hezayen *et al.* (2002)	*Htg. saccharevitans*	Xu *et al.* (2005c)
Hbf. haloterrestris	Hezayen *et al.* (2002)	*Htg. thermotolerans*	Montalvo-Rodríguez *et al.* (2000)
Hbf. lacisalsi	Xu *et al.* (2005b)	*Natrialba*	Kamekura & Dyall-Smith (1995)
Hbf. nitratireducens	Xin *et al.* (2001); Hezayen *et al.* (2002)	*Nab. asiatica*	Kamekura & Dyall-Smith (1995)
Halococcus	Schoop (1935); Grant (2001b)	*Nab. aegyptia*	Hezayen *et al.* (2001)

Hcc. morrhuae	Kocur & Hodgkiss (1973)	*Nab. chahannaoensis*	Xu *et al.* (2001)
Hcc. dombrowskii	Stan-Lotter *et al.* (2002)	*Nab. hulunbeirensis*	Xu *et al.* (2001)
Hcc. saccharolyticus	Montero *et al.* (1989)	*Nab. magadii*	Tindall *et al.* (1984); Kamekura *et al.* (1997)
Hcc. salifodinae	Denner *et al.* (1994)	*Nab. taiwanensis*	Hezayen *et al.* (2001)
Haloferax	Torreblanca *et al.* (1986)	*Natrinema*	McGenity *et al.* (1998)
Hfx. volcanii	Mullakhanbhai & Larsen (1975); Torreblanca *et al.* (1986)	*Nnm. pellirubrun*	McGenity *et al.* (1998)
Hfx. alexandrinus	Asker & Ohta (2002)	*Nnm. altunense*	Xu *et al.* (2005a)
Hfx. denitrificans	Tomlinson *et al.* (1986); Tindall *et al.* (1989)	*Nnm. pallidum*	McGenity *et al.* (1998)
Hfx. gibbonsii	Juez *et al.* (1986)	*Nnm. versiforme*	Xin *et al.* (2000)
Hfx. lucentense	Gutierrez *et al.* (2002)	*Natronobacterium*	Tindall *et al.* (1984)
Hfx. mediterranei	Rodriguez-Valera *et al.* (1983); Torreblanca *et al.* (1986)	*Nbt. gregoryi*	Tindall *et al.* (1984)
Hfx. sulfurifontis	Elshahed *et al.* (2004)	*Natronococcus*	Tindall *et al.* (1984)
Halogeometricum	Montalvo-Rodríguez *et al.* (1998)	*Ncc. occultus*	Tindall *et al.* (1984)
Hgm. borinquense	Montalvo-Rodríguez *et al.* (1998)	*Ncc. amylolyticus*	Kanai *et al.* (1995)
Halomicrobium	Oren *et al.* (2002a)	*Natronolimnobius*	Itoh *et al.* (2005)
Hmc. mukohataei	Ihara *et al.* (1997); Oren *et al.* (2002a)	*Nln. baerhuensis*	Itoh *et al.* (2005)
Halorhabdus	Wainø *et al.* (2000)	*Nln. innermongolicus*	Itoh *et al.* (2005)
Hrd. utahensis	Wainø *et al.* (2000)	*Natronomonas*	Kamekura *et al.* (1997)
Halorubrum	McGenity & Grant (1995)	*Nmn. pharaonis*	Soliman & Trüper (1982); Kamekura *et al.* (1997)
Hrr. saccharovorum	Tomlinson & Hochstein (1976); McGenity & Grant (1995)	*Natronorubrum*	Xu *et al.* (1999)
Hrr. alkaliphilum	Feng *et al.* (2005)	*Nrr. bangense*	Xu *et al.* (1999)
Hrr. coriense	Kamekura & Dyall-Smith (1995); Oren & Ventosa (1996)	*Nrr. tibetense*	Xu *et al.* (1999)

Their absolute requirement for NaCl and their optimal growth at high NaCl concentrations are the most typical features of haloarchaea. The question of their mechanisms of haloadaptation has attracted much attention and, in contrast to most other prokaryotes, which accumulate intracellular organic compounds called compatible solutes, haloarchaea compensate for the high salt concentration in the environment by accumulating mainly KCl, up to 5 M (Grant *et al.*, 2001).

As mentioned previously, haloarchaea are found in many hypersaline environments; they are the predominant microbial biota of saturated ponds of salterns and salt lakes. Several species are haloalkaliphiles, being able to grow optimally at alkaline pH, that also inhabit soda lakes. Ecological studies demonstrate that they may reach high cell densities (> 10^7 cells ml^{-1}). Traditional studies based on cultivation of viable cells suggested that the predominant species found in most neutrophilic hypersaline environments were related to the genera *Halobacterium*, *Halorubrum*, *Haloferax* and *Haloarcula* (Rodriguez-Valera *et al.*, 1985; Rodriguez-Valera, 1988; Benlloch *et al.*, 2001). However, more recent molecular ecological studies based on cultivation-independent methods indicate that, at least in most environments studied, members of these genera constitute small proportions of the microbial community. Several recent studies carried out in hypersaline environments have allowed some general conclusions to be drawn. Square haloarchaea are very abundant, but they have not been isolated until recently; many environmental clones within the haloarchaeal group are also obtained that are not phylogenetically closely related to previously described species. Some clones related to *Halorubrum*, *Haloarcula*, *Natronobacterium* and *Natronomonas* are observed (Benlloch *et al.*, 2001, 2002; Burns *et al.*, 2004b). Recent studies on a saltern in Slovenia showed that the haloarchaeal community in the crystallizer was strongly dominated by two groups of *Halorubrum*-related environmental phylotypes. In addition, members of four other haloarchaeal genera and two groups of environmental phylotypes were observed. However, the square haloarchaeal morphotype was not observed (Pasic *et al.*, 2005). Another recent study of a saltern in San Diego showed that the predominant haloarchaea were members of *Halobacterium*; the presence of *Haloarcula* and *Halorubrum* was also detected as well as two novel lineages not closely related to any haloarchaeal species (Bidle *et al.*, 2005).

Haloarchaea are used in many studies as archaeal models since they can be easily grown under laboratory conditions. In contrast to other archaeal extremophiles, which require special culture conditions, haloarchaea grow well in complex media and in some cases in minimal media under aerobic conditions by using the standard procedures for growing other prokaryotes. They can also be genetically manipulated; genetic exchange mechanisms are well established and methods for their genetic manipulation in the laboratory are available (Robb *et al.*, 1995). The complete genome sequences of quite a few haloarchaea have been reported, including *Halobacterium salinarum* NRC-1 (2571 kb) (Ng *et al.*, 2000), *Haloarcula marismortui* ATCC 43049^T (4274 kb) (Baliga *et al.*, 2004) and *Natronomonas pharaonis* DSM 2160^T (2749 kb) (Falb *et al.*, 2005). Several other genome sequencing projects are in progress: a second strain of *Halobacterium salinarum* and strains of *Halobaculum gomorrense*, *Halobiforma lacisalsi*, *Haloferax volcanii*, *Halorubrum lacusprofundi* and *Natrialba asiatica*. Clearly, more effort would be necessary in this field in order to have genomic information available that reflects the diversity found within the haloarchaea.

Besides their use as excellent models to study the molecular biology of archaea and their mechanisms of adaptation to extreme environments and the important ecological roles that they play in hypersaline habitats, haloarchaea have very interesting biotechnological applications. Bacteriorhodopsin in the form of purple membrane patches produced by *Halobacterium salinarum* is commercially available; other compounds of industrial interest produced by haloarchaea are extracellular hydrolytic enzymes, exopolysaccharides, polyhydroxyalkanoates (PHAs) used as bioplastics and the halocins (antimicrobial compounds produced by haloarchaea). These and many other potential applications have been reviewed in detail elsewhere (Ventosa & Nieto, 1995; Margesin & Schinner, 2001; Mellado & Ventosa, 2003).

FROM 'SQUARE BACTERIA' TO 'SQUARE HALOARCHAEA'

In 1980, A. E. Walsby described the abundant presence of 'square bacteria' in a small, saturated brine pool, or sabkha, in the Sinai Peninsula (Walsby, 1980). These micro-organisms were collected from the surface of the pool, had a large number of gas vesicles and presented unique morphologies never observed previously in the microbial world. The cells are squares and very thin, with sizes from 1·5 to 11 μm and a thickness of about 0·2 μm (Fig. 1). Division planes were observed, with an arrangement indicating that division occurred in two planes alternating at right angles, so that each square grows to a rectangle which then divides into two equal squares, producing sheets divided like postage stamps (Walsby, 1980; Parkes & Walsby, 1981). Also, different gas vesicles were observed, from spindle-shaped to cylindrical with conical ends and, in many cases, they were concentrated at the cell periphery. Both forms may occur within the same cell, as had been observed previously in *Halobacterium* (Parkes & Walsby, 1981). Similar observations were made from other hypersaline environments in different geographical areas, from salt ponds located in Baja California, Mexico, or San Francisco Bay, California (Stoeckenius, 1981). Ultrastructure studies confirmed that the square cells observed by Walsby were micro-organisms with a typical prokaryote structure (Stoeckenius, 1981; Kessel & Cohen, 1982).

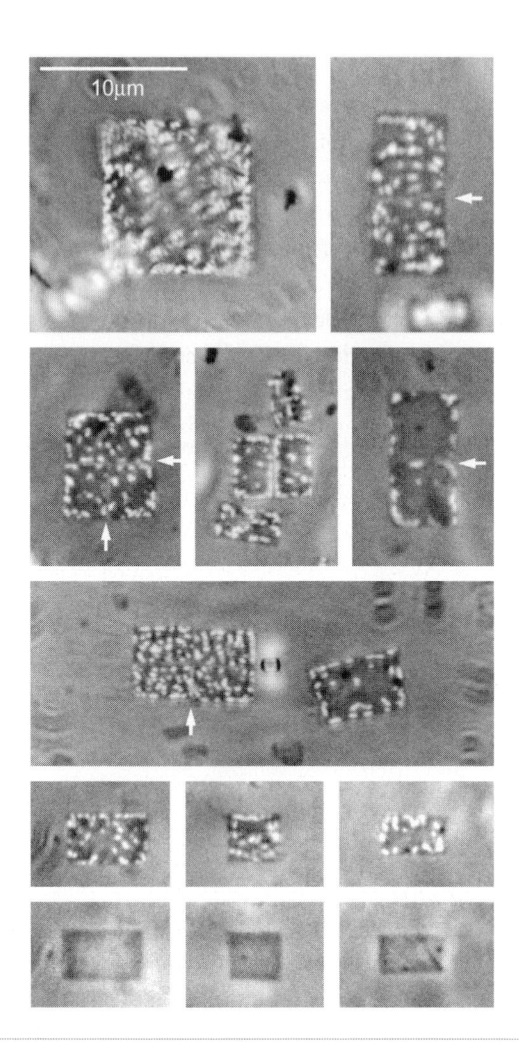

Fig. 1. The square archaeon from the Sinai. Phase-contrast light micrographs of the original square haloarchaeon discovered in a brine sample from a sabkha near Nabq, Sinai. Division lines (arrows) are visible in some of the cells but not in the largest (top left). Gas vesicles show as bright refractile granules in all cells except those in the bottom row, which have been exposed to pressure (the same cells as shown in the row above). Bar, 10 μm. Reproduced from Walsby (2005) with permission from Elsevier.

Direct observations of brines of salterns and other hypersaline environments suggested that these square cells were very abundant in such habitats, especially in the most concentrated ponds with salinities higher than 3–4 M NaCl, and they have been reported to occur in many geographical locations (Oren, 1999). Oren *et al.* (1996) suggested on the basis of polar lipid analysis that they could be archaea not related to the genera *Halobacterium* or *Haloarcula*. When samples from a crystallizer pond of a saltern in Eliat, Israel, with large numbers of gas-vacuolated square cells (representing about

55 % of the total population) were studied with respect to polar lipid composition, Oren *et al.* (1996) observed that the most frequent lipids present were typical of halophilic archaea, members of the family *Halobacteriaceae*, but that the square archaea may belong to a new genus. More recently, molecular techniques revealed that they may represent a large proportion of the microbial population in many hypersaline environments (Benlloch *et al.*, 1996, 2001, 2002; Antón *et al.*, 1999; Burns *et al.*, 2004b).

Many attempts have been made to isolate these square micro-organisms, but only recently have two independent studies reported their isolation and cultivation (Bolhuis *et al.*, 2004; Burns *et al.*, 2004a). Previous studies based on the isolation of pure cultures from different hypersaline environments reported the presence of halophilic archaea with morphologies ranging from squares to discs and triangles, and not the typical rod or spherical cell shapes that are observed within other bacterial groups. An early study describing 'box-shaped' halophiles (Javor *et al.*, 1982) did not permit the isolation of square archaea, since the cells were pleomorphic and lacked the gas vesicles observed in Walsby's square cells. Most isolates studied in recent years that show pleomorphic cells have been placed in the genera *Haloferax* (Ventosa, 2001b) or *Haloarcula* (Ventosa, 2001a). Most species of the genus *Haloarcula* have unique morphologies, such as *Haloarcula japonica*, which typically shows many triangular and rhomboid cells (Takashina *et al.*, 1990), or *Haloarcula quadrata* (Oren *et al.*, 1999). The latter species was described based on a motile, square prokaryote isolated from the Gavish Sabkha in the Sinai Peninsula. The square morphology and interesting mode of motility were reported by Alam *et al.* (1984). The cells have a single or several flagella anchored from a single or several locations that form a bundle. As in the case of other haloarchaea, this species requires high salt concentrations for growth and for stabilization of the cells. Optimal growth is obtained in the presence of 3·4–4·3 M NaCl and 0·1–0·5 M Mg^{2+}. It is interesting to note that at least 50–100 mM Mg^{2+} was required for maintenance of the square morphology (Oren *et al.*, 1999). However, its lack of gas vesicles and a more variable morphology suggest that it does not correspond to the 'square bacteria' described by Walsby (1980).

The recent isolation of Walsby's square haloarchaea, 25 years after their original description, by two independent groups of microbiologists can be considered as an important event. It required the use of new approaches that differed from traditional culture methods, as well as extraordinary persistence.

Burns *et al.* (2004b) studied water samples from a crystallizer pond of a saltern located in Geelong, Victoria, Australia, in which they observed by microscopy the abundance of square cells. They used a combination of serial dilutions of the samples in liquid media with different compositions (extinction cultures), and growth of the square haloarchaea was determined by PCR screening. When conventional media were used, cells were similar to *Halorubrum* species. However, when a medium composed of filtered and sterilized natural saltern water supplemented with 50 μM amino acids mixture and 0·5 % pyruvate was used, square cells were detected both by light microscopy and PCR screening after 3 weeks of incubation in three dilution cultures. The best-growing isolate, culture C23, was grown further and reached cell densities of about 10^7 cells ml^{-1}, at which point cultures were slightly turbid by eye. However, it does not grow on solid media. Further studies with isolate C23 determined its doubling time at around 1·5 days. A faster growth rate was observed when a rich medium (composed of 23 % salt water plus 0·1 % yeast extract and 0·5 % peptone) was supplemented with 0·5 % pyruvate, obtaining final cell densities of 10^8 cells ml^{-1}. Microscopy studies showed that the cells of isolate C23 were thin, square (mean of 2·6 μm) or rectangular (mean 2·3 × 3 μm). Apart from occasional triangular cells, isolate C23 showed a homogeneous morphology. Furthermore, it contained gas vesicles that were easily collapsed by pressure and poly-*β*-hydroxybutyrate granules of about 0·3 μm diameter; in both cases, they were often found near the periphery of the cells (Fig. 2). Comparison of the complete 16S rRNA gene sequence with sequences of all members of the *Halobacteriaceae* and clone sequences available from the databases confirmed that isolate C23 is closely related to the 'square haloarchaea group' clones, showing more than 99 % sequence similarity. However, the closest 16S rRNA gene sequence similarity with any of the cultured species was about 91 %, with the sequence of *Halogeometricum borinquense*, a very pleomorphic haloarchaeon, showing different shapes (from rods to triangles, squares or oval cells) but not the typical square of Walsby's archaea (Montalvo-Rodríguez *et al.*, 1998).

Bolhuis *et al.* (2004) studied samples obtained from salt-saturated crystallizer ponds of a solar saltern in Alicante, Spain, in which square cells were observed microscopically, and they were inoculated into different culture media. When media composed of hypersaline artificial sea water supplemented with high concentrations of yeast extract, peptone or tryptone and glycerol or other alternative carbon sources were used, rapid growth of halophilic prokaryotes was obtained but square cells were not observed under the microscope. Only in media with very low concentrations of yeast extract and glycerol (less than 0·5 g l^{-1} in both cases) and after long incubation periods of 1–2 months were the square cells enriched, but they did not grow on solid media when agar was added as a solidifying agent. After several further enrichments in these media for a period of 2 years, they obtained a culture of square cells that was not pure. Finally, in a medium with a low concentration of yeast extract (0·1 g l^{-1}) supplemented with sodium pyruvate (1 g l^{-1}) they obtained an enrichment with square cells that were plated in the same medium with agarose (10 g l^{-1}) instead of agar as solidifying agent. The first pure culture of a square isolate that contained gas vesicles was designated strain HBSQ001.

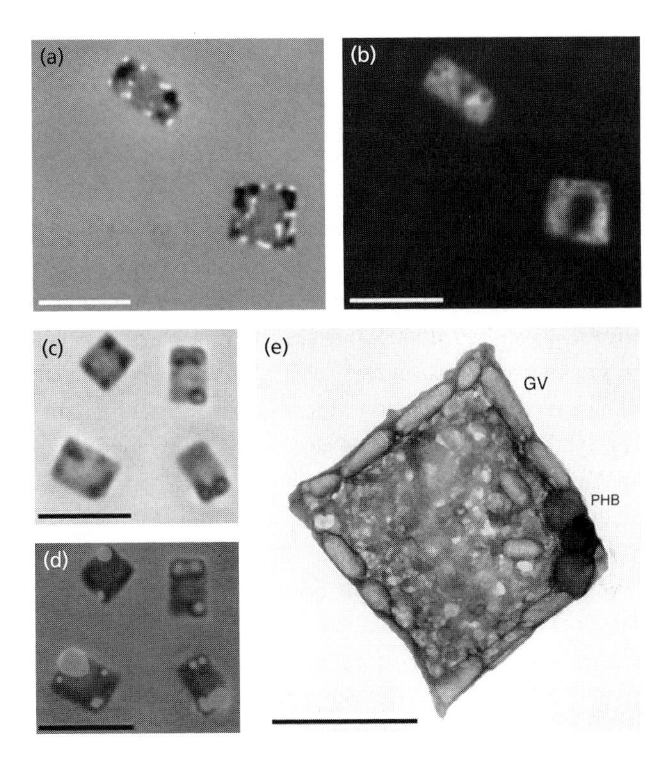

Fig. 2. Cells of isolate C23 viewed by light and electron microscopy. (a) Cells examined by phase-contrast microscopy; (b) the same field viewed by epifluorescence microscopy after acridine orange staining; (c) cells fixed and stained with Nile blue A (a poly-β-hydroxybutyrate-specific fluorescent stain) and viewed by phase-contrast microscopy; (d) the same field as in (c) viewed by epifluorescence, and the image superimposed over that of (c) to show the cell borders relative to the red-fluorescing granules. Bars, 5 μm (a–d). (e) Uranyl acetate- (2 %) stained preparation examined by transmission electron microscopy. Bar, 1 μm. Reproduced from Burns *et al.* (2004a) with permission from Elsevier.

The cells were 2–5 μm wide and about 0·1–0·5 μm thick; in fresh cultures they observed very large cells (20 × 20 to 40 × 40 μm and 0·1–0·5 μm thick) in which division walls were not observed. This isolate required at least 3 M NaCl and showed an elevated tolerance of high concentrations of $MgCl_2$, being able to grow in media containing over 2 M $MgCl_2$ and 3·3 M NaCl.

When the 16S rRNA gene sequence of strain HBSQ001 was determined, the phylogenetic study showed that it was closely related to the SPhT clone sequence (99 % similarity) that had been observed previously as related to Walsby's square archaeon. Only 93 % 16S rRNA gene sequence similarity was obtained with the haloarchaeon T1.3, a strain isolated from an ancient salt deposit (McGenity *et al.*, 2000), and lower values were seen with haloarchaeal species. This low 16S rRNA gene sequence similarity suggests that the new isolate would represent a new genus of the family *Halobacteriaceae*.

Bolhuis and co-workers proposed the new name '*Haloquadratum walsbyi*' if this isolate were to constitute a new taxon (Bolhuis *et al.*, 2004). However, a complete taxonomic characterization would be required, including phenotypic as well as chemotaxonomic data (such as polar lipid analysis). Furthermore, it would be interesting to determine the sequence similarity between the two novel square archaea, strains HBSQ001 and C23, in order to determine whether they are closely related, and to carry out a comparative study to determine whether they are members of the same genus and species.

The reasons for the difficulties in isolating the square archaea can be attributed to several factors, such as their sensitivity to agar and to high concentrations of yeast extract, their slow growth rate (doubling time of 1–2 days under optimal conditions), which permits rapid growth of other extremely halophilic micro-organisms in the enriched cultures, and probably their requirement for relatively high concentrations of $MgCl_2$ to maintain the square morphology.

The isolation by two independent research groups of the square archaea introduces a new methodology that permits the isolation of new fresh isolates of square archaea, allowing us to establish whether they are represented by a single or several novel species. Furthermore, it will be possible to determine their roles in the natural hypersaline environments in which, on the basis of molecular ecology techniques, they represent a large proportion of the microbial population, and to study other physiological or molecular aspects such as their high tolerance of $MgCl_2$ or the unique square morphology. The complete sequence of the genome of these micro-organisms will also constitute a highlight for comparative purposes with other organisms.

Recently, Walsby (2005) reviewed the features of the square archaea and the biological significance of the square shape. This shape depends on the lack of turgor; most cells are distended by turgor pressure, generated by the difference in water potential between the external medium and the cell. In spherical or cylindrical cells, the pressure produces stress in the cell wall. Square cells, like other haloarchaea, have no turgor pressure (Walsby, 1980) and are free of stress in the cell wall, thus they need less complex cell walls than other prokaryotes and might, in theory, adopt any shape. The shape of the cell might be dependent on the shape and arrangement of the wall subunits. The cell wall of square archaea is composed of regularly arranged subunits, forming a dominantly hexagonal lattice (Walsby, 1980; Parkes & Walsby, 1981; Stoeckenius, 1981; Kessel & Cohen, 1982). The advantages of the thinness of the square cells could be related to the large surface area, which would permit the uptake of nutrients as well as more efficient light absorption, since it has been shown that the square archaea contain purple membranes that are used for phototrophic growth. Furthermore, the gas vesicles offer another clear advantage: they permit the cells to concentrate and rest parallel in a horizontal position on the water surface. This movement to the surface has less energetic cost in cells with gas vesicles than with flagella and could explain the slow growth strategy and their dominance in saturated hypersaline environments (Walsby, 2005).

SALINIBACTER AND OTHER EXTREMELY HALOPHILIC BACTERIA

The genus *Salinibacter*, with the single species *Salinibacter ruber*, was proposed by Antón *et al.* (2002) for red-pigmented, motile, rod-shape extremely halophilic bacteria isolated from saltern crystallizer ponds in Alicante and Mallorca (Spain). It grows optimally between 20 and 30 % salt and requires at least 15 % salts. Therefore it is as halophilic and salt-dependent as most extremely halophilic archaea belonging to the family *Halobacteriaceae* (Grant *et al.*, 2001). It is phylogenetically related to the phylum *Bacteroidetes* but the closest cultivated relative is *Rhodothermus marinus* (showing only about 89 % 16S rRNA gene sequence similarity), a thermophilic and slightly halophilic bacterium. This bacterium constitutes a major component of the community of hypersaline environments and was previously known only by its phylotype (designated '*Candidatus* Salinibacter') (Antón *et al.*, 2000). By using fluorescent oligonucleotide probes, this organism has been shown to be abundant in the crystallizer ponds of salterns, representing between 5 and 25 % of the total prokaryotic community (Antón *et al.*, 2000). They may be also responsible for the red coloration of saltern crystallizer ponds (Oren & Rodriguez-Valera, 2001). However, their presence has not been observed in other salterns studied recently (Bidle *et al.*, 2005).

Despite its relatively recent isolation and description, new data about this interesting bacterium, which can be considered as one of the most halophilic organisms belonging to the *Bacteria*, have been accumulated in recent years. *S. ruber* has an extremely high intracellular potassium content but very low concentrations of organic osmotic solutes (glutamate, glycine betaine and *N*-α-acetyl lysine) (Oren *et al.*, 2002b). Thus, it uses a similar mode of haloadaptation to that of the haloarchaea, which also have high intracellular K^+ concentrations, and does not accumulate organic osmotic solutes such as those used by most halophilic or halotolerant bacteria (Ventosa *et al.*, 1998; Oren *et al.*, 2002b). Amino acid analysis of bulk protein of *S. ruber* showed a high content of acidic amino acids, a low proportion of basic amino acids, a low content of hydrophobic amino acids and a high abundance of serine (Oren & Mana, 2002). When these authors determined the levels of activity of several cytoplasmic enzymes at different KCl and NaCl concentrations, they showed that these enzymes are adapted to function in the presence of high salt concentrations (Oren & Mana, 2002). However, it seems that the behaviour of each enzyme towards salt varies considerably. Thus, while some enzymes are truly halophilic and require salt for activity and stability, such as the NAD-dependent isocitrate dehydrogenase (Oren & Mana, 2002) or the fatty acid synthetase complex (Oren *et al.*, 2004), other enzymes may function just as well in the presence of high salt or in its absence, like the NADP-dependent isocitrate dehydrogenase, or may even be inhibited by high salt concentrations (Oren & Mana, 2002). Two glutamate dehydrogenases have also been detected in *S. ruber*; they show differences in their affinity for substrates, in their pH and salt dependence for activity and stability and in their regulation by different effectors. While one glutamate dehydrogenase depends on high salt concentrations for both activity and stability, the other shows a strong dependence on high salt concentrations for stability but not for activity (Bonete *et al.*, 2003).

In a recent study, the malate dehydrogenase of *S. ruber* was purified and characterized in detail, comparing it with those produced by extremely halophilic archaea or non-halophilic micro-organisms (Madern & Zaccai, 2004). In contrast to most other halophilic enzymes, which unfold when incubated at low salt concentrations, this enzyme is completely stable in the absence of salt; its amino acid composition does not display the strong acidic character specific to halophilic proteins. Its activity is reduced by high salt concentrations, but remains sufficient for the enzyme to sustain catalysis at approximately 30 % of its maximal rates in 3 M KCl (Madern & Zaccai, 2004). The differences in the behaviour of the enzymes of *S. ruber* pose questions about the origins of the genes encoding such proteins and the possibility of horizontal gene transfer from other bacteria or archaea that might inhabit the same environments. Sequencing of the complete genomes of two different strains of *S. ruber* is in progress and will allow comparative analysis at the molecular level of the features of this interesting micro-organism, not only concerning its physiological and biochemical behaviour but also other basic and applied aspects in comparison with the extremely halophilic archaea, with which it must compete in natural environments.

The bright-red pigmentation of *S. ruber* is due to a novel carotenoid that has been designated salinixanthin, which constitutes more than 96 % of the total carotenoid content. The chemical structure of this C_{40} carotenoid has been reported (Lutnaes *et al.*, 2002), being different from the C_{50} carotenoids of the bacterioruberin group present in haloarchaea. It has been proposed that salinixanthin would provide protection from photodamage and stabilize the cell membrane (Lutnaes *et al.*, 2002). However, it has recently been established that salinixanthin is not the only pigment in the *S. ruber* cell membrane; it also contains an unusual retinal protein that uses salinixanthin to harvest light energy and then uses it for transmembrane proton transport. This light-driven proton pump, similar to bacteriorhodopsin or the archaerhodopsins but with two chromophores, has been designated xanthorhodopsin (Balashov *et al.*, 2005). The membrane lipids of *S. ruber* have also been analysed recently. Apart from the phospholipids typical of the bacterial domain, a novel sulfonolipid that represents about 10 % of total cellular lipids has been identified (Corcelli *et al.*, 2004). It has been suggested that this novel sulfonolipid could be used as a chemotaxonomic marker for the detection of *Salinibacter* within the halophilic microbial community in hypersaline environments (Corcelli *et al.*, 2004).

S. ruber is not the only extremely halophilic bacterium that has been described. Several other examples of extremely halophilic bacteria are *Halorhodospira halophila* (formerly *Ectothiorhodospira halophila*), a phototrophic bacterium that grows optimally in media with about 25 % salts (Hirschler-Rea *et al.*, 2003), *Halanaerobium lacusrosei*, a strictly anaerobic bacterium isolated from hypersaline sediments of Lake Retba in Senegal (Cayol *et al.*, 1995), *Acetohalobium arabaticum* (Zhilina & Zavarzin, 1990) and *Halobacteroides lacunaris* (Zhilina *et al.*, 1991), two anaerobic organisms isolated from hypersaline lakes, and the actinomycete *Actinopolyspora halophila* (Gochnauer *et al.*, 1975).

MODERATELY HALOPHILIC BACTERIA

Moderately halophilic bacteria are a group of micro-organisms that grow optimally in media containing 0·5 M (3 %) to 2·5 M (15 %) NaCl (Ventosa *et al.*, 1998). They are represented by a large number of species belonging to many different genera. In 1980, when the Approved Lists of Bacterial Names (Skerman *et al.*, 1980) were published, only six species were described: *Salinivibrio costicola* (formerly *Vibrio costicola*), a vibrio motile by a single sheathed polar flagellum, that is frequently isolated from salted meats and other food products and is present in salt lakes and in ponds of salterns of intermediate salinities (Garcia *et al.*, 1987), *Nesterenkonia* (*Micrococcus*) *halobia*, *Halomonas* (*Paraccoccus*) *halodenitrificans*, *Halomonas* (*Flavobacterium*) *halmophila*, *Marinococcus* (*Planococcus*) *halophilus* and *Spirochaeta halophila*. Most of them were isolated from salted foods or as laboratory contaminants and their taxonomic descriptions were very incomplete. The role(s) that these bacteria played in natural habitats was unknown. However, extensive studies carried out over the last 25 years have given us a better knowledge of their biodiversity, phylogenetic relationships, physiological and haloadaptative mechanisms and, more recently, biotechnological potential. Currently, the number of bacterial species with a moderately halophilic response is very large and they are represented in many different bacterial phyla, especially the Gram-positives (low- and high-G+C content), spirochaetes and *Bacteroidetes*, but most of them belong to the *Proteobacteria*. Table 2 shows some representative genera that include moderately halophilic bacterial species. In many cases they are members of genera that include both moderately halophilic and other non-halophilic or halotolerant species, but in most cases they are included in genera with only moderately halophilic representatives. However, it cannot be disputed that other non-halophilic species could also be included in the near future within these genera. The

Table 2. Representative genera that include moderately halophilic species

Gram-positive	Gram-negative	
Alkalibacillus	*Algoriphagus*	*Psychrobacter*
Bacillus	*Alteromonas*	*Rhodospirillum*
Clostridium	*Arhodomonas*	*Rhodothalassium*
Gracilibacillus	*Chromohalobacter*	*Rhodovibrio*
Halobacillus	*Desulfocella*	*Roseisalinus*
Lentibacillus	*Desulfohalobium*	*Salegentibacter*
Marinococcus	*Desulfovibrio*	*Salinimonas*
Nocardiopsis	*Halanaerobacter*	*Salinisphaera*
Pontibacillus	*Halanaerobium*	*Salinivibrio*
Prauserella	*Halochromatium*	*Salipiger*
Saccharomonospora	*Haloincola*	*Selenihalanaerobacter*
Salinibacillus	*Halomonas*	*Spirochaeta*
Streptomonospora	*Halorhodospira*	*Sporohalobacter*
Tenuibacillus	*Halospina*	*Staleya*
Tetragenococcus	*Halothermothrix*	*Sulfitobacter*
Thalassobacillus	*Halothiobacillus*	*Thiohalocapsa*
Virgibacillus	*Halovibrio*	
Yania	*Idiomarina*	
Microbacterium	*Marinicola*	
Filobacillus	*Marinobacter*	
Dietzia	*Methylarcula*	
Marinibacillus	*Methylohalobius*	
Desulfotomaculum	*Muricauda*	
Salinicoccus	*Natroniella*	
Nesterenkonia	*Nitrincola*	
Jeotgalicoccus	*Orenia*	
Jeotgalibacillus	*Palleronia*	
Natronincola	*Pseudoalteromonas*	
Sporosarcina	*Psychroflexus*	

close phylogenetic relationship of many moderately halophilic and non-halophilic bacteria indicates that they are not evolutionary descendents of a single lineage and that they evolved as extremophilic micro-organisms adapted to high saline environments by developing haloadaptation mechanisms that are similar to those of halotolerant bacteria.

The biology of moderately halophilic bacteria has been reviewed elsewhere (Ventosa *et al.*, 1998), with special emphasis on their taxonomy, ecology, physiology and metabolism, genetics, as well as biotechnological applications. Most studies carried out with laboratory cultures have used a limited number of species, especially members of the genera *Halomonas*, *Chromohalobacter* and *Salinivibrio* (Ventosa *et al.*, 1998). However, most species have been described very recently and, besides their taxonomic description and in some cases a few other data concerning their physiological adaptation to hypersaline environments, little is known about them. It would be necessary to carry out extensive studies over the next few years in order to know in more detail whether the moderately halophilic bacteria constitute a homogeneous group with respect to their mechanisms of haloadaptation and other molecular-biological aspects, or there may be interesting new aspects to be explored. To date, only one genome has been sequenced, from the heterotrophic Gram-negative bacterium *Chromohalobacter salexigens*, but this sequence is still not completely available, and this is a field in which more efforts are clearly necessary.

The most interesting features of moderately halophilic bacteria are their haloadaptation mechanisms and, since they may grow over a wide range of salt concentrations, their ability to adapt to changes in environmental osmolarity. Their main strategy for coping with osmotic stress caused by changing salinity of the surrounding medium is based on the intracellular accumulation of organic compounds, compatible solutes, which function as osmoprotectants. These compatible solutes can be taken up from the external medium if they are present, as in the cases of the betaines or choline, or can be synthesized by the cells, like ectoine or β-hydroxyectoine (Ventosa *et al.*, 1998; Nieto & Vargas, 2002).

The ecological distribution of moderately halophilic bacteria based on culture-dependent methods was already established (Rodriguez-Valera *et al.*, 1985; Rodriguez-Valera, 1988; Ventosa *et al.*, 1998). Recent studies based on molecular ecology techniques have been carried out (Bowman *et al.*, 2000; Benlloch *et al.*, 2002; Ma *et al.*, 2004; Rees *et al.*, 2004; Bidle *et al.*, 2005). However, new approaches would be required in order to define their ecological role and relationships with haloarchaea and halotolerant bacteria in hypersaline environments.

The moderately halophilic bacteria have very interesting industrial applications and they show many potential biotechnological uses, but most of them need to be studied in more detail. Their ability to produce organic compounds as compatible solutes is one of the best-known industrial applications. Several compatible solutes such as ectoine or hydroxyectoine are commercially available and are used in the cosmetics industry. Since many of them can grow well under low (0·5 % NaCl) or high (20–25 % NaCl) salinities, they are used for the biological treatment of saline industrial waste effluents and for the biodegradation of toxic compounds (Ventosa & Nieto, 1995; Mellado & Ventosa, 2003). More recently, they are being explored for the production of several extracellular enzymes (amylases, proteases, lipases) (Mellado *et al.*, 2004).

VIRUSES

Few studies have been carried out with respect to the presence of viruses in hypersaline environments and their ecological roles. Most studies have focused on haloarchaeal viruses (haloviruses). Surprisingly, considering the presence of bacteria in such environments, information concerning bacteriophages is almost non-existent. Three phages, designated F9-11, F5-4 and F12-9, from lysogenic strains of *Halomonas halophila* (formerly *Deleya halophila*) have been isolated and characterized (Calvo *et al.*, 1988, 1994); they have isometric heads and non-contractile tails. The salt concentration for optimal adsorption and phage production is between 2·5 and 7·5 %. Another bacteriophage specific for a moderately halophilic bacterium was described by Kauri *et al.* (1991). The first haloarchaeal virus was reported by Torsvik & Dundas (1974), but the most extensive studies were carried out by Wolfram Zillig and colleagues at the Max Planck Institute in Munich. They described and studied several haloviruses, particularly the halovirus ϕH, that infect *Halobacterium salinarum* (Schnabel *et al.*, 1982). More recently, Mike Dyall-Smith and co-workers, from the University of Melbourne, have greatly contributed to our knowledge of the diversity and molecular characteristics of haloviruses. Several recent articles have reviewed them in detail and can be consulted for information in more depth (Dyall-Smith *et al.*, 2003, 2005; Tang *et al.*, 2004a).

Considering that virus populations in natural environments are assumed to be greater than those of their prokaryotic hosts and that each host species is susceptible to infection by several different viruses, it may be deduced that a great diversity of such viruses must exist in hypersaline environments. However, to date, only about 15 haloviruses have been described and very few have been studied in detail. Few attempts have been made to isolate viruses from hypersaline environments, and most studies used *Halobacterium salinarum* as host archaeon, a species that is not predominant in most hypersaline habitats studied.

The halovirus ϕH is the most intensively studied halovirus. It is a temperate, head–tail virus with a genome of about 59 kb linear double-stranded (ds) DNA. Apart from a few methylase genes, it shows little sequence similarity to the genomes of bacteria, bacteriophages or eukaryotic viruses. At the level of replication, morphology and control of lysogeny, it shows similarities to P-type coliphages, such as P1 (Stolt & Zillig, 1994a, b). Another temperate, head–tail halovirus, ϕCh1, with a linear dsDNA genome of 55 kb, has been studied. Its host is the haloalkaliphilic archaeon *Natrialba magadii*. It shows extensive similarity to halovirus ϕH, but they have several differences that reflect their different hosts and ecosystems (Klein *et al.*, 2002).

Two head–tail haloviruses, HF1 and HF2, isolated from crystallization ponds of an Australian saltern, have also been studied (Tang *et al.*, 2004a). Unlike the previously reported viruses, they are lytic and infect a wide range of haloarchaea (*Halobacterium*, *Haloferax*, *Haloarcula*, *Natrialba*, *Haloterrigena* and *Halorubrum*). The two haloviruses have similar features, like a head-and-tail morphology and linear dsDNA genomes with similar sizes. The complete sequences of the HF1 and HF2 genomes have recently been completed (Tang *et al.*, 2002, 2004b). They are the largest archaeal virus genomes sequenced; HF1 is 75·9 kb (Tang *et al.*, 2004b) and HF2 is 77·7 kb (Tang *et al.*, 2002). Comparison of the two genomes showed that they are 94·4 % identical. Except for a single base change, HF1 and HF2 are identical in sequence over the first 48 kb, but they are very different in the other 28 kb region, suggesting a recent recombination event between either HF1 or HF2 and other HF-like haloviruses. This example suggests that there is a high level of recombination among viruses that live in hypersaline environments (Tang *et al.*, 2004b).

Very recently, a new halovirus, designated SH1, has been isolated from a hypersaline lake in Australia. It has a spherical morphology and infects *Haloarcula hispanica* and a *Halorubrum* isolate. It is an icosahedral, dsDNA virus with morphology that suggests the presence of a membrane underneath the protein capsid (Porter *et al.*, 2005). Bamford *et al.* (2005) have determined the complete sequence of the SH1 genome (31 kb) and identified genes for 11 structural proteins. The SH1 genome is unique and, except for a few open reading frames, shows no detectable similarity to other sequences from databases, but the overall structure of the SH1 virion and its linear genome with inverted terminal repeats is reminiscent of lipid-containing dsDNA bacteriophages like PRD1 (Bamford *et al.*, 2005).

In summary, knowledge of the diversity of haloviruses has increased, particularly in recent years, but is still in its infancy when compared to knowledge of bacteriophages. Their isolation from hypersaline waters was probably biased by the use of host strains that perhaps are not the predominant microbiota in natural environments. Direct electron microscopy of hypersaline waters shows that they maintain high levels of virus-like particles (about 10 times higher than the cell population), with recognizable morphotypes including head–tail and lemon-shaped particles (Guixa-Boixereu *et al.*, 1996; Oren *et al.*, 1997; Pedrós-Alió, 2004). Guixa-Boixereu *et al.* (1996) determined the abundance of viruses in two different salterns in Spain and they observed that the number of viruses increased in parallel to that of prokaryotes, from 10^7 ml^{-1} in the lowest salinity ponds to $10^9\,ml^{-1}$ in the most concentrated ponds (crystallizers), thus maintaining a proportion of 10 virions per prokaryotic cell throughout the salinity gradient. A lemon-shaped virus was found infecting square archaea; its abundance increased along the salinity gradient together with the abundance of the square archaea. In addition, many square cells were infected by viruses with other morphologies. Two additional studies (Díez *et al.*, 2000; Sandaa *et al.*, 2003) carried out in the salterns of Alicante (Spain) determined the genetic diversity of the viruses by pulsed-field gel electrophoresis. Sandaa *et al.* (2003) detected an increase in diversity from sea water to intermediate salinity (15 %), followed by a decrease at higher salinities. Oren *et al.* (1997) observed the presence of large numbers of virus-like particles by electron-microscopy techniques in water samples of the Dead Sea. Up to 10^7 virus-like particles ml^{-1} were detected, showing a variety of morphologies, from spindle-shaped to polyhedral and tailed phages. The recent culture under laboratory conditions of the square haloarchaea will permit the isolation of haloviruses that infect them and will increase our knowledge of virus diversity in hypersaline environments. Further studies are necessary in order to understand their role in the ecology and evolution of haloarchaea.

OTHER HALOPHILIC ORGANISMS

According to the information given earlier in this chapter, the organisms that inhabit hypersaline environments are predominantly prokaryotes, but other eukaryotic organisms may also be present, especially in habitats with lower salinities. They include different species that are adapted to the high salt concentrations or may just survive under these extreme conditions, such as algae, diatoms, protozoa or fungi. Several publications have reviewed their biodiversity in hypersaline environments (Rodriguez-Valera, 1988; Javor, 1989; Pedrós-Alió, 2004; Gunde-Cimerman *et al.*, 2005a, b). In saltern ponds, the change in species composition is associated with salinity and species halotolerance or salt requirements. Thus, halotolerant species can be expected to be progressively replaced by halophilic organisms as salinity increases; the presence of different eukaryotic species and their abundance decrease continuously with increasing salinity. Primary production is due to cyanobacteria and green algae; *Dunaliella* is the unicellular green algae responsible for most of the primary production in hypersaline environments. It is considered an excellent model organism for the study of salt adaptation in algae. Besides, some *Dunaliella* strains can accumulate very large amounts of β-carotene; this is produced commercially and constitutes a good example of biotechnological applications of halophilic micro-organisms (Oren, 2005). Rodriguez-Valera *et al.* (1985) and Rodriguez-Valera (1988) reported high productivity values in saltern ponds between 10 and 30 % salinity, with a maximum at around 25 % salts, which corresponded to the highest densities of *Dunaliella*. In ponds of salterns and saline lakes with salinities higher than 10–15 %, large organisms disappear and the brine shrimp *Artemia salina* and the larvae of the brine fly *Ephydra* are the only macroscopic organisms that are observed.

Gunde-Cimerman and colleagues, from the University of Ljubljana, Slovenia, have carried out several studies focused on the isolation and characterization of fungi isolated from salterns in Slovenia and other geographical locations (Gunde-Cimerman *et al.*, 2004, 2005b). They observed a surprisingly rich diversity of fungi. Enumeration of fungi in these habitats revealed their presence in relatively large numbers (up to 4×10^4 ml^{-1}), but the biodiversity appears to be limited to a small number of fungal genera. The melanized fungi were represented by a yeast-like fungus called black yeast and the related genus *Cladosporium*. Among the non-melanized fungi most frequently observed were species of the genera *Aspergillus*, *Penicillium* with teleomorphic stages and *Wallemia*, *Scupolariopsis* and *Alternaria*. Most species are halotolerant, but recent data support the existence of halophilic species (Gunde-Cimerman *et al.*, 2004; Kogej *et al.*, 2005). The black yeasts that were detected with the highest frequency just before the increase in NaCl for the crystallization process were *Hortaea werneckii*, *Phaeotheca triangularis*, *Trimmatostroma salinum* and *Aureobasidium pullulans*. Since *Hortaea werneckii*, *Phaeotheca triangularis* and *Trimmatostroma salinum* are not known outside saline environments, it is assumed that hypersaline waters are their natural habitat (Gunde-Cimerman *et al.*, 2000, 2004). More recently, non-melanized yeasts have been isolated from several salterns worldwide and some salt lakes (Dead Sea, Great Salt Lake). Among the species isolated from these environments were *Pichia guilliermondii*, *Debaryomyces hansenii*, *Yarrowia lipolytica* and *Candida parapsilosis*, well-known contaminants of low-water-activity food products, as well as other species not previously related to hypersaline habitats nor known for their halotolerance, such as *Rhodosporidium sphaerocarpum*, *Rhodosporidium babjevae* and *Rhodotorula larynges* (Butinar *et al.*, 2005).

Independently, an exhaustive study carried out by Buchalo *et al.* (2000) permitted the taxonomic characterization of filamentous fungi isolated from the Dead Sea. They included 26 species representing 13 different genera of Zygomycotina (*Absidia glauca*), Ascomycotina (most representative were species of *Aspergillus*, *Chaetomium*, *Cladosporium*, *Penicillium* and *Eurotium*, as well as a new species of the genus *Gymnascella*, designated *Gymnascella marismortui*) and mitosporic fungi (four species belonging to the genera *Acremonium*, *Stachybotrys* and *Ulocladium*). It must be pointed out that most fungal species isolated from the Dead Sea waters are common soil fungi and are probably contaminant halotolerant organisms that may be adapted to live in this hypersaline environment or are present as dormant spores (Buchalo *et al.*, 2000).

ACKNOWLEDGEMENTS

I thank Cristina Sánchez-Porro for her assistance and critical reading of the manuscript and Niall A. Logan for the review and suggested changes in the article. I also thank Mike Dyall-Smith and Tony Walsby for the supply of the original figures used in this article. The author's research was supported by grants from the Quality of Life and Management of Living Resources Programme of the European Commission (QLK3-CT-2002-01972), Spanish Ministerio de Educación y Ciencia (BMC2003-1344 and BIO2002-11399-E) and Junta de Andalucía.

REFERENCES

Alam, M., Claviez, M., Oesterhelt, D. & Kessel, M. (1984). Flagella and motility behaviour of square bacteria. *EMBO J* **3**, 2899–2903.

Antón, J., Llobet-Brossa, E., Rodriguez-Valera, F. & Amann, R. (1999). Fluorescence *in situ* hybridization analysis of the prokaryotic community inhabiting crystallizer ponds. *Environ Microbiol* **1**, 517–523.

Antón, J., Rosselló-Mora, R., Rodríguez-Valera, F. & Amann, R. (2000). Extremely halophilic bacteria in crystallizer ponds from solar salterns. *Appl Environ Microbiol* **66**, 3052–3057.

Antón, J., Oren, A., Benlloch, S., Rodríguez-Valera, F., Amann, R. & Rosselló-Mora, R. (2002). *Salinibacter ruber* gen. nov., sp. nov., a novel, extremely halophilic member of the *Bacteria* from saltern crystallizer ponds. *Int J Syst Evol Microbiol* **52**, 485–491.

Asker, D. & Ohta, Y. (2002). *Haloferax alexandrinus* sp. nov., an extremely halophilic canthaxanthin-producing archaeon from a solar saltern in Alexandria (Egypt). *Int J Syst Evol Microbiol* **52**, 729–738.

Balashov, S. P., Imasheva, E. S., Boichenko, V. A., Anton, J., Wang, J. M. & Lanyi, J. K. (2005). Xanthorhodopsin: a proton pump with a light-harvesting carotenoid antenna. *Science* **309**, 2061–2064.

Baliga, N. S., Bonneau, R., Facciotti, M. T. & 12 other authors (2004). Genome sequence of *Haloarcula marismortui*: a halophilic archaeon from the Dead Sea. *Genome Res* **14**, 2221–2234.

Bamford, D. H., Ravantti, J. J., Ronnholm, G., Laurinavicius, S., Kukkaro, P., Dyall-Smith, M., Somerharju, P., Kalkkinen, N. & Bamford, J. K. (2005). Constituents of SH1, a novel lipid-containing virus infecting the halophilic euryarchaeon *Haloarcula hispanica*. *J Virol* **79**, 9097–9107.

Benlloch, S., Acinas, S. G., Martinez-Murcia, A. J. & Rodriguez-Valera, F. (1996). Description of prokaryotic biodiversity along the salinity gradient of a multipond solar saltern by direct PCR amplification 16S rDNA. *Hydrobiologia* **329**, 19–31.

Benlloch, S., Acinas, S. G., Antón, J., Lopez-Lopez, A., Luz, S. P. & Rodriguez-Valera, F. (2001). Archaeal biodiversity in crystallizer ponds from a solar saltern: culture versus PCR. *Microb Ecol* **41**, 12–19.

Benlloch, S., Lopez-Lopez, A., Casamayor, E. O. & 9 other authors (2002). Prokaryotic genetic diversity throughout the salinity gradient of a coastal solar saltern. *Environ Microbiol* **4**, 349–360.

Bidle, K., Amadio, W., Oliveira, P., Paulish, T., Hicks, S. & Earnest, C. (2005). A phylogenetic analysis of haloarchaea found in a solar saltern. *BIOS* **76**, 89–96.

Bolhuis, H., te Poele, E. M. & Rodriguez-Valera, F. (2004). Isolation and cultivation of Walsby's square archaeon. *Environ Microbiol* **6**, 1287–1291.

Bonete, M. J., Perez-Pomares, F., Diaz, S., Ferrer, J. & Oren, A. (2003). Occurrence of two different glutamate dehydrogenase activities in the halophilic bacterium *Salinibacter ruber*. *FEMS Microbiol Lett* **226**, 181–186.

Bowman, J. P., McCammon, S. A., Rea, S. M. & McMeekin, T. A. (2000). The microbial composition of three limnologically disparate hypersaline Antarctic lakes. *FEMS Microbiol Lett* **183**, 81–88.

Brock, T. D. (1979). Ecology of saline lakes. In *Strategies of Microbial Life in Extreme Environments*, pp. 29–47. Edited by M. Shilo. Weinheim: Verlag Chemie.

Buchalo, A. S., Nevo, E., Wasser, S. P. & Volz, P. A. (2000). Newly discovered halophilic fungi in the Dead Sea (Israel). In *Journey to Diverse Microbial Worlds. Adaptation to Exotic Environments*, pp. 241–252. Edited by J. Seckbach. Dordrecht: Kluwer Academic.

Burns, D. G., Camakaris, H. M., Janssen, P. H. & Dyall-Smith, M. L. (2004a). Cultivation of Walsby's square haloarchaeon. *FEMS Microbiol Lett* **238**, 469–473.

Burns, D. G., Camakaris, H. M., Janssen, P. H. & Dyall-Smith, M. L. (2004b). Combined use of cultivation-dependent and cultivation-independent methods indicates that members of most haloarchaeal groups in an Australian crystallizer pond are cultivable. *Appl Environ Microbiol* **70**, 5258–5265.

Butinar, L., Santos, S., Spencer-Martins, I., Oren, A. & Gunde-Cimerman, N. (2005). Yeast diversity in hypersaline habitats. *FEMS Microbiol Lett* **244**, 229–234.

Calvo, C., Garcia de la Paz, A. M., Bejar, V., Quesada, E. & Ramos-Cormenzana, A. (1988). Isolation and characterization of phage F9-11 from a lysogenic *Deleya halophila* strain. *Curr Microbiol* **17**, 49–53.

Calvo, C., Garcia de la Paz, A. M., Martinez-Checa, F. & Caba, M. A. (1994). Behaviour of two *D. halophila* bacteriophages with respect to salt concentrations and other environmental factors. *Toxicol Environ Chem* **43**, 85–93.

Castillo, A. M., Gutiérrez, M. C., Kamekura, M., Ma, Y., Cowan, D. A., Jones, B. E., Grant, W. D. & Ventosa, A. (2006). *Halovivax asiaticus* gen. nov., sp. nov., a novel, extremely halophilic archaeon isolated from Inner Mongolia, China. *Int J Syst Evol Microbiol* **56** (in press).

Cayol, J. L., Ollivier, B., Patel, B. K. C., Ageron, E., Grimont, P. A. D., Prensier, G. & Garcia, J. L. (1995). *Haloanaerobium lacusroseus* sp. nov., an extremely halophilic fermentative bacterium from the sediments of a hypersaline lake. *Int J Syst Bacteriol* **45**, 790–797.

Corcelli, A., Lattanzio, V. M., Mascolo, G., Babudri, F., Oren, A. & Kates, M. (2004). Novel sulfonolipid in the extremely halophilic bacterium *Salinibacter ruber*. *Appl Environ Microbiol* **70**, 6678–6685.

Denner, E. B. M., McGenity, T. J., Busse, H.-J., Grant, W. D., Wanner, G. & Stan-Lotter, H. (1994). *Halococcus salifodinae* sp. nov., an archaeal isolate from an Austrian salt mine. *Int J Syst Bacteriol* **44**, 774–780.

Díez, B., Antón, J., Guixa-Boixereu, N., Pedrós-Alió, C. & Rodríguez-Valera, F. (2000). Pulsed-field gel electrophoresis analysis of virus assemblages present in a hypersaline environment. *Int Microbiol* **3**, 159–164.

Dyall-Smith, M., Tang, S. L. & Bath, C. (2003). Haloarchaeal viruses: how diverse are they? *Res Microbiol* **154**, 309–313.

Dyall-Smith, M. L., Burns, D. G., Camakaris, H. M., Janssen, P. H., Russ, B. E. & Porter, K. (2005). Haloviruses and their hosts. In *Adaptation to Life at High Salt Concentrations in Archaea, Bacteria, and Eukarya*, pp. 553–564. Edited by N. Gunde-Cimerman, A. Oren & A. Plemenitaš. Berlin: Springer.

Elazari-Volcani, B. (1957). Genus XII. *Halobacterium*. In *Bergey's Manual of Determinative Bacteriology*, 7th edn, pp. 207–212. Edited by R. S. Breed, E. G. D. Murray & N. R. Smith. Baltimore: Williams & Wilkins.

Elshahed, M. S., Savage, K. N., Oren, A., Gutierrez, M. C., Ventosa, A. & Krumholz, L. R. (2004). *Haloferax sulfurifontis* sp. nov., a halophilic archaeon isolated from a sulfide- and sulfur-rich spring. *Int J Syst Evol Microbiol* **54**, 2275–2279.

Falb, M., Pfeiffer, F., Palm, P., Rodewald, K., Hickmann, V., Tittor, J. & Oesterhelt, D. (2005). Living with two extremes: conclusions from the genome sequence of *Natronomonas pharaonis*. *Genome Res* **15**, 1336–1343.

Fan, H., Xue, Y., Ma, Y., Ventosa, A. & Grant, W. D. (2004). *Halorubrum tibetense* sp. nov., a novel haloalkaliphilic archaeon from Lake Zabuye in Tibet, China. *Int J Syst Evol Microbiol* **54**, 1213–1216.

Feng, J., Zhou, P. J. & Liu, S. J. (2004). *Halorubrum xinjiangense* sp. nov., a novel halophile isolated from saline lakes in China. *Int J Syst Evol Microbiol* **54**, 1789–1791.

Feng, J., Zhou, P., Zhou, Y. G., Liu, S. J. & Warren-Rhodes, K. (2005). *Halorubrum alkaliphilum* sp. nov., a novel haloalkaliphile isolated from a soda lake in Xinjiang, China. *Int J Syst Evol Microbiol* **55**, 149–152.

Franzmann, P. D., Stackebrandt, E., Sanderson, K., Volkman, J. K., Cameron, D. E., Stevenson, P. L., McMeekin, T. A. & Burton, H. R. (1988). *Halobacterium lacusprofundi* sp. nov. a halophilic bacterium isolated from Deep Lake, Antarctica. *Syst Appl Microbiol* **11**, 20–27.

Garcia, M. T., Ventosa, A., Ruiz-Berraquero, F. & Kocur, M. (1987). Taxonomic study and emended description of *Vibrio costicola*. *Int J Syst Bacteriol* **37**, 251–256.

Gibbons, N. E. (1974). Family V. *Halobacteriaceae* fam. nov. In *Bergey's Manual of Determinative Bacteriology*, 8th edn, pp. 269–273. Edited by R. E. Buchanan & N. E. Gibbons. Baltimore: Williams & Wilkins.

Gochnauer, M. B., Leppard, G. G., Komaratat, P., Kates, M., Novitsky, T. & Kushner, D. J. (1975). Isolation and characterization of *Actinopolyspora halophila*, gen. et sp. nov., an extremely halophilic actinomycete. *Can J Microbiol* **21**, 1500–1511.

Gonzalez, C., Gutierrez, C. & Ramirez, C. (1979). *Halobacterium vallismortis* sp. nov. an amylolytic and carbohydrate-metabolizing, extremely halophilic bacterium. *Can J Microbiol* **24**, 710–715.

Grant, W. D. (1990). General view of halophiles. In *Superbugs. Microorganisms in Extreme Environments*, pp. 15–37. Edited by K. Horikoshi & W. D. Grant. Tokyo: Springer.

Grant, W. D. (2001a). Genus I. *Halobacterium* Elazari-Volcani 1957, 207,[AL] emend. Larsen and Grant 1989, 2222. In *Bergey's Manual of Systematic Bacteriology*, 2nd edn, vol. 1, pp. 301–305. Edited by D. R. Boone, R. W. Castenholz & G. M. Garrity. New York: Springer.

Grant, W. D. (2001b). Genus IV. *Halococcus* Schoop 1935, 817[AL]. In *Bergey's Manual of Systematic Bacteriology*, 2nd edn, vol. 1, pp. 311–314. Edited by D. R. Boone, R. W. Castenholz & G. M. Garrity. New York: Springer.

Grant, W. D., Kamekura, M., McGenity, T. J. & Ventosa, A. (2001). Class III. *Halobacteria* class. nov. In *Bergey's Manual of Systematic Bacteriology*, 2nd edn, vol. 1, pp. 294–301. Edited by D. R. Boone, R. W. Castenholz & G. M. Garrity. New York: Springer.

Gruber, C., Legat, A., Pfaffenhuemer, M., Radax, C., Weidler, G., Busse, H. J. & Stan-Lotter, H. (2004). *Halobacterium noricense* sp. nov., an archaeal isolate from a bore core of an alpine Permian salt deposit, classification of *Halobacterium* sp. NRC-1 as a strain of *H. salinarum* and emended description of *H. salinarum*. *Extremophiles* **8**, 431–439.

Guixa-Boixereu, N., Calderón-Paz, J. I., Heldal, M., Bratbak, G. & Pedrós-Alió, C. (1996). Viral lysis and bacteriovory as prokaryotic loss factors along a salinity gradient. *Aquat Microb Ecol* **11**, 215–227.

Gunde-Cimerman, N., Zalar, P., de Hoog, S. & Plemenitaš, A. (2000). Hypersaline waters in salterns – natural ecological niches for halophilic black yeasts. *FEMS Microbiol Ecol* **32**, 235–240.

Gunde-Cimerman, N., Zalar, P., Petrovic, U., Turk, M., Kogej, T., de Hoog, G. S. & Plemenitaš, A. (2004). Fungi in salterns. In *Halophilic Microorganisms*, pp. 103–113. Edited by A. Ventosa. Berlin: Springer.

Gunde-Cimerman, N., Oren, A. & Plemenitaš, A. (editors) (2005a). *Mikrosafari. The Beautiful World of Microorganisms in the Salterns*. Ljubljana: Državna Založba Slovenije (in English and Slovenian).

Gunde-Cimerman, N., Oren, A. & Plemenitaš, A. (editors) (2005b). *Adaptation to Life at High Salt Concentrations in Archaea, Bacteria, and Eukarya*. Dordrecht: Springer.

Gutierrez, M. C., Kamekura, M., Holmes, M. L., Dyall-Smith, M. L. & Ventosa, A. (2002). Taxonomic characterization of *Haloferax* sp. ("*H. alicantei*") strain Aa 2.2: description of *Haloferax lucentensis* sp. nov. *Extremophiles* **6**, 479–483.

Hezayen, F. F., Rehm, B. H. A., Tindall, B. J. & Steinbuechel, A. (2001). Transfer of *Natrialba asiatica* B1T to *Natrialba taiwanensis* sp. nov. and description of *Natrialba aegyptiaca* sp. nov., a novel extremely halophilic, aerobic, non-pigmented member of the *Archaea* from Egypt that produces extracellular poly(glutamic acid). *Int J Syst Evol Microbiol* **51**, 1133–1142.

Hezayen, F. F., Tindall, B. J., Steinbüchel, A. & Rehm, B. H. A. (2002). Characterization of a novel halophilic archaeon, *Halobiforma haloterrestris* gen. nov., sp. nov., and transfer of *Natronobacterium nitratireducens* to *Halobiforma nitratireducens* comb. nov. *Int J Syst Evol Microbiol* **52**, 2271–2280.

Hirschler-Rea, A., Matheron, R., Riffaud, C., Moune, S., Eatock, C., Herbert, R. A., Willison, J. C. & Caumette, P. (2003). Isolation and characterization of spirilloid purple phototrophic bacteria forming red layers in microbial mats of Mediterranean salterns: description of *Halorhodospira neutriphila* sp. nov. and emendation of the genus *Halorhodospira*. *Int J Syst Evol Microbiol* **53**, 153–163.

Ihara, K., Watanabe, S. & Tamura, T. (1997). *Haloarcula argentinensis* sp. nov. and *Haloarcula mukohataei* sp. nov., two new extremely halophilic archaea collected in Argentina. *Int J Syst Bacteriol* **47**, 73–77.

Itoh, T., Yamaguchi, T., Zhou, P. & Takashina, T. (2005). *Natronolimnobius baerhuensis* gen. nov., sp. nov. and *Natronolimnobius innermongolicus* sp. nov., novel haloalkaliphilic archaea isolated from soda lakes in Inner Mongolia, China. *Extremophiles* **9**, 111–116.

Javor, B. J. (1989). *Hypersaline Environments. Microbiology and Biogeochemistry*. New York: Springer.

Javor, B., Requadt, C. & Stoeckenius, W. (1982). Box-shaped halophilic bacteria. *J Bacteriol* **151**, 1532–1542.

Juez, G., Rodriguez-Valera, F., Ventosa, A. & Kushner, D. J. (1986). *Haloarcula hispanica* spec. nov. and *Haloferax gibbonsii* spec. nov., two new species of extremely halophilic archaebacteria. *Syst Appl Microbiol* **8**, 75–79.

Kamekura, M. & Dyall-Smith, M. L. (1995). Taxonomy of the family *Halobacteriaceae* and the description of two new genera *Halorubrobacterium* and *Natrialba*. *J Gen Appl Microbiol* **41**, 333–350.

Kamekura, M., Dyall-Smith, M. L., Upasani, V., Ventosa, A. & Kates, M. (1997). Diversity of alkaliphilic halobacteria: proposals for transfer of *Natronobacterium vacuolatum*, *Natronobacterium magadii*, and *Natronobacterium pharaonis* to *Halorubrum*, *Natrialba*, and *Natronomonas* gen. nov., respectively, as *Halorubrum vacuolatum* comb. nov., *Natrialba magadii* comb. nov., and *Natronomonas pharaonis* comb. nov., respectively. *Int J Syst Bacteriol* **47**, 853–857.

Kanai, H., Kobayashi, T., Aono, R. & Kudo, T. (1995). *Natronococcus amylolyticus* sp. nov., a haloalkaliphilic archaeon. *Int J Syst Bacteriol* **45**, 762–766.

Kates, M. (1993). Membrane lipids of extreme halophiles: biosynthesis, function and evolutionary significance. *Experientia (Basel)* **49**, 1027–1036.

Kauri, T., Ackerman, H.-W., Goel, U. & Kushner, D. J. (1991). A bacteriophage of a moderately halophilic bacterium. *Arch Microbiol* **156**, 435–438.

Kessel, M. & Cohen, Y. (1982). Ultrastructure of square bacteria from a brine pool in Southern Sinai. *J Bacteriol* **150**, 851–860.

Klein, R., Baranyi, U., Rossler, N., Greineder, B., Scholz, H. & Witte, A. (2002). *Natrialba magadii* virus *φ*Ch1: first complete nucleotide sequence and functional organization of a virus infecting a haloalkaliphilic archaeon. *Mol Microbiol* **45**, 851–863.

Kocur, M. & Hodgkiss, W. (1973). Taxonomic status of the genus *Halococcus* Schoop. *Int J Syst Bacteriol* **23**, 151–156.

Kogej, T., Ramos, J., Plemenitas, A. & Gunde-Cimerman, N. (2005). The halophilic fungus *Hortaea werneckii* and the halotolerant fungus *Aureobasidium pullulans* maintain low intracellular cation concentrations in hypersaline environments. *Appl Environ Microbiol* **71**, 6600–6605.

Kushner, D. J. & Kamekura, M. (1988). Physiology of halophilic eubacteria. In *Halophilic Bacteria*, vol. I, pp. 109–138. Edited by F. Rodriguez-Valera. Boca Raton, FL: CRC Press.

Lanyi, J. K. (1995). Bacteriorhodopsin as a model for proton pumps. *Nature* **375**, 461–463.

Lizama, C., Monteoliva-Sánchez, M., Suárez-García, A., Rosselló-Mora, R., Aguilera, M., Campos, V. & Ramos-Cormenzana, A. (2002). *Halorubrum tebenquichense* sp. nov., a novel halophilic archaeon isolated from the Atacama saltern, Chile. *Int J Syst Evol Microbiol* **52**, 149–155.

Lutnaes, B. F., Oren, A. & Liaaen-Jensen, S. (2002). New C_{40}-carotenoid acyl glycoside as principal carotenoid in *Salinibacter ruber,* an extremely halophilic eubacterium. *J Nat Prod* **65**, 1340–1343.

Ma, Y., Zhang, W., Xue, Y., Zhou, P., Ventosa, A. & Grant, W. D. (2004). Bacterial diversity of the Inner Mongolian Baer Soda Lake as revealed by 16S rRNA gene sequence analyses. *Extremophiles* **8**, 45–51.

Madern, D. & Zaccai, G. (2004). Molecular adaptation: the malate dehydrogenase from the extreme halophilic bacterium *Salinibacter ruber* behaves like a non-halophilic protein. *Biochimie* **86**, 295–303.

Margesin, R. & Schinner, F. (2001). Potential of halotolerant and halophilic microorganisms for biotechnology. *Extremophiles* **5**, 73–83.

McGenity, T. J. & Grant, W. D. (1995). Transfer of *Halobacterium saccharovorum*, *Halobacterium sodomense*, *Halobacterium trapanicum* NRC 34021 and *Halobacterium lacusprofundi* to the genus *Halorubrum* gen. nov., as *Halorubrum saccharovorum* comb. nov., *Halorubrum sodomense* comb. nov., *Halorubrum trapanicum* comb. nov., and *Halorubrum lacusprofundi* comb. nov. *Syst Appl Microbiol* **18**, 237–243.

McGenity, T. J., Gemmell, R. T. & Grant, W. D. (1998). Proposal of a new halobacterial genus *Natrinema* gen. nov., with two species *Natrinema pellirubrum* nom. nov. and *Natrinema pallidum* nom. nov. *Int J Syst Bacteriol* **48**, 1187–1196.

McGenity, T. J., Gemmell, R. T., Grant, W. D. & Stan-Lotter, H. (2000). Origins of halophilic microorganisms in ancient salt deposits. *Environ Microbiol* **2**, 243–250.

Mellado, E. & Ventosa, A. (2003). Biotechnological potential of moderately and extremely halophilic microorganisms. In *Microorganisms for Health Care, Food and Enzyme Production*, pp. 233–256. Edited by J. L. Barredo. Kerala: Research Signpost.

Mellado, E., Sanchez-Porro, C., Martin, S. & Ventosa, A. (2004). Extracellular hydrolytic enzymes produced by moderately halophilic bacteria. In *Halophilic Microorganisms*, pp. 285–295. Edited by A. Ventosa. Berlin: Springer.

Montalvo-Rodríguez, R., Vreeland, R. H., Oren, A., Kessel, M., Betancourt, C. & López-Garriga, J. (1998). *Halogeometricum borinquense* gen. nov., sp. nov., a novel halophilic archaeon from Puerto Rico. *Int J Syst Bacteriol* **48**, 1305–1312.

Montalvo-Rodríguez, R., López-Garriga, J., Vreeland, R. H., Oren, A., Ventosa, A. & Kamekura, M. (2000). *Haloterrigena thermotolerans* sp. nov., a halophilic archaeon from Puerto Rico. *Int J Syst Evol Microbiol* **50**, 1065–1071.

Montero, C. G., Ventosa, A., Rodriguez-Valera, F., Kates, M., Moldoveanu, N. & Ruiz-Berraquero, F. (1989). *Halococcus saccharolyticus* sp. nov., a new species of extremely halophilic non-alkaliphilic cocci. *Syst Appl Microbiol* **12**, 167–171.

Mullakhanbhai, M. F. & Larsen, H. (1975). *Halobacterium volcanii* spec. nov., a Dead Sea halobacterium with a moderate salt requirement. *Arch Microbiol* **104**, 207–214.

Mwatha, W. E. & Grant, W. D. (1993). *Natronobacterium vacuolata* sp. nov., a haloalkaliphilic archaeon isolated from Lake Magadi, Kenya. *Int J Syst Bacteriol* **43**, 401–404.

Ng, W. V., Kennedy, S. P., Mahairas, G. G. & 40 other authors (2000). Genome sequence of *Halobacterium* species NRC-1. *Proc Natl Acad Sci U S A* **97**, 12176–12181.

Nieto, J. J. & Vargas, C. (2002). Synthesis of osmoprotectants by moderately halophilic bacteria: genetic and applied aspects. *Recent Res Devel Microbiol* **6**, 403–418.

Oren, A. (1983). *Halobacterium sodomense* sp. nov., a Dead Sea *Halobacterium* with an extremely high magnesium requirement. *Int J Syst Bacteriol* **33**, 381–386.

Oren, A. (1999). The enigma of square and triangular halophilic archaea. In *Enigmatic Microorganisms and Life in Extreme Environments*, pp. 337–355. Edited by J. Seckbach. Dordrecht: Kluwer.

Oren, A. (2005). A hundred years of *Dunaliella* research: 1905–2005. *Saline Systems* **1**, 2.

Oren, A. & Mana, L. (2002). Amino acid composition of bulk protein and salt relationships of selected enzymes of *Salinibacter ruber*, an extremely halophilic bacterium. *Extremophiles* **6**, 217–223.

Oren, A. & Rodriguez-Valera, F. (2001). The contribution of halophilic bacteria to the red coloration of saltern crystallizer ponds. *FEMS Microbiol Ecol* **36**, 123–130.

Oren, A. & Ventosa, A. (1996). A proposal for the transfer of *Halorubrobacterium distributum* and *Halorubrobacterium coriense* to the genus *Halorubrum* as *Halorubrum distributum* comb. nov. and *Halorubrum coriense* comb. nov., respectively. *Int J Syst Bacteriol* **46**, 1180.

Oren, A., Ginzburg, M., Ginzburg, B. Z., Hochstein, L. I. & Volcani, B. E. (1990). *Haloarcula marismortui* (Volcani) sp. nov., nom. rev., an extremely halophilic bacterium from the Dead Sea. *Int J Syst Bacteriol* **40**, 209–210.

Oren, A., Gurevich, P., Gemmell, R. T. & Teske, A. (1995). *Halobaculum gomorrense* gen. nov., sp. nov., a novel extremely halophilic archaeon from the Dead Sea. *Int J Syst Bacteriol* **45**, 747–754.

Oren, A., Duker, S. & Ritter, S. (1996). The polar lipid composition of Walsby's square bacterium. *FEMS Microbiol Lett* **138**, 135–140.

Oren, A., Bratbak, G. & Heldal, M. (1997). Occurrence of virus-like particles in the Dead Sea. *Extremophiles* **1**, 143–149.

Oren, A., Ventosa, A., Gutiérrez, M. C. & Kamekura, M. (1999). *Haloarcula quadrata* sp. nov., a square, motile archaeon isolated from a brine pool in Sinai (Egypt). *Int J Syst Bacteriol* **49**, 1149–1155.

Oren, A., Elevi, R., Watanabe, S., Ihara, K. & Corcelli, A. (2002a). *Halomicrobium mukohataei* gen. nov., comb. nov., and emended description of *Halomicrobium mukohataei*. *Int J Syst Evol Microbiol* **52**, 1831–1835.

Oren, A., Heldal, M., Norland, S. & Galinski, E. A. (2002b). Intracellular ion and organic solute concentrations of the extremely halophilic bacterium *Salinibacter ruber*. *Extremophiles* **6**, 491–498.

Oren, A., Rodriguez-Valera, F., Antón, J., Benlloch, S., Rosselló-Mora, R., Amann, R., Coleman, J. & Russell, N. J. (2004). Red, extremely halophilic, but not archeal: the physiology and ecology of *Salinibacter ruber*, a bacterium isolated from saltern crystallizer ponds. In *Halophilic Microorganisms*, pp. 63–74. Edited by A. Ventosa. Berlin: Springer.

Parkes, K. & Walsby, A. E. (1981). Ultrastructure of a gas-vacuolate square bacterium. *J Gen Microbiol* **126**, 503–506.

Pasic, L., Bartual, S. G., Ulrih, N. P., Grabnar, M. & Velikonja, B. H. (2005). Diversity of halophilic archaea in the crystallizers of an Adriatic solar saltern. *FEMS Microbiol Ecol* **54**, 491–498.

Pedrós-Alió, C. (2004). Trophic ecology of solar salterns. In *Halophilic Microorganisms*, pp. 33–48. Edited by A. Ventosa. Berlin: Springer.

Petter, H. F. M. (1931). On bacteria of salted fish. *Proc K Ned Akad Wet Amsterdam* **34**, 1417–1423.

Porter, K., Kukkaro, P., Bamford, J. K., Bath, C., Kivela, H. M., Dyall-Smith, M. L. & Bamford, D. H. (2005). SH1: a novel, spherical halovirus isolated from an Australian hypersaline lake. *Virology* **335**, 22–33.

Purdy, K. J., Cresswell-Maynard, T. D., Nedwell, D. B., McGenity, T. J., Grant, W. D., Timmis, K. N. & Embley, T. M. (2004). Isolation of haloarchaea that grow at low salinities. *Environ Microbiol* **6**, 591–595.

Rees, H. C., Grant, W. D., Jones, B. E. & Heaphy, S. (2004). Diversity of Kenyan soda lake alkaliphiles assessed by molecular methods. *Extremophiles* **8**, 63–71.

Robb, F. T., Place, A. R., Sowers, K. R., Schreier, H. J., Dasarma, S. & Fleischmann, E. M. (editors) (1995). *Archaea. a Laboratory Manual*. Cold Spring Harbor, NY: Cold Spring Harbor Laboratory.

Rodriguez-Valera, F. (1988). Characteristics and microbial ecology of hypersaline environments. In *Halophilic Bacteria*, vol. 1, pp. 3–30. Edited by F. Rodriguez-Valera. Boca Raton, FL: CRC Press.

Rodriguez-Valera, F., Juez, G. & Kushner, D. J. (1983). *Halobacterium mediterranei* spec. nov., a new carbohydrate-utilizing extreme halophile. *Syst Appl Microbiol* **4**, 369–381.

Rodriguez-Valera, F., Ventosa, A., Juez, G. & Imhoff, J. F. (1985). Variation of environmental features and microbial populations with salt concentrations in a multi-pond saltern. *Microb Ecol* **11**, 107–115.

Sandaa, R. A., Foss Skjoldal, E. & Bratbak, G. (2003). Virioplankton community structure along a salinity gradient in a solar saltern. *Extremophiles* **7**, 347–351.

Schnabel, H., Zillig, W., Pfaffle, M., Schnabel, R., Michel, H. & Delius, H. (1982). *Halobacterium halobium* phage *ϕ*H. *EMBO J* **1**, 87–92.

Schoop, G. (1935). *Halococcus litoralis*, ein obligat halphiler Farbstoffbildner. *Dtsch Tierarztl Wochenschr* **43**, 817–820 (in German).

Skerman, V. B. D., McGowan, V. & Sneath, P. H. A. (1980). Approved lists of bacterial names. *Int J Syst Bacteriol* **30**, 225–420.

Soliman, G. S. H. & Trüper, H. G. (1982). *Halobacterium pharaonis* sp. nov., a new extremely haloalkaliphilic archaebacterium with a low magnesium requirement. *Zentralbl Bakteriol Mikrobiol Hyg 1 Abt Orig C* **3**, 318–329.

Stan-Lotter, H., Pfaffenhuemer, M., Legat, A., Busse, H. J., Radax, C. & Gruber, C. (2002). *Halococcus dombrowskii* sp. nov., an archaeal isolate from a Permian alpine salt deposit. *Int J Syst Evol Microbiol* **52**, 1807–1814.

Stoeckenius, W. (1981). Walsby's square bacterium: fine structure of an orthogonal procaryote. *J Bacteriol* **148**, 352–360.

Stolt, P. & Zillig, W. (1994a). Transcription of the halophage *ϕ*H repressor gene is abolished by transcription from an inversely oriented lytic promoter. *FEBS Lett* **344**, 125–128.

Stolt, P. & Zillig, W. (1994b). Gene regulation in halophage *ϕ*H; more than promoters. *Syst Appl Microbiol* **16**, 591–596.

Takashina, T., Hamamoto, T., Otozai, K., Grant, W. D. & Horikoshi, K. (1990). *Haloarcula japonica* sp. nov., a new triangular halophilic archaebacterium. *Syst Appl Microbiol* **13**, 177–181.

Tang, S. L., Nuttall, S., Ngui, K., Fisher, C., Lopez, P. & Dyall-Smith, M. (2002). HF2: a double-stranded DNA tailed haloarchaeal virus with a mosaic genome. *Mol Microbiol* **44**, 283–296.

Tang, S. L., Fisher, C., Ngui, K., Nuttall, S. D. & Dyall-Smith, M. (2004a). Genome sequences of the head-tail haloviruses HF1 and HF2. In *Halophilic Microorganisms*, pp. 255–262. Edited by A. Ventosa. Berlin: Springer.

Tang, S. L., Nuttall, S. & Dyall-Smith, M. (2004b). Haloviruses HF1 and HF2: evidence for a recent and large recombination event. *J Bacteriol* **186**, 2810–2817.

Tindall, B. J., Ross, H. N. M. & Grant, W. D. (1984). *Natronobacterium* gen. nov. and *Natronococcus* gen. nov., two new genera of haloalkaliphilic archaebacteria. *Syst Appl Microbiol* **5**, 41–57.

Tindall, B. J., Tomlinson, G. A. & Hochstein, L. I. (1989). Transfer of *Halobacterium denitrificans* (Tomlinson, Jahnke, and Hochstein) to the genus *Haloferax* as *Haloferax denitrificans* comb. nov. *Int J Syst Bacteriol* **39**, 359–360.

Tomlinson, G. A. & Hochstein, L. I. (1976). *Halobacterium saccharovorum* sp. nov., a carbohydrate-metabolizing, extremely halophilic bacterium. *Can J Microbiol* **22**, 587–591.

Tomlinson, G. A., Jahnke, L. L. & Hochstein, L. I. (1986). *Halobacterium denitrificans* sp. nov., an extremely halophilic denitrifying bacterium. *Int J Syst Bacteriol* **36**, 66–70.

Torreblanca, M., Rodriguez-Valera, F., Juez, G., Ventosa, A., Kamekura, M. & Kates, M. (1986). Classification of non-alkaliphilic halobacteria based on numerical taxonomy and polar lipid composition, and description of *Haloarcula* gen. nov. and *Haloferax* gen. nov. *Syst Appl Microbiol* **8**, 89–99.

Torsvik, T. & Dundas, I. D. (1974). Bacteriophage of *Halobacterium salinarium*. *Nature* **248**, 680–681.

Ventosa, A. (2001a). Genus II. *Haloarcula*. In *Bergey's Manual of Systematic Bacteriology*, 2nd edn, vol. 1, pp. 305–309. Edited by D. R. Boone, R. W. Castenholz & G. M. Garrity. New York: Springer.

Ventosa, A. (2001b). Genus V. *Haloferax*. In *Bergey's Manual of Systematic Bacteriology*, 2nd edn, vol. 1, pp. 315–318. Edited by D. R. Boone, R. W. Castenholz & G. M. Garrity. New York: Springer.

Ventosa, A. & Nieto, J. J. (1995). Biotechnological applications and potentialities of halophilic microorganisms. *World J Microbiol Biotechnol* **11**, 85–94.

Ventosa, A. & Oren, A. (1996). *Halobacterium salinarum* nom. corrig., a name to replace *Halobacterium salinarium* (Elazari-Volcani) and to include *Halobacterium halobium* and *Halobacterium cutirubrum*. *Int J Syst Bacteriol* **46**, 347.

Ventosa, A., Nieto, J. J. & Oren, A. (1998). Biology of moderately halophilic aerobic bacteria. *Microbiol Mol Biol Rev* **62**, 504–544.

Ventosa, A., Gutierrez, M. C., Kamekura, M. & Dyall-Smith, M. L. (1999). Proposal to transfer *Halococcus turkmenicus*, *Halobacterium trapanicum* JCM 9743 and strain GSL-11 to *Haloterrigena turkmenica* gen. nov., comb. nov. *Int J Syst Bacteriol* **49**, 131–136.

Ventosa, A., Gutierrez, M. C., Kamekura, M., Zvyagintseva, I. S. & Oren, A. (2004). Taxonomic study of *Halorubrum distributum* and proposal of *Halorubrum terrestre* sp. nov. *Int J Syst Evol Microbiol* **54**, 389–392.

Vreeland, R. H., Straight, S., Krammes, J., Dougherty, K., Rosenzweig, W. D. & Kamekura, M. (2002). *Halosimplex carlsbadense* gen. nov., sp. nov., a unique halophilic archaeon, with three 16S rRNA genes, that grows only in defined medium with glycerol and acetate or pyruvate. *Extremophiles* **6**, 445–452.

Wainø, M., Tindall, B. J. & Ingvorsen, K. (2000). *Halorhabdus utahensis* gen. nov., sp. nov., an aerobic, extremely halophilic member of the *Archaea* from Great Salt Lake, Utah. *Int J Syst Evol Microbiol* **50**, 183–190.

Walsby, A. E. (1980). A square bacterium. *Nature* **283**, 69–71.

Walsby, A. E. (2005). Archaea with square cells. *Trends Microbiol* **13**, 193–195.

Xin, H., Itoh, T., Zhou, P., Suzuki, K., Kamekura, M. & Nakase, T. (2000). *Natrinema versiforme* sp. nov., an extremely halophilic archaeon from Aibi salt lake, Xinjiang, China. *Int J Syst Evol Microbiol* **50**, 1297–1303.

Xin, H., Itoh, T., Zhou, P., Suzuki, K. & Nakase, T. (2001). *Natronobacterium nitratireducens* sp. nov. a haloalkaliphilic archaeon isolated from a soda lake in China. *Int J Syst Evol Microbiol* **51**, 1825–1829.

Xu, Y., Zhou, P. & Tian, X. (1999). Characterization of two novel haloalkaliphilic archaea, *Natronorubrum bangense* gen. nov., sp. nov. and *Natronorubrum tibetense* gen. nov., sp. nov. *Int J Syst Bacteriol* **49**, 261–266.

Xu, Y., Wang, Z., Xue, Y., Zhou, P., Ma, Y., Ventosa, A. & Grant, W. D. (2001). *Natrialba hulunbeirensis* sp. nov. and *Natrialba chahannaoensis* sp. nov., novel haloalkaliphilic archaea from soda lakes in Inner Mongolia Autonomous Region, China. *Int J Syst Evol Microbiol* **51**, 1693–1698.

Xu, X. W., Ren, P. G., Liu, S. J., Wu, M. & Zhou, P. J. (2005a). *Natrinema altunense* sp. nov., an extremely halophilic archaeon isolated from a salt lake in Altun Mountain in Xinjiang, China. *Int J Syst Evol Microbiol* **55**, 1311–1314.

Xu, X. W., Wu, M., Zhou, P. J. & Liu, S. J. (2005b). *Halobiforma lacisalsi* sp. nov., isolated from a salt lake in China. *Int J Syst Evol Microbiol* **55**, 1949–1952.

Xu, X. W., Liu, S. J., Tohty, D., Oren, A., Wu, M. & Zhou, P. J. (2005c). *Haloterrigena* *saccharevitans* sp. nov., an extremely halophilic archaeon from Xin-Jiang, China. *Int J Syst Evol Microbiol* **55**, 2539–2542.

Xue, Y., Fan, H., Ventosa, A., Grant, W. D., Jones, B. D., Cowan, D. A. & Ma, Y. (2005). *Halalkalicoccus tibetensis* gen. nov., sp. nov., representing a novel genus of haloalkaliphilic archaea. *Int J Syst Evol Microbiol* **55**, 2501–2505.

Zhilina, T. N. & Zavarzin, G. A. (1990). A new extremely halophilic homoacetogen bacterium, *Acetohalobium arabaticum* gen. nov., sp. nov. *Dokl Akad Nauk SSSR* **311**, 745–747 (in Russian).

Zhilina, T. N., Miroshnikova, L. V., Osipov, G. A. & Zavarzin, G. A. (1991). *Halobacteroides lucunaris* sp. nov. – a new saccharolytic, anaerobic, extremely halophilic organism from the lagoon-like hypersaline Lake Chokrak. *Mikrobiologiia* **60**, 714–724 (in Russian).

Zvyagintseva, I. S. & Tarasov, A. L. (1987). Extreme halophilic bacteria from saline soils. *Mikrobiologiia* **56**, 839–844 (in Russian).

Genomic islands and evolution of catabolic pathways

Stéphan Lacour, Muriel Gaillard and Jan Roelof van der Meer

Department of Fundamental Microbiology, University of Lausanne, CH-1015 Lausanne, Switzerland

INTRODUCTION

Evolution of catabolic pathways in bacteria is most often equivalent to 'catabolic pathway expansion' or 'new acquisition of catabolic properties', although in essence it could also mean loss or deletion. Even if we limit ourselves to this narrow interpretation, changes in the repertoire of catabolic functions of a bacterium are largely attributable to the activity of mobile genetic elements (van der Meer, 1997, 2002). Typically, mobile genetic elements may create new recombinations between previously disconnected DNA fragments, even from different bacterial origins, thus assembling bits and pieces together. Such recombinations are not necessarily needed to result in a single, smoothly transcribed new operon; any workable gene or fragment of genes within the boundaries of the new host cell may contribute to the catabolic expansion of functions (Dogra *et al.*, 2004; Müller *et al.*, 2004). Classical examples of evolutionarily ancient pathway expansions are the formation of the operons for toluene and xylene degradation in pseudomonads (Harayama *et al.*, 1987; Harayama & Rekik, 1993; Greated *et al.*, 2002). Observations of (most likely) very recent pathway expansions include the formation of pathways for chlorobenzene degradation, such as in *Pseudomonas* sp. strain P51 (van der Meer *et al.*, 1991) and *Ralstonia* sp. strain JS705 (van der Meer *et al.*, 1998; Müller *et al.*, 2003), or pathways for 2,4-dichlorophenoxyacetic acid degradation in *Ralstonia eutropha* (now *Cupriavidus necator*) JMP134 (Laemmli *et al.*, 2000; Trefault *et al.*, 2004). These and other cases have shown that incoming gene fragments with catabolic gene functions can complement certain pathway deficiencies (such as allowing chlorinated as opposed to non-chlorinated intermediates to be metabolized), leading to a strong selective advantage for growth.

SGM symposium 66: Prokaryotic diversity: mechanisms and significance.
Editors N. A. Logan, H. M. Lappin-Scott & P. C. F. Oyston. Cambridge University Press. ISBN 0 521 86935 8 ©SGM 2006

On the other hand, the majority of recombinations may lead to redundancies or imperfect complementations, of which the selective advantage is doubtful or unclear.

Likewise, mobile genetic elements play an important role in bacterial evolution in general and in adaptation to changing conditions, in either clinical, industrial or natural environments, and the outcomes of these adaptations are obvious from the development of antibiotic resistances, of pathogenicity functions and, as explained above, of catabolic functions. Mobile genetic elements are collectively designated the *horizontal gene pool* and consist of transferable or mobilizable plasmids, transposons, integrons, genomic islands and phages (Thomas, 2000). Lateral or horizontal gene transfer, the exchange of DNA between bacterial strains not of the same species, has until recently been considered as perhaps a curious phenomenon limited to a number of very specific conjugative plasmids, phages or bacterial strains capable of natural transformation. However, now that more than 500 bacterial genomes have been completely sequenced or are available in raw annotation, microbiologists have recognized that horizontal gene transfer must have been (and probably is) more general and more abundant than previously assumed (Ochman *et al.*, 2000; Koonin & Galperin, 2003).

GENOMIC ISLANDS

Genomic islands (GEIs) are probably the latest addition to the types of elements comprising the horizontal gene pool. The concept of pathogenicity islands (PAIs) was developed in the late 1980s by J. Hacker and colleagues, who investigated the genetic basis of virulence of uropathogenic isolates of *Escherichia coli* (Knapp *et al.*, 1986; Hacker *et al.*, 1990). PAIs encode virulence factors and are frequently found to be unstable (Groisman & Ochman, 1996). However, at this time, identification of PAIs relied on classical genetic techniques such as pulsed-field gel electrophoresis and Southern hybridization, which were used to compare digested genomes of pathogenic versus non-pathogenic isolates of the same species. Nowadays, PAIs and other GEIs are discovered by systematic statistical screening of completely sequenced bacterial genomes (e.g. Greub *et al.*, 2004) or of genomes of very closely related isolates (e.g. *E. coli*; Perna *et al.*, 2001). Loci with an altered nucleotide profile (e.g. tetranucleotide usage or so-called residual skew analysis) from the rest of the bacterial genome can be detected and investigated in more detail in order to define the potential proportion of DNA exogenous to each genome (Dufraigne *et al.*, 2005; Regeard *et al.*, 2005). By using residual TA-skew analysis, for example, a novel GEI was discovered in *Parachlamydiaceae* UWE25 (Greub *et al.*, 2004).

Thus, different lines of evidence have demonstrated that bacterial genomes evolve in part by capturing or deleting very discrete, relatively large DNA regions (2–200 kb).

Although not all of those are GEIs in the more restricted definition, which will be discussed in more detail below, many genomes do contain clearly definable GEIs (Mohd-Zain *et al.*, 2004). Other fragments may more closely resemble prophages or have no trace of a recognizable mobile element (Perna *et al.*, 2001). In addition, it has become evident that GEIs not only consist of pathogenicity members, but also occur in non-pathogenic bacteria, conferring various traits, such as nitrogen fixation, symbiosis functions or degradation of aromatic compounds (Dobrindt *et al.*, 2004). It has also been possible to study the genealogy of GEIs themselves, detecting regions of higher sequence and functional conservation and other more variable regions in which auxiliary gene functions could accumulate. Dedicated databases now exist which try to organize the information on the different integrative islands (Mantri & Williams, 2004) and even on mobile genetic elements in general (Leplae *et al.*, 2004). A picture emerges of GEIs as a diverse family of genetic elements with a large variability in functional content, some of which are capable of conjugative self-transfer.

Here we will review in a bit more detail what is currently known on the structure, mode of evolution and 'life-style' of GEIs. Particular attention will be given to GEIs with catabolic gene functions, such as the *clc* element, which was discovered in the bacterium *Pseudomonas* sp. strain B13. This GEI was named after the presence of the *clc* genes, which enable the host bacterium to use chlorinated catechols as growth intermediates.

STRUCTURE OF GENOMIC ISLANDS

GEIs are genetic entities that range in size from about 15 kb (so-called 'islets'; rarely mobile) up to 500 kb (e.g. the symbiosis island in *Mesorhizobium loti*) (Sullivan & Ronson, 1998; Sullivan *et al.*, 2002). The definition of a GEI is somewhat diffuse, but GEIs are named as such if they contain any of the following components: integration site and site duplication, integrase, plasmid or phage-like functions (Dobrindt *et al.*, 2004). It should be noted that GEIs can be defective or missing some of these components. The most complete GEI is probably one that can excise and integrate into one or a few specific chromosomal sites, can mobilize itself to another recipient bacterium and contains recognizable auxiliary functions, such as pathogenicity, host-interaction or carbon-compound catabolism (Fig. 1). For this reason, GEIs are considered by some authors to be part of a larger group of conjugative and integrative elements, which would also include conjugative transposons and integrative plasmids (Burrus *et al.*, 2002).

For reasons that are only partially understood, GEIs are often (but not necessarily) inserted in the 3′ end of tRNA genes (Ravatn *et al.*, 1998b; Sullivan & Ronson, 1998; Larbig *et al.*, 2002; Williams, 2002). Insertion into the integration site is catalysed by site-specific phage-like recombinases called *integrases*, which are usually encoded

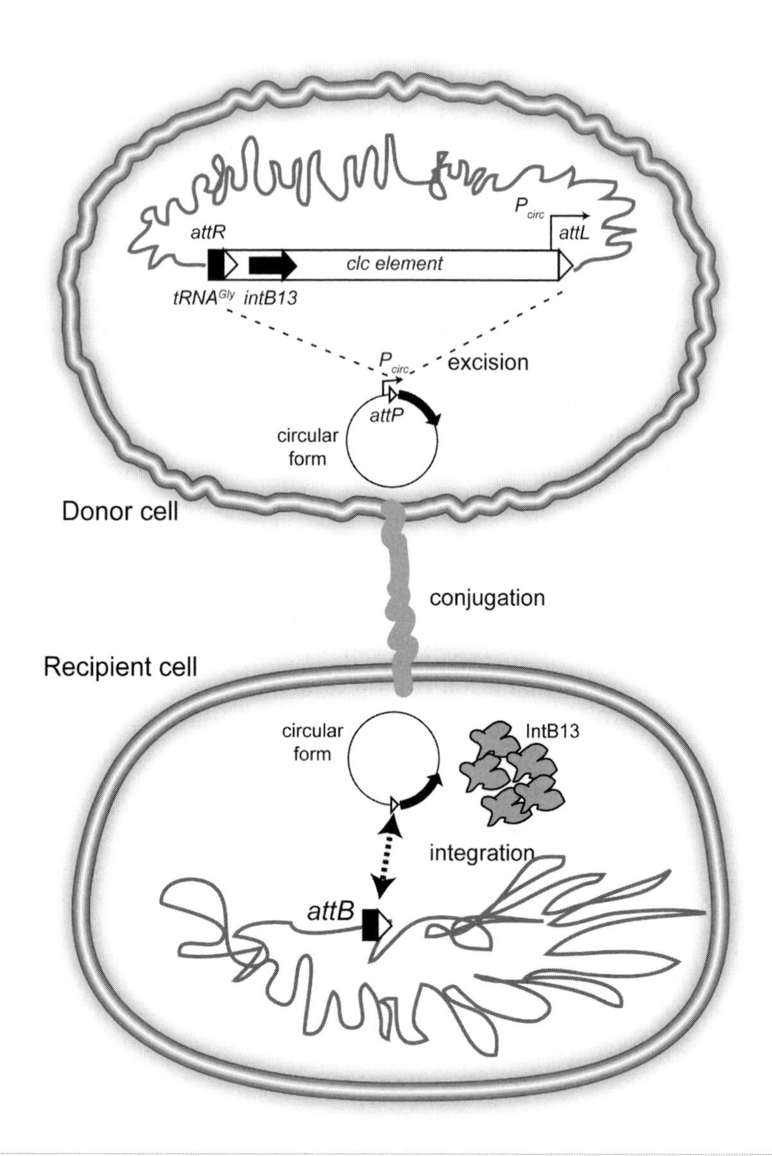

Fig. 1. Schematic outline of the 'life-style' of the *clc* element of *Pseudomonas* sp. strain B13 ('donor'). Only one integrated copy is drawn for simplicity, within borders of the gene for $tRNA^{Gly}$ (*attR*) and the 18 bp duplication (*attL*). The position of the integrase gene is indicated, as well as that of the P*circ* promoter, which faces out in the integrated form but drives integrase expression in the excised form. The excised form, in which the *attR* and *attL* ends are recombined (*attP*), is presumed to be the source for single-stranded conjugative transfer. In a new cell ('recipient'), the episomal form is replicated and reintegrates when a suitable integration site (*attB*, gene for $tRNA^{Gly}$) is present.

by the GEI itself. Integrases are not strictly conserved among all known GEIs, and various different tRNA genes can be used (Williams, 2002). Several integrases from GEIs relate to the lambda, P4 or XerD families (van der Meer *et al.*, 2001; Mohd-Zain *et al.*, 2004). The integrases are also implicated in the excision process of GEIs, where they are assisted by at least an auxiliary excisionase protein often called Xis (Nunes-Düby *et al.*, 1998; Burrus & Waldor, 2003; Lesic *et al.*, 2004). The *int* gene for the integrase is often situated at one extremity of the island, and adjacent to the tRNA gene in the integrated GEI form (*attR*; Fig. 1). This can have a profound effect on the regulation of the life-style of the element (see below). Integrases recognize specific motifs, 15–20 nucleotides long, which are also called directed repeats since they flank the GEI in its integrated chromosomal form (Hochhut & Waldor, 1999; van der Meer *et al.*, 2001; Williams, 2002). After excision, the two GEI ends form a closed link with a single copy of the recombination site (*attP*). Excision is mostly non-replicative and a single copy of the recombination site is again formed on the chromosome (*attB*) (Burrus & Waldor, 2003; V. Sentchilo and J. R. van der Meer, unpublished). The GEI can reintegrate from its excised form into the *attB* site or, when it is capable of self-transfer or can be mobilized by a co-residing GEI or phage, can be transferred into a new recipient cell. If that recipient carries a suitable *attB* site, the GEI can integrate into it. Integration is the result of a site-specific recombination between a 15–20 bp motif within the 3′ extremity of the tRNA gene (Ravatn *et al.*, 1998c; Hochhut & Waldor, 1999; Williams, 2002). For a few islands, the circular excised DNA molecule in which the two ends are connected has been detected by PCR using divergent primers annealing at the extremities of the island (Ravatn *et al.*, 1998c; Burrus & Waldor, 2003; Lesic *et al.*, 2004; Schubert *et al.*, 2004). Otherwise, it remains very difficult to isolate such an episomal molecule, which suggests that excised GEIs don't replicate independently from the host chromosome but rely on reintegration or lateral transfer to a new host to proliferate.

GEIs are mostly classified by the most prominent functional characteristic that they carry. As we have seen, PAIs contain genes that potentially provide the pathogenic character to the host and symbiosis islands enable infection of the host plant roots, whereas catabolic islands supply gene functions that are potentially important for degradation of carbon compounds (Hacker & Carniel, 2001; Burrus *et al.*, 2002; van der Meer & Sentchilo, 2003; Dobrindt *et al.*, 2004). In addition, GEIs carry one or more components devoted to the life-style of the element: integrase, transfer functions and/or regulatory elements, although it is not unusual to find fragmented GEIs and those lacking one or more of the 'life-style' functions. Some attempts have been made to define the minimal GEI module on the basis of genometric comparisons, with the result being some 33 identifiable core genes (Mohd-Zain *et al.*, 2004). Some GEIs have transfer functions clearly related to the relaxosome and type IV secretion system of conjugative plasmids (*tra* and *vir* genes) (Beaber *et al.*, 2002; Toussaint *et al.*, 2003; Greub *et al.*, 2004). Thus, these GEIs may have been the result of an ancient fusion between prophages and conjugative plasmids (Osborn & Boltner, 2002).

PECULIARITIES OF THE *clc* ELEMENT

The *clc* element is a GEI that would be classified as a 'catabolic' or 'ecological' island. Among other things, it harbours the *clcRABDE* cluster of genes that encode the inducible chlorocatechol oxidative pathway that enables its host to use 3-chlorobenzoate (3CBA) via 3- and 4-chlorocatechol (Dorn *et al.*, 1974; Frantz & Chakrabarty, 1987). This catabolic pathway was identified some 30 years ago in *Pseudomonas putida* pAC27 (Chatterjee *et al.*, 1981) and in the isolate *Pseudomonas* sp. strain B13 (Schmidt & Knackmuss, 1980). Different research groups observed that the *clc* genes could be transferred by bacterial matings into new recipients (Oltmanns *et al.*, 1988; Ravatn *et al.*, 1998a; Springael *et al.*, 2002). For that reason, they were thought to be encoded by a 110 kb conjugative plasmid, although this was never consistently isolated (Weisshaar *et al.*, 1987). In the late 1990s, we demonstrated by pulsed-field gel electrophoresis and Southern hybridization of genomic DNA from strain B13 and various transconjugants of *P. putida* F1 that the *clc* genes were actually present in the chromosome on a 105 kb mobile region (Ravatn *et al.*, 1998b). This raised the hypothesis that the *clc* genes were part of an integrative and conjugative element, perhaps an integrative plasmid, that enabled the transfer of the *clc* catabolic genes. Later, after analysing several different recipients of the *clc* element (Ravatn *et al.*, 1998c), it was discovered that the element occupies a specific integration site consisting of the 3′-end of a gene for $tRNA^{Gly}$ (see below). A distantly related bacteriophage P4-type integrase gene was situated close to the integration site (Ravatn *et al.*, 1998c). Finally, the integration process between the 3′ 18 bp of the $tRNA^{Gly}$ gene (*attB*) and a homologous fragment of the *clc* element (*attP*) mediated by the IntB13 integrase protein could be isolated and reproduced in *E. coli* (Ravatn *et al.*, 1998c). From this experiment, it was concluded that the *clc* element is not an integrative plasmid, but a conjugative GEI (van der Meer *et al.*, 2001).

To discover the nature of the other genes contained on the *clc* element and perhaps clarify the nature of the conjugative process, the complete sequence of the *clc* element was determined (GenBank accession no. AJ617740; M. Gaillard, T. Vallaeys, F.-J. Vorhoelter, C. Werlen, V. Sentchilo, A. Pühler and J. R. van der Meer, unpublished). The *clc* element has a total size of almost 103 kbp and was predicted to contain at least 107 open reading frames (ORFs). In addition to the *clc* gene cluster, two other potential catabolic pathways were encoded on the *clc* element. One of these might enable the cells to degrade 2-aminophenol, whereas the other might be important for oxidation of salicylate or anthranilate to catechol (M. Gaillard, T. Vallaeys, F.-J. Vorhoelter, C. Werlen, V. Sentchilo, A. Pühler and J. R. van der Meer, unpublished). Most strikingly, a region of about 50 kb of the *clc* element returned very little functional prediction, although the ORFs in this area were very highly conserved with regions from other bacterial genomes (with the result of a somewhat disappointing annotation as 'conserved hypothetical protein'). Statistically significant sequence similarity to protein motifs from the *tra*- and *vir*-type conjugative machinery was detected for only for a few ORFs (see below).

clc ELEMENT EXCISION AND INTEGRATION

As described above, the *clc* element inserts in a site-specific manner at the 3′ end of a $tRNA^{Gly}$ gene. Since most proteobacteria harbour three $tRNA^{Gly}$ genes, it was possible to find more than one integrated copy of the *clc* element in the recipient chromosome (Ravatn *et al.*, 1998b; Springael *et al.*, 2002). Strain B13 itself has two integrated copies, whereas transconjugants of *Ralstonia*, *Burkholderia* and various pseudomonads exhibited one, two or more copies (Ravatn *et al.*, 1998b; Springael *et al.*, 2002). By comparing the integration sites for the *clc* element in different host bacteria, it was found that a specific 18 bp motif (5′-GTCTCGTTTCCCGCTCCA-3′) was required for integration to take place and probably also a hairpin structure immediately downstream of the $tRNA^{Gly}$ gene (van der Meer *et al.*, 2001). This hairpin sequence has distinct but non-perfect sequence similarity to a similar region on the *clc* element itself and may therefore be important for the selection of the target site by the element through hybridization. The large inverted repeat may also have a functional role as a Rho-independent terminator, preventing high transcription activity coming from the $tRNA^{Gly}$ gene. No genetic evidence has yet been provided to elucidate the role of this DNA feature in integration of the *clc* element. However, its conservation in the different insertion sites argues for a specific involvement in the integration process. Perhaps the absence of an inverted repeat sequence downstream of the $tRNA^{Gly}$ of *E. coli* is the reason why it has not yet been possible to transfer the *clc* element into this species (S. Lacour, unpublished). It is also worth noticing that the *intB13* ORF is preceded by a 200 bp untranslated region which could also be involved in fine-tuning integrase accumulation, most probably by post-transcriptional modulation or RNA modulation (similarly to examples given by Johansson & Cossart, 2003). Although the integrase is essential for integration (Ravatn *et al.*, 1998c), it may also be involved in the excision processes. However, this has not yet been demonstrated, and the excisionase of the *clc* element is unfortunately not known. Excision of the *clc* element leads to a closed 'circular' intermediate DNA molecule in which the two ends (*attR* and *attL*) are closed up. This intermediate form can be detected by PCR (Ravatn *et al.*, 1998b) or Southern hybridization (Sentchilo *et al.*, 2003a). By PCR, it can also be shown that the chromosomal insertion site ligates after excision (V. Sentchilo and J. R. van der Meer, unpublished). Judging from quantitative Southern hybridizations, between 5 and 7 % of the *clc* elements in a population of strain B13 are present in the excised form (Sentchilo *et al.*, 2003b).

The process of excision and integration is subject to relatively tight regulation, as far as is currently understood. Expression of the integrase in the excised form is driven by a strong constitutive promoter present at and facing outward from the left end (*attL*) of the *clc* element (Sentchilo *et al.*, 2003a). By joining the two ends of the integrated form together in the circular excised form, this promoter is placed in front of the integrase gene. This promoter (named P*circ*) has σ^{70}-like promoter elements (so called –35 and –10 hexamers) and is constitutively active. This was demonstrated by placing the junction DNA in front of a promoterless *gfp* reporter gene (Sentchilo *et al.*, 2003a). The consequence of the strong promoter driving integrase expression in the circular excised form may be a temporary overexpression of integrase protein and an increased likelihood of reintegration. From the inability to isolate the circular form in larger quantities and because of the strong promoter in the circular form driving integrase expression, one might conclude that the integrated form is the preferred form for the *clc* element. Despite the absence of experimental data and of an identified origin of transfer, we assume that the element is transferred from the excised molecule as a single-stranded DNA molecule in the same way as conjugative plasmids. Loss of the element from strain B13 (as a consequence of conjugative transfer) has never been reported.

By assuming that integrase activity is also needed for excision, it becomes 'logical' that integrase expression is turned down in the integrated form to prevent loss of the element. By using primer extension analysis on successive deletions of the region upstream of the *int* gene and by promoter gene fusions to the *gfp* gene, another promoter region (called P*int*) was identified which lies between 70 and 150 bp upstream of the *intB13* start codon (Sentchilo *et al.*, 2003a). Because of the low level of mRNA transcription originating from this promoter, the promoter elements could not be precisely identified. However, there are two pairs of putative –35 and –10 boxes in this region (TTGAAA/TTTTTT and CGGAAA/TTTTTT) and the corresponding DNA fragments are bound by purified *E. coli* RNA polymerase (S. Lacour and K. Globig, unpublished). By using a P*int*: : *gfp* fusion, it was found that the P*int* promoter is a very weak promoter which leads to GFP expression in a very small proportion of the bacterial population when assayed in *Pseudomonas* sp. strain B13 (about 10 % for cultures in stationary phase) (Sentchilo *et al.*, 2003b). From promoter deletion studies and complementation analysis with plasmids containing parts of the left side of the *clc* element, it was concluded that integrase expression in the integrated form must be transcriptionally controlled. Mutants were isolated in genes at the left end of the *clc* element that up- or downregulated *gfp* expression from P*int*: : *gfp* fusions (Sentchilo *et al.*, 2003a). This suggested that a balance between two counteracting regulatory proteins might be responsible for controlling P*int* expression. One of these presumed factors, the *inrR* gene (ORF94689), was studied a little further, but its mode of action has not yet been elucidated. InrR does not share any sequence similarity with known transcriptional regulators and does not possess a DNA-binding motif. Purified His-tagged InrR did not bind the P*int* promoter DNA, suggesting that it is not a classical transcription activator (S. Lacour and K. Globig, unpublished).

The single-copy chromosomal P*int*–*gfp* reporter system in strain B13 was also used to study the conditions under which the *clc* element excises (Sentchilo *et al.*, 2003b). Although only a small proportion of cells in a B13 population induced GFP formation, it could be established that stationary phase conditions and previous growth on 3CBA resulted in a statistically significant increase in GFP expression from P*int*. In addition, the relative amount of excised form and the frequency of *clc* transfer in matings increased under those two conditions. It is curious that growth on 3CBA resulted in an enhanced frequency of excision and *clc* conjugation, since metabolism of 3CBA through chlorocatechols is only made possible by the presence of *clc*. On the other hand, cell density, UV irradiation and toxic or heat stress did not result in enhanced *clc* element excision, thus making an SOS-type phage response as was observed for the SXT element less likely (Beaber *et al.*, 2004). More recently, we have discovered that the same conditions also stimulate expression of *clc* element genes that are possibly implicated in unwinding the double-stranded circular form to the transferably active form (M. Gaillard and J. R. van der Meer, unpublished). However, the mechanism for enhanced induction by 3CBA is not understood, although it has become clear that 3CBA itself is not acting as some secret signalling molecule or effector. Interestingly, the presence of 3CBA in model membrane biofilm reactors seeded with another *clc* element donor strain, *P. putida* BN210, very rapidly selected for the appearance of spontaneous new recipients with the *clc* element and displacement of the donor strain (Springael *et al.*, 2002). This suggests that environmental pollutants, direct or indirect, can influence horizontal gene transfer rates.

GENOMIC ISLANDS STRONGLY RELATED TO THE *clc* ELEMENT

One might have argued that finding the *clc* element once, and in a strain maintained for a long time in the laboratory (strain B13), does not necessarily indicate that GEIs are important carriers for catabolic gene functions and evolution. However, at least two very similar islands have been detected in other environmental isolates, and very clear relationships between the *clc* element and other putative GEIs exist. An element very similar to the *clc* element, detected in the genome of *Burkholderia xenovorans* strain LB400 (http://genome.jgi-psf.org/finished_microbes/burfu/burfu.home.html), shares an overall identity of 99 % with the *clc* element, except for two regions not present in the *clc* element (M. Gaillard, T. Vallaeys, F.-J. Vorhoelter, C. Werlen, V. Sentchilo, A. Pühler and J. R. van der Meer, unpublished). One of these is a 20 kb insert flanked by two copies of an IS element related to IS*Ppu12* and which carries the genes for an *ortho*-halobenzoate dioxygenase system similar to that of *Pseudomonas aeruginosa* strain

JB2 (Hickey & Sabat, 2001). The presence of this region suggests that the Ppu12 insertion element mobilized a distinct set of genes into an ancestor *clc* element present in *Burkholderia*, which could potentially enlarge the range of metabolizable compounds in this organism. The *ortho*-halobenzoate dioxygenase system has been implicated in growth on 2-chlorobenzoate in *P. aeruginosa* 142 (Tsoi *et al.*, 1999), but metabolism of 2-chlorobenzoate in *B. xenovorans* is disputed (Maltseva *et al.*, 1999). The other natural variant was found in *Ralstonia* sp. strain JS705, which was isolated from a polluted groundwater aquifer (van der Meer *et al.*, 1998; Müller *et al.*, 2003). In this strain, an element similar to *clc* exists, which has captured an additional 10 kb region containing genes for a toluene and chlorobenzene multicomponent ring dioxygenase and *cis*-dihydrodiol dehydrogenase. The activity of these enzymes enables the host bacterium to metabolize monochloro- and 1,4-dichlorobenzene to chlorocatechols, which are further converted by the *clc* gene-encoded enzymes, likewise present on the *clc* element. It could also be shown, by a mating between strain JS705 and the recipient strain *Ralstonia eutropha* JMP289 and selection for the capacity to degrade chlorobenzenes (Müller *et al.*, 2003), that this element is self-transferable. Thus, one can conclude that the *clc* element is not just a singular unique finding, but is more widespread in the natural environment and is capable of acquiring additional DNA. This also indicates that catabolic islands are as dynamic as, for example, catabolic plasmids and contribute to the adaptation and evolution of catabolic properties in environmental bacteria (van der Meer, 2006).

In addition to having two highly related GEI partners, the *clc* element was found to be closely related to various PAIs in *P. aeruginosa* (PAGI-2, -3, -4 and PAPI-1) (Larbig *et al.*, 2002; Klockgether *et al.*, 2004), to genomic regions in the plant pathogen *Xylella fastidiosa* 9a5c (Nunes *et al.*, 2003), in *Xanthomonas campestris* and in *Azoarcus* (Rabus *et al.*, 2005) and to an unidentified GEI in *Ralstonia metallidurans* strain CH34 (Fig. 2). Interestingly, and in contrast to the elements in *B. xenovorans* and strain JS705, the relationships between *clc* and the other GEIs or genomic regions were confined to about half of the *clc* element, which did not contain the catabolic genes. Nucleotide similarities between this region of the *clc* element (plus the integrase) and the other genomic regions vary from roughly 70 to 95 % (Mohd-Zain *et al.*, 2004). As discussed further below, this large conserved region of nearly 50 kb might contain the DNA transfer functions and, thus, may also be present and perhaps functional in other related GEIs. Nevertheless, large-scale comparisons reveal several gaps in this region, most notably between the *clc* element and PAGI-4 and PAPI-1 from *P. aeruginosa* (Fig. 2).

Such comparisons further suggest that the 'relatives' of the *clc* element have a modular structure, consisting of the integrase gene at one end, a variable region in between and

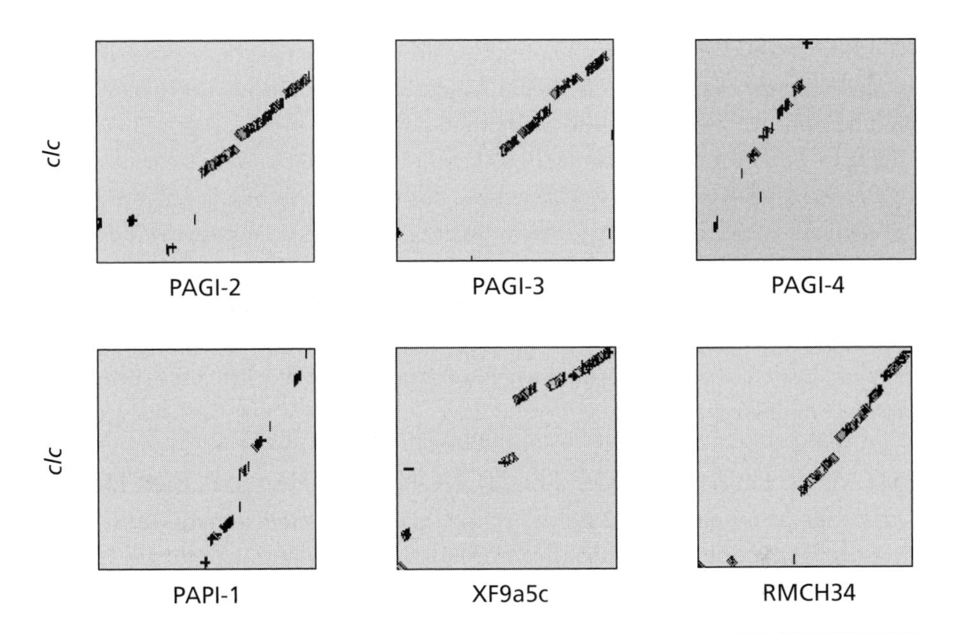

Fig. 2. Global analysis of relatedness between the *clc* element and other recognized GEIs or genomic regions, performed by BLAST 2 analysis (Tatusova & Madden, 1999). Default parameters were used for the comparisons. Diagrams show similarity hits (crosses), with the *clc* element sequence (~105 kb region) on the *y*-axis and the corresponding region of the targets (~100–150 kb elements) on the *x*-axis. Targets included *P. aeruginosa* GEIs PAGI-2 (GenBank accession no. AF440523), PAGI-3 (AF440524), PAGI-4 (AY258138) and PAPI-1 (AY273869), *X. fastidiosa* strain 9a5c (XF9a5c; NC_002488) and *Ralstonia metallidurans* strain CH34 (RMCH34; NZ_AAAI00000000).

some 50 kb at the other end, possibly involved in conjugative transfer. In a recent report, Mohd-Zain *et al.* (2004) have concatenated and aligned 15 GEIs and genome regions on the basis of similarities to a GEI in *Haemophilus influenzae* (ICE*Hin1056*). All these (potential) GEIs originated in beta- and gammaproteobacteria, such as *Haemophilus*, *Salmonella*, *P. aeruginosa*, *Pseudomonas fluorescens*, *Yersinia* and *Ralstonia*, including the *clc* element. All except one contained an integrase gene, either from the P4 family or from the XerD family. Based on phylogeny of the *int* gene, three major lineages were found, grouping the *int* from the *clc* element together with PAGI-3 and 4 and with the *X. fastidiosa* element. From detailed comparison of all the other component genes, Mohd-Zain *et al.* (2004) suggested that two contiguous regions of 13 and 20 genes could be found among all the elements, forming a virtual core island of approximately 18 kb. The two regions in the various GEIs were interrupted by other clusters of genes, however. Fifteen of the genes were conserved in all the GEIs and were organized in the same order, forming a congruent structure (with a few minor exceptions). This remarkable synteny suggests that the diverse GEIs shared a common evolutionary origin despite the fact that amino acid sequence similarities were found to be as low as 20–25 % for the conserved genes and despite the high divergence of fitness genes in islands with sizes ranging from 49 to 140 kb. This region of conserved genes had been recognized earlier (Sentchilo *et al.*, 2003a). Some of the homologues of the core genes have been recognized to play roles in plasmid replication (Gerdes *et al.*, 2000) or in conjugative DNA transfer (Adamczyk & Jagura-Burdzy, 2003). As for the rest of the GEI contents, a broad variety of functions was detected, including antibiotic and antiseptic resistances, type IV secretion systems, signalling systems, toxin production and a wide range of metabolic and catabolic functions. These results suggest again that the GEIs are not evolutionary 'novelties' per se, because such a widespread distribution and divergence already exists, and that they themselves change by incorporating and deleting discrete clusters of genes. Much of this concept has been recognized for plasmid evolution as well.

PUTATIVE DNA TRANSFER FUNCTIONS OF THE *clc* ELEMENT

One of the major questions with regard to GEIs is their mode of (suspected) self-transfer. For most of the GEIs detected by genome sequencing and database searches, no experimental evidence exists at present for self-transfer, with the exception of the *clc* element, ICE*Hin1056* and SXT of *Vibrio cholerae*. From the *clc* sequence information, it is possible to conclude (by exclusion) that the 3′ half of the *clc* element (starting around position 46777) must encode functions necessary for the conjugative transfer of the island. We deduce this from the fact that the *clc* element can transfer from strain B13 to other hosts and from those again to others, even in hosts (such as *R. eutropha*) which themselves do not encode conjugative proteins. However, a role for certain host proteins in the transfer process cannot be excluded without further experimental evidence. In addition, the role of proteins in determining the host range for successful integration is unknown. As mentioned earlier, 70 % of this region of the *clc* element is composed of ORFs with a hypothetical function because of low sequence similarity to known proteins. The fact that half of the putative genes are highly conserved in other, related GEIs would argue for an essential role of this region for the life-style of this type of island (M. Gaillard, T. Vallaeys, F.-J. Vorhoelter, C. Werlen, V. Sentchilo, A. Pühler and J. R. van der Meer, unpublished; Mohd-Zain *et al.*, 2004).

Among the numerous hypothetical proteins are a few predicted membrane-associated (or possibly periplasmic) proteins which may constitute components of uncharacterized export, secretion or transduction systems. At least four ORFs show evidence for putative secretion signals and might be exported outside the host cell. Two ORFs encode proteins which share significant but low sequence similarity with phage-related proteins and at least six others with plasmid proteins, four being homologous to *tra* and two to *par* genes. The predicted polypeptides of six ORFs, listed in Table 1, hint at a role in conjugative transfer of the *clc* element. On the other hand, it was impossible to identify a full group of *tra* or *vir* genes homologous to conjugative DNA transfer

Table 1. Suspected conjugative functions encoded on the *clc* element

ORFs are numbered according to *clc* element ORF numbering given in GenBank accession no. AJ617740.

ORF	Homologue(s)	Conserved domain	Putative function
59888	VirB4	COG3451 VirB4	Energy generation through NTP hydrolysis
68987	TraG, TrwB, VirD4	Pfam02534 TraG/TraD family, COG3505 VirD4	Coupling protein, potential pump
73676	PilL	None	Lipoprotein; pilus constituent
75419	DNA helicase	HELICc (helicase superfamily), DEXH-box (ATP binding)	Unwinding of DNA in relaxosome
91884	TraE	COG0550 TOP1Ac (DNA and ATP binding) and TOPRIM (nucleotidyl transferase)	DNA topoisomerase, acting at *oriT*
94175	SSB (PriB)	COG0629, SSB	Single-stranded DNA-binding protein; replication
98147	ParA	Pfam0991 (ATPase domain)	Regulatory role in conjugation
100033	ParBc	Pfam02195 (nuclease domain)	Regulatory role in conjugation

functions. This suggests that the conjugation system encoded by the *clc* element is at best very distantly related to the known relaxosome and mating-pair formation systems of the *tra*, *trw*, *trb* and *vir* types (Gomis-Rüth *et al.*, 2002; Adamczyk & Jagura-Burdzy, 2003). Interestingly, some GEIs encode clearly identifiable *tra*-related conjugative systems, e.g. SXT (Beaber *et al.*, 2002) and CTn*4371*, the biphenyl transposon of *Cupriavidus oxalatica* (Toussaint *et al.*, 2003), which might point to a plasmid origin in the distant past (Osborn & Boltner, 2002). When assuming that the *clc* element, after initial excision to a circular intermediate, would behave like a conjugative plasmid of which only a single-stranded DNA is transferred, one could 'fit' the possible function of those hypothetical proteins. The DNA topoisomerase (putative peptide from ORF91884) would be needed to nick the double strand at *oriT*, whereas the helicase (ORF75419) could be part of a large protein complex called the relaxosome and unwind the DNA to single strands, to which single-stranded binding protein (SSB; ORF94175) could attach prior to its transfer. The DNA–protein complex could be presented to a conjugation machinery, during which the TraG (TrwB, VirD4) homologue (ORF68987) might act as a coupling and transport protein (Tato *et al.*, 2005) and the VirB4 homologue (ORF59888) might provide energy. The predicted peptides from both ORF68987 and 59888 were found to possess domains for NTP or ATP hydrolysis which could energize DNA transport. Recent studies revealed that VirB4, the largest component of the *Agrobacterium* T-DNA transporter, forms a hexameric complex and acts as a docking site at the entrance of the secretion system (Middleton *et al.*, 2005). In the studied models, it acts in concert with VirD4 and VirB11 to transport substrates through the channel (Tato *et al*., 2005), but those components were not identified in the *clc* island. The PilL lipoprotein (encoded by ORF73676), whose precise function is not yet clear, could also be one of the anchor proteins of the mosaic conjugative pilus encoded by the *clc* element. The *pilL* gene is always present in the *pil* set of genes found in *Yersinia* or *Salmonella* PAIs and is conserved in various GEIs related to the *clc* element (Mohd-Zain *et al*., 2004). We assume so far that the evidence for a conjugative system on the *clc* element is strong enough, but the nature of this system may be different from the known type IV secretion systems and relaxosomes. Finally, the InrR protein, for integrase regulator (Sentchilo *et al*., 2003a), is also present in all the GEIs related to the *clc* element and therefore is thought to be essential for regulation of the excision process, eventually coupled to the expression of the DNA transfer system.

GENOMIC ISLANDS AND EVOLUTION OF CATABOLIC PATHWAYS

The *clc* element is a conjugative catabolic island found in bacteria isolated from soil and sewage water. Analysis of the *clc* element has shown that self-transfer of catabolic gene functions and catabolic expansion through additional modular integration is not confined to the classical catabolic plasmids (Assinder & Williams, 1990; Sayler *et al*., 1990; van der Meer *et al*., 1992) or catabolic transposons (Tan, 1999). On the other hand, many more catabolic plasmids have been discovered than GEIs with catabolic properties, the current exceptions being the *clc* element and its relatives, the CTn*4371* biphenyl transposon (Toussaint *et al*., 2003) and further mobile elements that are suspected but not yet characterized in detail (Hickey *et al*., 2001; Nishi *et al*., 2000). Therefore, we currently have little idea of what might be the different selective advantages of maintaining catabolic pathways on plasmids or on GEIs.

ACKNOWLEDGEMENTS

The authors would like to thank the Swiss National Science Foundation and the Roche Research Foundation for their support to M. G. and S. L. (grants 3100-067229 and 230-2004, respectively).

REFERENCES

Adamczyk, M. & Jagura-Burdzy, G. (2003). Spread and survival of promiscuous IncP-1 plasmids. *Acta Biochim Pol* **50**, 425–453.

Assinder, S. J. & Williams, P. A. (1990). The TOL plasmids: determinants of the catabolism of toluene and the xylenes. *Adv Microb Physiol* **31**, 1–69.

Beaber, J. W., Hochhut, B. & Waldor, M. K. (2002). Genomic and functional analyses of SXT, an integrating antibiotic resistance gene transfer element derived from *Vibrio cholerae*. *J Bacteriol* **184**, 4259–4269.

Beaber, J. W., Hochhut, B. & Waldor, M. K. (2004). SOS response promotes horizontal dissemination of antibiotic resistance genes. *Nature* **427**, 72–74.

Burrus, V. & Waldor, M. K. (2003). Control of SXT integration and excision. *J Bacteriol* **185**, 5045–5054.

Burrus, V., Pavlovic, G., Decaris, B. & Guédon, G. (2002). Conjugative transposons: the tip of the iceberg. *Mol Microbiol* **46**, 601–610.

Chatterjee, D. K., Kellogg, S. T., Hamada, S. & Chakrabarty, A. M. (1981). Plasmid specifying total degradation of 3-chlorobenzoate by a modified *ortho* pathway. *J Bacteriol* **146**, 639–646.

Dobrindt, U., Hochhut, B., Hentschel, U. & Hacker, J. (2004). Genomic islands in pathogenic and environmental microorganisms. *Nat Rev Microbiol* **2**, 414–424.

Dogra, C., Raina, V., Pal, R., Suar, M., Lal, S., Gartemann, K.-H., Holliger, C., van der Meer, J. R. & Lal, R. (2004). Organization of *lin* genes and IS*6100* among different strains of hexachlorocyclohexane-degrading *Sphingomonas paucimobilis*: evidence for horizontal gene transfer. *J Bacteriol* **186**, 2225–2235.

Dorn, E., Hellwig, M., Reineke, W. & Knackmuss, H.-J. (1974). Isolation and characterization of a 3-chlorobenzoate degrading pseudomonad. *Arch Microbiol* **99**, 61–70.

Dufraigne, C., Fertil, B., Lespinats, S., Giron, A. & Deschavanne, P. (2005). Detection and characterization of horizontal transfers in prokaryotes using genomic signature. *Nucleic Acids Res* **33**, e6.

Frantz, B. & Chakrabarty, A. M. (1987). Organization and nucleotide sequence determination of a gene cluster involved in 3-chlorocatechol degradation. *Proc Natl Acad Sci U S A* **84**, 4460–4464.

Gerdes, K., Moller-Jensen, J. & Bugge Jensen, R. (2000). Plasmid and chromosome partitioning: surprises from phylogeny. *Mol Microbiol* **37**, 455–466.

Gomis-Rüth, F. X., de la Cruz, F. & Coll, M. (2002). Structure and role of coupling proteins in conjugal DNA transfer. *Res Microbiol* **153**, 199–204.

Greated, A., Lambertsen, L., Williams, P. A. & Thomas, C. M. (2002). Complete sequence of the IncP-9 TOL plasmid pWW0 from *Pseudomonas putida*. *Environ Microbiol* **4**, 856–871.

Greub, G., Collyn, F., Guy, L. & Roten, C. A. (2004). A genomic island present along the bacterial chromosome of the *Parachlamydiaceae* UWE25, an obligate amoebal endosymbiont, encodes a potentially functional F-like conjugative DNA transfer system. *BMC Microbiol* **4**, 48.

Groisman, E. A. & Ochman, H. (1996). Pathogenicity islands: bacterial evolution in quantum leaps. *Cell* **87**, 791–794.

Hacker, J. & Carniel, E. (2001). Ecological fitness, genomic islands and bacterial pathogenicity. A Darwinian view of the evolution of microbes. *EMBO Rep* **2**, 376–381.

Hacker, J., Bender, L., Ott, M., Wingender, J., Lund, B., Marre, R. & Goebel, W. (1990). Deletions of chromosomal regions coding for fimbriae and hemolysins occur in vitro and in vivo in various extraintestinal *Escherichia coli* isolates. *Microb Pathog* **8**, 213–225.

Harayama, S. & Rekik, M. (1993). Comparison of the nucleotide sequences of the *meta*-cleavage pathway genes of TOL plasmid pWW0 from *Pseudomonas putida* with other *meta*-cleavage genes suggests that both single and multiple nucleotide substitutions contribute to enzyme evolution. *Mol Gen Genet* **239**, 81–89.

Harayama, S., Rekik, M., Wasserfallen, A. & Bairoch, A. (1987). Evolutionary relationships between catabolic pathways for aromatics: conservation of gene order and nucleotide sequences of catechol oxidation genes of pWW0 and NAH7 plasmids. *Mol Gen Genet* **210**, 241–247.

Hickey, W. J. & Sabat, G. (2001). Integration of matrix-assisted laser desorption ionization-time of flight mass spectrometry and molecular cloning for the identification and functional characterization of mobile *ortho*-halobenzoate oxygenase genes in *Pseudomonas aeruginosa* strain JB2. *Appl Environ Microbiol* **67**, 5648–5655.

Hickey, W. J., Sabat, G., Yuroff, A. S., Arment, A. R. & Pérez-Lesher, J. (2001). Cloning, nucleotide sequencing, and functional analysis of a novel, mobile cluster of biodegradation genes from *Pseudomonas aeruginosa* strain JB2. *Appl Environ Microbiol* **67**, 4603–4609.

Hochhut, B. & Waldor, M. K. (1999). Site-specific integration of the conjugal *Vibrio cholerae* SXT element into *prfC*. *Mol Microbiol* **32**, 99–110.

Johansson, J. & Cossart, P. (2003). RNA-mediated control of virulence gene expression in bacterial pathogens. *Trends Microbiol* **11**, 280–285.

Klockgether, J., Reva, O., Larbig, K. & Tümmler, B. (2004). Sequence analysis of the mobile genome island pKLC102 of *Pseudomonas aeruginosa* C. *J Bacteriol* **186**, 518–534.

Knapp, S., Hacker, J., Jarchau, T. & Goebel, W. (1986). Large, unstable inserts in the chromosome affect virulence properties of uropathogenic *Escherichia coli* O6 strain 536. *J Bacteriol* **168**, 22–30.

Koonin, E. V. & Galperin, M. Y. (2003). *Sequence – Evolution – Function: Computational Approaches in Comparative Genomics*. Boston: Kluwer Academic.

Laemmli, C. M., Leveau, J. H. J., Zehnder, A. J. B. & van der Meer, J. R. (2000). Characterization of a second *tfd* gene cluster for chlorophenol and chlorocatechol metabolism on plasmid pJP4 in *Ralstonia eutropha* JMP134(pJP4). *J Bacteriol* **182**, 4165–4172.

Larbig, K. D., Christmann, A., Johann, A., Klockgether, J., Hartsch, T., Merkl, R., Wiehlmann, L., Fritz, H. J. & Tummler, B. (2002). Gene islands integrated into $tRNA^{Gly}$ genes confer genome diversity on a *Pseudomonas aeruginosa* clone. *J Bacteriol* **184**, 6665–6680.

Leplae, R., Hebrant, A., Wodak, S. J. & Toussaint, A. (2004). ACLAME: a CLAssification of Mobile genetic Elements. *Nucleic Acids Res* **32**, D45–D49.

Lesic, B., Bach, S., Ghigo, J. M., Dobrindt, U., Hacker, J. & Carniel, E. (2004). Excision of the high-pathogenicity island of *Yersinia pseudotuberculosis* requires the combined actions of its cognate integrase and Hef, a new recombination directionality factor. *Mol Microbiol* **52**, 1337–1348.

Maltseva, O. V., Tsoi, T. V., Quensen, J. F., III, Fukuda, M. & Tiedje, J. M. (1999). Degradation of anaerobic reductive dechlorination products of Aroclor 1242 by four aerobic bacteria. *Biodegradation* **10**, 363–371.

Mantri, Y. & Williams, K. P. (2004). Islander: a database of integrative islands in prokaryotic genomes, the associated integrases and their DNA site specificities. *Nucleic Acids Res* **32**, D55–D58.

Middleton, R., Sjolander, K., Krishnamurthy, N., Foley, J. & Zambryski, P. (2005). Predicted hexameric structure of the *Agrobacterium* VirB4 C terminus suggests VirB4 acts as a docking site during type IV secretion. *Proc Natl Acad Sci U S A* **102**, 1685–1690.

Mohd-Zain, Z., Turner, S. L., Cerdeño-Tárraga, A. M. & 7 other authors (2004). Transferable antibiotic resistance elements in *Haemophilus influenzae* share a common evolutionary origin with a diverse family of syntenic genomic islands. *J Bacteriol* **186**, 8114–8122.

Müller, T. A., Werlen, C., Spain, J. & van der Meer, J. R. (2003). Evolution of a chlorobenzene degradative pathway among bacteria in a contaminated groundwater mediated by a genomic island in *Ralstonia*. *Environ Microbiol* **5**, 163–173.

Müller, T. A., Byrde, S. M., Werlen, C., van der Meer, J. R. & Kohler, H.-P. E. (2004). Genetic analysis of phenoxyalkanoic acid degradation in *Sphingomonas herbicidovorans* MH. *Appl Environ Microbiol* **70**, 6066–6075.

Nishi, A., Tominaga, K. & Furukawa, K. (2000). A 90-kilobase conjugative chromosomal element coding for biphenyl and salicylate catabolism in *Pseudomonas putida* KF715. *J Bacteriol* **182**, 1949–1955.

Nunes, L. R., Rosato, Y. B., Muto, N. H. & 9 other authors (2003). Microarray analyses of *Xylella fastidiosa* provide evidence of coordinated transcription control of laterally transferred elements. *Genome Res* **13**, 570–578.

Nunes-Düby, S. E., Kwon, H. J., Tirumalai, R. S., Ellenberger, T. & Landy, A. (1998). Similarities and differences among 105 members of the Int family of site-specific recombinases. *Nucleic Acids Res* **26**, 391–406.

Ochman, H., Lawrence, J. G. & Groisman, E. A. (2000). Lateral gene transfer and the nature of bacterial innovation. *Nature* **405**, 299–304.

Oltmanns, R. H., Rast, H. G. & Reineke, W. (1988). Degradation of 1,4-dichlorobenzene by enriched and constructed bacteria. *Appl Microbiol Biotechnol* **28**, 609–616.

Osborn, A. M. & Boltner, D. (2002). When phage, plasmids, and transposons collide: genomic islands, and conjugative- and mobilizable-transposons as a mosaic continuum. *Plasmid* **48**, 202–212.

Perna, N. T., Plunkett, G., III, Burland, V. & 25 other authors (2001). Genome sequence of enterohaemorrhagic *Escherichia coli* O157 : H7. *Nature* **409**, 529–533.

Rabus, R., Kube, M., Hieder, J., Beck, A., Heitmann, K., Widdel, F. & Reinhardt, R. (2005). The genome sequence of an anaerobic aromatic-degrading denitrifying bacterium, strain EbN1. *Arch Microbiol* **183**, 27–36.

Ravatn, R., Zehnder, A. J. B. & van der Meer, J. R. (1998a). Low-frequency horizontal transfer of an element containing the chlorocatechol degradation genes from *Pseudomonas* sp. strain B13 to *Pseudomonas putida* F1 and to indigenous bacteria in laboratory-scale activated-sludge microcosms. *Appl Environ Microbiol* **64**, 2126–2132.

Ravatn, R., Studer, S., Springael, D., Zehnder, A. J. B. & van der Meer, J. R. (1998b). Chromosomal integration, tandem amplification, and deamplification in *Pseudomonas putida* F1 of a 105-kilobase genetic element containing the chlorocatechol degradative genes from *Pseudomonas* sp. strain B13. *J Bacteriol* **180**, 4360–4369.

Ravatn, R., Studer, S., Zehnder, A. J. B. & van der Meer, J. R. (1998c). Int-B13, an unusual site-specific recombinase of the bacteriophage P4 integrase family, is responsible for chromosomal insertion of the 105-kilobase *clc* element of *Pseudomonas* sp. strain B13. *J Bacteriol* **180**, 5505–5514.

Regeard, C., Maillard, J., Dufraigne, C., Deschavanne, P. & Holliger, C. (2005). Indications for acquisition of reductive dehalogenase genes through horizontal gene transfer by *Dehalococcoides ethenogenes* strain 195. *Appl Environ Microbiol* **71**, 2955–2961.

Sayler, G. S., Hooper, S. W., Layton, A. C. & King, J. M. H. (1990). Catabolic plasmids of environmental and ecological significance. *Microb Ecol* **19**, 1–20.

Schmidt, E. & Knackmuss, H.-J. (1980). Chemical structure and biodegradability of halogenated aromatic compounds. Conversion of chlorinated muconic acids into maleoylacetic acid. *Biochem J* **192**, 339–347.

Schubert, S., Dufke, S., Sorsa, J. & Heesemann, J. (2004). A novel integrative and conjugative element (ICE) of *Escherichia coli*: the putative progenitor of the *Yersinia* high-pathogenicity island. *Mol Microbiol* **51**, 837–848.

Sentchilo, V. S., Zehnder, A. J. B. & van der Meer, J. R. (2003a). Characterization of two alternative promoters for integrase expression in the *clc* genomic island of *Pseudomonas* sp. strain B13. *Mol Microbiol* **49**, 93–104.

Sentchilo, V. S., Ravatn, R., Werlen, C., Zehnder, A. J. B. & van der Meer, J. R. (2003b). Unusual integrase gene expression on the *clc* genomic island in *Pseudomonas* sp. strain B13. *J Bacteriol* **185**, 4530–4538.

Springael, D., Peys, K., Ryngaert, A. & 8 other authors (2002). Community shifts in a seeded 3-chlorobenzoate degrading membrane biofilm reactor: indications for involvement of in situ horizontal transfer of the *clc*-element from inoculum to contaminant bacteria. *Environ Microbiol* **4**, 70–80.

Sullivan, J. T. & Ronson, C. W. (1998). Evolution of rhizobia by acquisition of a 500-kb symbiosis island that integrates into a phe-tRNA gene. *Proc Natl Acad Sci U S A* **95**, 5145–5149.

Sullivan, J. T., Trzebiatowski, J. R., Cruickshank, R. W. & 11 other authors (2002). Comparative sequence analysis of the symbiosis island of *Mesorhizobium loti* strain R7A. *J Bacteriol* **184**, 3086–3095.

Tan, H.-M. (1999). Bacterial catabolic transposons. *Appl Microbiol Biotechnol* **51**, 1–12.

Tato, I., Zunzunegui, S., de la Cruz, F. & Cabezon, E. (2005). TrwB, the coupling protein involved in DNA transport during bacterial conjugation, is a DNA-dependent ATPase. *Proc Natl Acad Sci U S A* **102**, 8156–8161.

Tatusova, T. A. & Madden, T. L. (1999). BLAST 2 Sequences, a new tool for comparing protein and nucleotide sequences. *FEMS Microbiol Lett* **174**, 247–250.

Thomas, C. M. (editor) (2000). *The Horizontal Gene Pool: Bacterial Plasmids and Gene Spread*. Amsterdam: Harwood Academic.

Toussaint, A., Merlin, C., Monchy, S., Benotmane, M. A., Leplae, R., Mergeay, M. & Springael, D. (2003). The biphenyl- and 4-chlorobiphenyl-catabolic transposon Tn*4371*, a member of a new family of genomic islands related to IncP and Ti plasmids. *Appl Environ Microbiol* **69**, 4837–4845.

Trefault, N., de la Iglesia, R., Molina, A. M., Manzano, M., Ledger, T., Pérez-Pantoja, D., Sanchez, M. A., Stuardo, M. & Gonzalez, B. (2004). Genetic organization of the catabolic plasmid pJP4 from *Ralstonia eutropha* JMP134 (pJP4) reveals mechanisms of adaptation to chloroaromatic pollutants and evolution of specialized chloroaromatic degradation pathways. *Environ Microbiol* **6**, 655–668.

Tsoi, T. V., Plotnikova, E. G., Cole, J. R., Guerin, W. F., Bagdasarian, M. & Tiedje, J. M. (1999). Cloning, expression, and nucleotide sequence of the *Pseudomonas aeruginosa* 142 *ohb* genes coding for oxygenolytic *ortho* dehalogenation of halobenzoates. *Appl Environ Microbiol* **65**, 2151–2162.

van der Meer, J. R. (1997). Evolution of novel metabolic pathways for the degradation of aromatic compounds. *Antonie van Leeuwenhoek* **71**, 159–178.

van der Meer, J. R. (2002). Evolution of metabolic pathways for degradation of environmental pollutants. In *Encyclopedia of Environmental Microbiology*, pp. 1194–1207. Edited by G. Bitton. New York: Wiley.

van der Meer, J. R. (2006). Evolution of catabolic pathways in *Pseudomonas* through gene transfer. In *Pseudomonas*, vol. 4. Edited by J. L. Ramos & R. C. Levesque. Dordrecht: Kluwer Academic (in press).

van der Meer, J. R. & Sentchilo, V. S. (2003). Genomic islands and the evolution of catabolic pathways in bacteria. *Curr Opin Biotechnol* **14**, 248–254.

van der Meer, J. R., Zehnder, A. J. B. & de Vos, W. M. (1991). Identification of a novel composite transposable element, Tn*5280*, carrying chlorobenzene dioxygenase genes of *Pseudomonas* sp. strain P51. *J Bacteriol* **173**, 7077–7083.

van der Meer, J. R., de Vos, W. M., Harayama, S. & Zehnder, A. J. B. (1992). Molecular mechanisms of genetic adaptation to xenobiotic compounds. *Microbiol Rev* **56**, 677–694.

van der Meer, J. R., Werlen, C., Nishino, S. & Spain, J. C. (1998). Evolution of a pathway for chlorobenzene metabolism leads to natural attenuation in a contaminated groundwater. *Appl Environ Microbiol* **64**, 4185–4193.

van der Meer, J. R., Ravatn, R. & Sentchilo, V. S. (2001). The *clc* element of *Pseudomonas* sp. strain B13 and other mobile degradative elements employing phage-like integrases. *Arch Microbiol* **175**, 79–85.

Weisshaar, M.-P., Franklin, F. C. H. & Reineke, W. (1987). Molecular cloning and expression of the 3-chlorobenzoate-degrading genes from *Pseudomonas* sp. strain B13. *J Bacteriol* **169**, 394–402.

Williams, K. P. (2002). Integration sites for genetic elements in prokaryotic tRNA and tmRNA genes: sublocation preference of integrase subfamilies. *Nucleic Acids Res* **30**, 866–875.

Horizontal gene transfer and its role in the emergence of new phenotypes

A. Mark Osborn

Department of Animal and Plant Sciences, The University of Sheffield, Western Bank, Sheffield S10 2TN, UK

INTRODUCTION

Introducing horizontal (or lateral) gene transfer (HGT) to first-year undergraduates used to be straightforward. A whistle-stop tour through the basic mechanisms of transformation, transduction and conjugation (with an honourable mention of homologous recombination) and descriptions of rough and smooth mutants of *Streptococcus pneumoniae*, the life cycles of phages lambda and T4 and the sexual proclivities of the 'F' factor was, until recently, little changed over the intervening years since their discovery. These processes (described in more detail below) also form the cornerstone of much of bacterial genetics, albeit from a predominantly *Escherichia coli*-centric perspective. This year celebrates the sixtieth anniversary of Joshua Lederburg and Edward Tatum's groundbreaking discovery of sexual transfer in bacteria (Lederburg & Tatum, 1946), yet to many the concept of HGT as a driving force in evolution has only come of age with the advent of comparative genomics (Kurland, 2000), which has revealed that far more is involved in HGT than plasmid-mediated conjugation. This review begins by providing an overview of the significance of HGT, defined here as 'the acquisition and stable maintenance of foreign DNA into the genome of a recipient cell' and introduces the key mobile genetic elements (MGEs) that are the agents of HGT. The roles of comparative genomics and bioinformatics and of the emerging field of metagenomics are discussed in relation to changes in perception of the significance of HGT in prokaryotic evolution, and the review concludes by highlighting key examples of recently recognized HGT-mediated phenotypes in the prokaryotic world.

SGM symposium 66: Prokaryotic diversity: mechanisms and significance.
Editors N. A. Logan, H. M. Lappin-Scott & P. C. F. Oyston. Cambridge University Press. ISBN 0 521 86935 8

THE SIGNIFICANCE OF HORIZONTAL GENE TRANSFER IN PROKARYOTIC ADAPTATION AND EVOLUTION

That HGT is rampant in the prokaryotic world is a message with which the scientific literature is awash, to such an extent that it is perhaps surprising to the untrained eye that distinct species of the *Bacteria* and *Archaea* can be identified. Yet a universal 'prokaryotome' of an all-powerful superbug exists only in the mind of newspaper headline writers, and prokaryotes do have their own specific identities and ecological niches. HGT provides significant challenges to the definition of a species concept in prokaryotes (Cohan, 2002), with Arber (1991) first suggesting that evolutionary trees should be drawn as networks instead of traditional phylogenies, to take account of the effects of HGT. Nevertheless, distinct species can be identified, although in some cases, such as the closely related *Bacillus cereus* and *Bacillus anthracis*, the major discriminatory features are plasmid-encoded functions (Ivanova *et al.*, 2003). That HGT has not prevented the evolution of distinct species is a consequence of the physical transfer of DNA not being the only criterion for evolution to occur. Additionally, there are clear barriers that limit the potential for DNA transfer to occur both between individual species and between more diverse taxonomic lineages, in particular due to differing cell-wall structures in Gram-positive and Gram-negative bacteria, although it is recognized that some broad-host-range plasmids, e.g. the IncP plasmid RK2 and other MGEs including the conjugative transposon Tn*916*, can cross this divide and, indeed, that gene transfer from a range of bacteria including *Agrobacterium*, *Rhizobium* and *Mesorhizobium* to plants can also occur (Broothaerts *et al.*, 2005). Barriers to transfer can be considered at four levels: (i) entry into the transfer process by the donor, (ii) selection of the recipient, (iii) uptake of DNA by the recipient and finally (iv) successful establishment of the incoming DNA within the recipient. These processes are reviewed in more detail by Thomas & Nielsen (2005), but it is highlighted here that, at each stage of the transfer process, bacteria have evolved mechanisms that can limit the successful transfer of DNA. Perhaps of most importance are systems that actively select against incoming DNA either by surface exclusion of DNA or by restriction following uptake. Even if such systems can be overcome for DNA which is to be integrated into the recipient chromosome, there will be an additional requirement for homologous or site-specific recombination. Self-replicating plasmids on the other hand may be limited in their host ranges either by a requirement for host-encoded factors and/or by competition with plasmids that are already present in the cell, in particular as a consequence of plasmid incompatibility (Novick, 1987). Together, such mechanisms serve to limit the extent of HGT in prokaryotes.

In addition to such controls upon transfer and maintenance, it is apparent that genes differ in their frequency of transfer. Some genes, such as those residing on plasmids and other MGEs, and which typically confer obvious adaptive phenotypes on recipient cells,

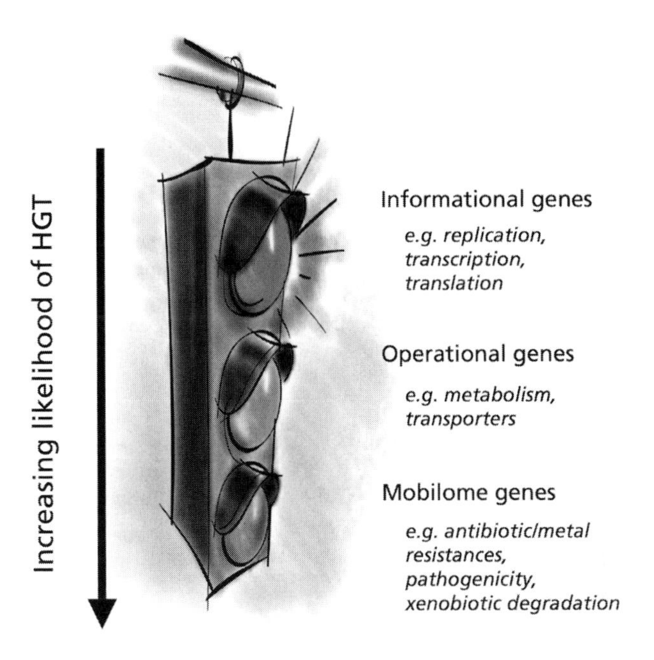

Fig. 1. Differential mobility of genes via HGT.

can be considered as members of a 'mobilome' (Fig. 1). In contrast, genes conferring key informational content within the genome, e.g. encoding DNA and RNA processing functions, appear to be less susceptible to DNA transfer (Fig. 1) and have been proposed by Peter Young of the University of York to represent a "Stay at 'ome". This assumed reduction in transfer of such informational genes, e.g. of the rRNA genes, underpins our current molecular phylogenies, in that such sequences serve as taxonomic markers which are not subject to HGT, although an increasing number of examples of the transfer of 16S rRNA genes have been reported (see Schouls *et al.*, 2003 and references therein). In between the extremes of the 'mobilome' and the 'stayatome', we find an increasing number of chromosomally located operational genes which are susceptible to transfer, in particular via transformation and transduction, but that are not as widely disseminated as those genes typically located on plasmids and other MGEs. Recent bioinformatics analysis of 116 prokaryotic genomes by Nakamura *et al.* (2004) has further suggested such gene transfer biases exist, in which they suggest that genes (excluding those carried directly on MGEs) that are involved in determining DNA-binding or pathogenicity or which confer functions linked to the cell envelope (e.g. encoding fimbriae or biosynthesis/degradation of surface polysaccharides) are most frequently transferred between bacteria. Fig. 1 shows an analogy for such variation in increasing likelihood of transfer in terms of a traffic light system.

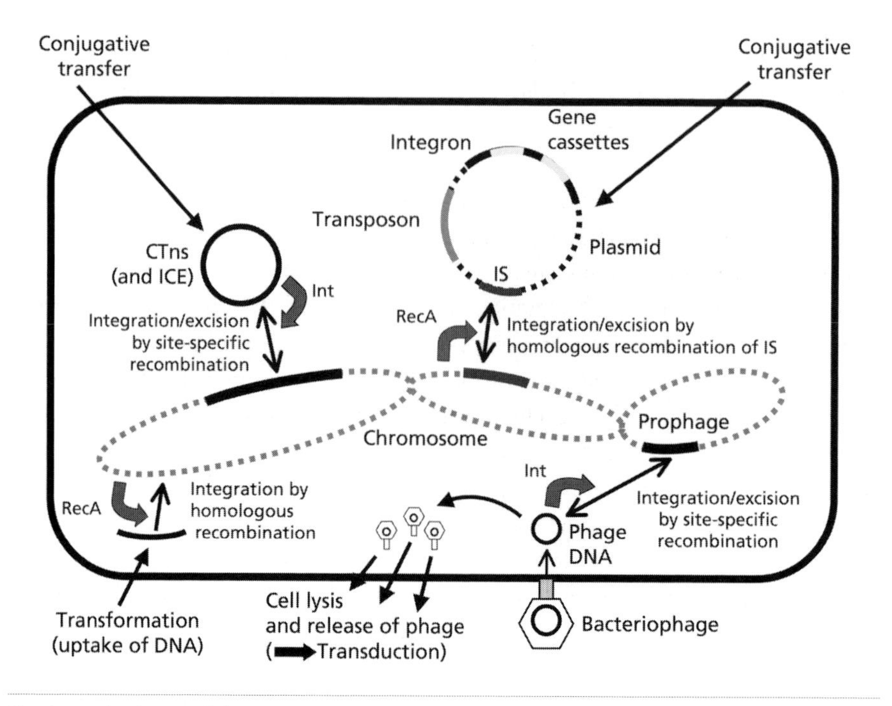

Fig. 2. Mechanisms and elements involved in bacterial HGT. Key HGT processes and MGEs are indicated. Integration of MGEs into the bacterial chromosome is mediated by MGE-encoded integrases (Int) or host chromosome-encoded RecA systems (RecA). See text for further details. CTns, Conjugative transposons.

TRANSFORMATION, TRANSDUCTION AND CONJUGATION: DNA ON THE MOVE

As discussed in the introduction, there are three primary mechanisms of HGT, namely transformation, transduction and conjugation. The discovery of transformation (Griffith, 1928), defined here as 'the uptake and incorporation of naked DNA, from the extracellular environment, into the recipient genome', and the subsequent evidence provided by Avery *et al.* (1944) that DNA was the carrier of genetic material, represent not only a key milestone in the discovery of HGT, but also in the wider field of genetics. Yet, despite these distinguished beginnings, transformation may be considered as the poor relation of HGT, in comparison to research on transduction and conjugation. This is primarily because transformation does not require the presence of a self-transmissible MGE such as a plasmid or phage genome. In this respect transformation represents a more stochastic process than transduction and conjugation, whereby the key requirement for transformation is the presence of naked DNA (of any sequence) in the same location as a cell which has the machinery to take up and incorporate the DNA molecule. Such bacteria are termed competent, though a distinction should be made between those bacteria that are naturally competent such as *Bacillus* and

Neisseria and those such as *E. coli* upon which competence can be conferred (e.g. by calcium chloride treatment). The key stages in transformation are firstly the uptake of DNA into the cell, during which the DNA is converted into a single-stranded molecule which is resistant to DNase degradation, and the subsequent stable maintenance of the DNA molecule. DNA uptake is mediated by a competence system, encoded by the recipient genome, including gene products that share homology to proteins from both type IV and type II secretion systems (reviewed in detail by Chen & Dubnau, 2004). Following successful uptake, integration of single-stranded DNA into the host genome is mediated by RecA-dependent homologous recombination (Fig. 2), necessitating sequence similarity between the incoming DNA and a region of the host genome. Alternatively, if the newly introduced DNA molecule encodes its own replication function (e.g. plasmid DNA), this molecule may also be stably maintained in the recipient. However, genetic exchange and adaptation via transformation-mediated processes is arguably most important when members of the same bacterial genus, though including different species, strains or clones, are present in large numbers in a particular environment, such as in the widely studied human commensal *Neisseria*, in which the transfer and evolution of penicillin-resistance genes has been shown to be a direct consequence of transformation (Spratt *et al.*, 1992). In such systems, cell mortality and turnover lead to a high availability of exogenous DNA for transformation by closely related but divergent bacteria, promoting genetic exchange.

As opposed to natural transformation, which is an active process requiring host-encoded machinery, the application of chemical treatments (i.e. salt additions) to naturally non-competent species such as *E. coli* to promote competence, as is used widely in recombinant DNA approaches, is a passive process whereby cell membrane structure is affected, facilitating DNA uptake. Similarly, electroporation, developed initially to transfer DNA into mouse lyoma cells (Neumann *et al.*, 1982) but now used widely as a more efficient alternative to chemically mediated transformation in molecular research, is also a passive process. Intriguingly, recent research has suggested that such passive DNA uptake as a consequence of electric pulses may also be a relevant mechanism for HGT in the environment; Cérémonie *et al.* (2004) have recently isolated *Pseudomonas* spp. from soil that they define as lightning-competent bacteria, in which they have demonstrated, *in situ*, using soil microcosms, that transformation of these bacteria can be stimulated by artificial lightning and hence that lightning can act as a potential promoter of bacterial HGT via transformation in soils.

The second recognized mechanism of HGT is transduction, first identified in *Salmonella* (Zinder & Lederberg, 1952) and defined here as 'the bacteriophage-mediated transfer of DNA between bacteria'. Our understanding of transduction processes is considerably more advanced than that of transformation, and has co-developed alongside research investigating the life cycles of bacteriophage (phage), following infection of a bacterial cell. Of most significance in HGT are the lytic and lysogenic life cycles, for which T4 and phage lambda, respectively, are the experimental paradigms (see Ackermann & DuBow, 1987). Briefly, for lytic phages within an infected bacterial cell, during virus production, bacterial chromosomal DNA which has been partially fragmented by phage-encoded enzymes can be randomly packaged into phage heads. Following phage-promoted cell lysis, the DNA carried within such so-called generalized transducing phage can then, following injection of their DNA into another bacterial cell, be incorporated into the genome of the recipient cell via homologous recombination, as in transformation. This non-specific transfer of DNA is termed generalized transduction. Alternatively, in lysogenic phages, phage DNA can be incorporated and stably maintained into the host chromosome in the form of a prophage. This integration is mediated by phage-encoded integrases via site-specific recombination (Fig. 2). However, following the switch to the lytic cycle (Ptashne, 2004), aberrant excision of the phage genome can result in mosaic molecules consisting of phage DNA and adjacent bacterial chromosomal DNA regions, which can then be packaged into phage heads, prior to lysis. Bacterial DNA within these resultant specialized transducing phage particles can again be injected into other bacterial cells and incorporated into bacterial genomes by homologous recombination. The significance of phage-mediated transduction is arguably of greatest importance in aquatic and in particular marine environments, in which virus numbers are estimated at 10^{10} l^{-1}, which is 5 to 25 times greater than bacterial numbers (Fuhrman, 1999). Such large numbers of virus afford tremendous potential for transduction-mediated transfer, supported by the discovery of phage-mediated transfer of photosynthesis genes in *Prochlorococcus*, a numerically important marine photoautotrophic genus (Lindell *et al.*, 2004). The role of marine viruses as drivers of HGT will be discussed in more detail later.

The third central mechanism of HGT is conjugation, classically defined as plasmid-mediated gene transfer following cell-to-cell contact. However, on the basis of the now widespread discovery of other conjugative MGEs such as conjugative transposons and integrative and conjugative elements (ICEs) (discussed later), conjugation is defined more broadly here as 'intercellular gene transfer, following cell-to-cell contact, mediated by transfer functions encoded on an MGE'. Conjugation was originally demonstrated in *E. coli* K-12, which is now known to carry the F plasmid, by Lederberg & Tatum (1946). In classical bacterial genetics experiments, the transfer of the F plasmid from the F^+ cells (carrying the plasmid) to F^- cells (lacking the plasmid) enables the self-transfer of the plasmid genome between cells. It is now established that the F plasmid (and other self-conjugative plasmids) carry two sets of genes that promote transfer that can be divided into mobilization (*mob* genes) and transfer (*tra* genes) functions (reviewed by Zechner *et al.*, 2000). The key component of the mobilization system is a relaxase protein that nicks DNA at a specific recognition site (*nic* or *oriT*) and, via a rolling-circle replication mechanism, generates a single-stranded DNA molecule suitable for transfer. In plasmids of Gram-negative organisms, the *tra* genes encode a hair-like surface appendage called the sex pilus, which consists of a membrane-located type IV secretion system (Cascales & Christie, 2003) through which DNA is transported across the membrane prior to being pumped into the recipient cell. Some Gram-positive bacteria also utilize type IV secretion systems, although in some genera, e.g. in plasmids of some members of *Streptomyces*, alternative transfer systems are present (reviewed by Grohmann *et al.*, 2003). In addition to enabling their own transfer, if a self-conjugative plasmid integrates into the host cell chromosome, typically via homologous recombination between insertion sequence (IS) elements found on the plasmid and the chromosome (Fig. 2), either parts of or indeed the entire chromosome can be transferred, in what is known as mobilization, by the plasmid *tra* system (*in cis*). Additionally, extrachromosomal plasmids are also able to mobilize other plasmids, which themselves carry *mob* genes and an origin of transfer (*oriT*), but lack their own *tra* genes, i.e. by *trans*-complementation. The genetics and cellular mechanisms that control conjugation have, to a large extent, been established in plasmid systems, yet the last 20 years have seen the identification of a diverse array of other MGEs that are also able to mediate conjugation events. These MGEs are discussed in the next section.

THE MOBILE GENETIC ELEMENTS: THE MOVERS AND SHAKERS OF HORIZONTAL GENE TRANSFER

The classically recognized agents of HGT as discussed above are the bacteriophage and the bacterial plasmids. More recently, we have seen the emergence of a plethora of other MGEs that either mediate intragenomic gene rearrangements (within a cell) or enable intergenomic transfer (between cells) via conjugation (and mobilization). The key feature of the MGEs is their mosaic nature, with the assortment of different MGEs being a direct consequence of their modular evolution, during which different functional modules (Toussaint & Merlin, 2002) such as replicons (rolling circle or theta), recombinases (tyrosine, serine or DDE recombinases), maintenance systems and transfer systems are combined (Fig. 2), together with the carriage of an assortment of accessory functional genes that encode a diverse range of phenotypic traits. A classification of the different functional modules is provided on the ACLAME web site (see Table 1).

Of those MGEs that catalyse intragenomic rearrangements, IS elements and transposons are the best characterized. IS elements, reviewed by Mahillon & Chandler (1998) and catalogued on the web site IS finder (Table 1), are typically small (< 1 to 3 kb) and consist of terminal inverted repeats flanking a transposase gene. These elements typically do not encode any function other than to enable their own intragenomic

Table 1. HGT and MGE web resources

Resource	Web address	Description	Reference
ACLAME	http://aclame.ulb.ac.be/	Catalogue and classification system for MGEs and their modular components	Leplae *et al.* (2004)
GOLD (Genomes OnLine Databases)	http://www.genomesonline.org/	Comprehensive database of completed and ongoing genome sequences	Bernal *et al.* (2001)
Horizstory and Lumbermill	http://coffee.biochem.dal.ca/	Source code for identification of HGT events	MacLeod *et al.* (2005)
IS finder	http://www-is.biotoul.fr/	Catalogue of IS from prokaryotes	Mahillon & Chandler (1998)
Microbes Online Database	http://www.microbesonline.org/	Database enabling comparative genomics	Alm *et al.* (2005)
Plasmid Genome Database	http://www.genomics.ceh.ac.uk/plasmiddb/	Catalogue of complete plasmid genome sequences	Mølbak *et al.* (2004)

transfer. Transposons are both more diverse and more complex than IS elements and, in addition to encoding self-transposition, carry other functional genes, e.g. antibiotic resistance or biodegradative functions. In addition to being found on the chromosome, both IS elements and transposons may also be located on plasmids or other MGEs, enabling their intergenomic transfer. Transposons themselves often contain integrons which themselves contain gene cassettes (Fig. 2), in a manner reminiscent of Russian dolls. Integrons consist of a site-specific recombination system coupled to a gene expression system carrying a strong promoter, downstream of which gene cassettes can be inserted in series at a specific attachment site, *attI*, catalysed by the integron-encoded recombinase (Hall *et al.*, 1999). Individual gene cassettes encode an array of different antibiotic-resistance genes, and recent analysis of PCR-amplified gene cassettes from soils suggests a wealth of other unknown functions that are proposed to represent a 'gene cassette metagenome' that serves as a resource for evolution (Holmes *et al.*, 2003).

In contrast to the above MGEs, an assortment of other mosaic MGEs are capable of conjugative transfer to other cells. These transfer-proficient MGEs include the conjugative transposons and mobilizable transposons and a diverse group of ICEs (Burrus *et al.*, 2002) (Fig. 2) that also share some similarities to genomic islands (GEIs). Together these have been proposed to represent a mosaic continuum (Osborn & Böltner, 2002) and can be subdivided in terms of both their mechanism of transfer and also the type of recombination system that they possess to enable integration into chromosomal (or plasmid) DNA. The first of these MGEs to be recognized were the conjugative transposons (reviewed by Salyers *et al.*, 1995), most notably Tn*916* (Franke & Clewell, 1981). Conjugative transposons have typically been isolated from Gram-positive bacteria and have been shown to consist of a backbone consisting of a plasmid-related conjugative transfer system together with a tyrosine-recombinase integration system related to those from integrative phages. Following transfer into a recipient cell, integration into the host chromosomes is semi-random. Conjugative transposons also typically carry a range of other MGEs and/or accessory functional genes. In contrast to the conjugative transposons, the emerging group of ICEs, which are more commonly found in Gram-negative bacteria, e.g. the related elements SXT from *Vibrio cholerae* (Beaber *et al.*, 2002) and R391 from *Providencia rettgeri* (Böltner *et al.*, 2002), whilst also comprising a mosaic of a phage-related integration system and plasmid-related conjugative transfer functions, tend to demonstrate site-specific integration into only one or a limited number of chromosomal integration sites, e.g. the *prfC* gene for R391 and SXT (Hochhut & Waldor, 1999). In this respect, these ICEs mirror the integration of GEIs, although the latter typically integrate into tRNA genes and can be distinguished from ICEs in that they are not self-conjugative. Further discussion of GEIs is provided in a review by Dobrindt *et al.* (2004) and also in the chapters by Klockgether *et al.* (2006) and Lacour *et al.* (2006) elsewhere in this volume.

The final group of MGEs are the mobilizable transposons. These elements consist of a recombinase integration system, enabling intragenomic transposition, but in addition carry mobilization functions enabling conjugative transfer mediated by transfer functions, supplied *in trans* (by *tra* genes on a conjugative plasmid, conjugative transposon or ICE). The majority of such mobilizable transposons identified to date are from the genus *Bacteroides*, e.g. the non-replication *Bacteroides* units (NBU) and Tn*4555*, which have a tyrosine recombinase that mediates transposition, and a *mob* gene and *oriT* sequence that enable mobilization (Shoemaker *et al.*, 1996; Tribble *et al.*, 1997). Alternatively, other mobilizable transposons have been found to include a serine recombinase and plasmid-related *mob* functions, e.g. Tn*4551* and Tn*4553* (Lyras & Rood, 2000), whilst a third type of mobilizable transposon represented by Tn*5398* from *Clostridium difficile* 630 carries no recognizable integration or mobilization genes, but does contain an *oriT* sequence, with both integration and mobilization functions believed to be supplied *in trans* by the co-resident conjugative transposon Tn*5397* (Farrow *et al.*, 2001).

Together, this diverse collection of MGEs affords tremendous potential for MGE-mediated intergenomic transfer and intragenomic rearrangements which, along with transformation, offers in principle the opportunity for rampant HGT, as suggested by some authors. However, until the advent of high-throughput genome sequencing and the subsequent emergence of the field of comparative genomics, our appreciation of the extent of HGT was limited partly to those MGEs and their conferred phenotypes which we could readily recognize as demonstrating an obvious adaptive trait such as the readily recognized examples of antibiotic and metal resistances.

COMPARATIVE GENOMICS, METAGENOMICS AND HORIZONTAL GENE TRANSFER

Evidence for the presence of horizontally transferred genes within chromosomes significantly predates the introduction of whole-genome sequencing approaches, for example the identification of pathogenicity islands within uropathogenic *E. coli* on the basis of differing G+C content between the pathogenicity determinant and the surrounding chromosomal DNA (Hacker *et al.*, 1983). Subsequently, following the determination of the first few bacterial genomes in the mid-1990s, the pioneering analysis by Lawrence & Ochman (1998) used both sequences with atypical G+C content and codon usage to identify putative horizontally transferred sequences and suggested that approximately 17 % of the open reading frames (ORFs) identified in the genome sequence had been acquired by HGT since the divergence of *E. coli* from *Salmonella*. As an increasing number of genome sequences became available, such approaches were extended to a wider group of prokaryotes (Ochman *et al.*, 2000) and suggested considerable variation in the extent of horizontally acquired sequences between genomes with, in particular, no or very little evidence of HGT identified in some small genomes such as *Mycobacterium genitalium* and *Borrelia burgdorferi*. Such analysis qualifies assertions that HGT is rampant and highlights the belief that factors such as restricted ecological niche will limit HGT in some species.

With the generation of over 280 complete prokaryotic genomes and a further 800 in progress according to the Genomes Online Database (see Table 1), there is now tremendous scope for genomics- and comparative genomics-based assessment of the role of HGT in adaptation and evolution. Availability of such sequences has driven the development of a series of analysis tools to identify horizontally transferred genes within genomes. Such approaches include the determination of how individual genes within a genome differ in their nucleotide composition in terms of codon usage in comparison with the prevalent codon usage of the entire genome (reviewed in more detail by Tsirigos & Rigoutsos, 2005). Alternatively, the relative abundance of individual dinucleotide sequences (i.e. AC, AG, AT, etc.) in comparison with that of individual nucleotides can also be used to identify regions that are dissimilar from the typical genomic content, either by searching blocks of sequence (e.g. 50 kb regions) or by analysis of individual ORFs (Karlin *et al.*, 1997). Improved derivations of such methods that consider both the abundance of multiple polynucleotide sequences (e.g. di-, tri- and tetranucleotides) and exclude reference to codon boundaries (Tsirigos & Rigoutsos, 2005) and alternatives based on phylogenetic approaches that identify and delineate both vertical and horizontal gene transmissions (MacLeod *et al.*, 2005) using Horizstory and Lumbermill (Table 1), or that specifically target pathogenicity (and other genomic) islands within genomes (Yoon *et al.*, 2005), will further extend our knowledge of the significance of HGT within individual genomes. From such studies to date, the evidence for a significant role in HGT in bacterial genome evolution is compelling. Nevertheless, such *in silico* inferences should not be considered as absolute arbiters of whether the presence of particular genes within a genome is the result of HGT. For example, Omelchenko *et al.* (2003) have proposed that mosaic operons evolve via HGT coupled to gene displacement *in situ* via the so-called selfish operon model (Lawrence & Roth, 1996), yet counter-arguments to the selfish operon model suggest that the formation of new operons using genes from within a genome is as common as the appearance of new operons in which formation is due to gene acquisition from another genome by HGT (Price *et al.*, 2005). This distinction between such operons is in fact a division between intra- and intergenomic gene acquisition.

Whilst many of the above approaches rely primarily on identification of regions or genes within individual chromosomes that are atypical compared with the chromosome as a whole and are thus predicted to have been acquired by HGT, arguably more powerful evidence for HGT can be determined by direct comparison of multiple genomes. For a number of species, several genome sequences have been completed, e.g. *E. coli*, *Helicobacter pylori*, *Prochlorococcus marinus* (see GOLD database; Table 1), whilst additionally some genera have complete genomes for a number of different species (e.g. *Bacillus*, *Pseudomonas*, *Staphylococcus*, *Streptococcus*). Direct comparison of such genomes either by pairwise comparison or via multiple comparisons is facilitated using bioinformatics packages such as the MicrobesOnline website for comparative genomics (Table 1).

The emerging field of metagenomics is also beginning to offer additional insights into the contribution of HGT to prokaryotic evolution. Metagenomics (reviewed by Handelsman, 2004) involves the isolation of DNA directly from an environmental sample, e.g. soil or water, followed by the construction of clone libraries, of either large insert libraries in bacterial artificial chromosomes or fosmids (~100 or ~30 kb inserts, respectively) or small fragment (<2 kb) shotgun libraries, as were used in the groundbreaking study by Venter *et al.* (2004), in which the Sargasso Sea metagenome was cloned. Irrespective of the choice of library construction strategy, metagenomic approaches circumvent the requirement for selective isolation of microbial strains and enable sequence information to be generated from the total as opposed to the culturable microbial community. For studying HGT, this allows the analysis of genomic information from environmentally relevant bacteria that are not in culture, and such approaches are beginning to yield information about classical MGEs and also the acquisition via

HGT of genes on the chromosomes of uncultured bacteria. In the analysis of the Sargasso Sea metagenome library, in addition to the identification of an estimated ~150 novel bacterial taxa, sequence assembly also enabled the construction from shotgun fractions of ten putative plasmids ranging in size from under 10 kb to over 100 kb. These plasmids contain sequences that are related to known plasmid replication and transfer genes, including *rep* gene sequences related to the IncN plasmid R46 and to the exogenously isolated plasmid pEMT1. In addition, a variety of genes encoding typical plasmid-conferred phenotypes including mercury, copper and UV resistance, together with multiple transposon sequences, demonstrates the value of such approaches to access MGEs from uncultured organisms. It should be noted, however, that these sequences assembled from shotgun fragments should not be considered as the definitive MGE, as some of the 'assembled' plasmids were lacking in readily identified replication regions (A. M. Osborn, unpublished data), although this may also be a consequence of divergence from previously characterized elements. An alternative metagenome-based investigation of HGT was undertaken using a fosmid library from DNA from anaerobic sediments, in which 12 fosmid clones, each containing a 23S rRNA gene to enable predictive species identification, were sequenced (Nesbø *et al.*, 2005). Subsequent phylogenetic comparison of the other ORFs identified on each fosmid to related sequences in the database enabled the identification of ORFs that were assessed to have different phylogenetic origins on individual fosmid clones. In this manner, for the 12 fosmid clones studied, the majority of ORFs on each clone were found to be of a similar taxonomic background to that of the rRNA gene. However, a significant fraction of the ORFs were nevertheless estimated to have been acquired by HGT (7–44 %, mean 17 %). These ORFs included acyl-CoA synthase genes and additionally integrase and transposase genes and, in three separate clones, there was evidence for transfer of both GEIs and, in one fosmid, a conjugative transposon that had integrated adjacent to a tRNA gene.

EMERGING PHENOTYPES CONVEYED BY HORIZONTAL GENE TRANSFER

Metagenomic approaches are not limited to the analysis of bacteria (or larger micro-organisms) and have also been used to investigate diversity within viruses. Such an approach has in particular been used to study bacteriophage communities in marine environments. Breitbart *et al.* (2002) have shown that over 65 % of the cloned sequences in a library constructed from DNA extracted from a 100 kDa filter to generate a phage concentrate (following sequential filtration of sea-water samples) were unrelated to previously reported viral or indeed other genomic sequences, suggesting considerable undiscovered variability in marine phage genomes. Given the considerable numbers of phage in marine environments (see above), this represents an enormous resource for increasing genomic variation in prokaryotic genomes, in particular from lysogenic phage. Whilst at face value this virus metagenome will consist mostly of genes considered essential for phage replication and survival, the subsequent discovery that some of these marine phage are found to carry genes, derived from bacteria, that encode components of photosynthetic systems suggests that marine phage and HGT (via transduction) play a key role in the evolution of major biogeochemical cycling pathways. Phage genomes from viruses infecting the two key photoautotrophic cyanobacteria (*Synechococcus* and *Prochlorococcus*) have been found to carry genes derived from these cyanobacteria that play a central role in photosynthesis, namely *psbA* and *psbD*, respectively encoding the photosystem II core reaction centre proteins D1 and D2 (Mann *et al.*, 2003; Lindell *et al.*, 2004). Whilst such genes are likely to have been acquired following aberrant excision following lysogeny, there remains the possibility that the maintenance of such genes on the cyanophage may promote photosynthetic activity of the host cell following infection and phage genome integration into the chromosome and thus select for the presence of cells containing the prophage. In addition, in the genomes of phage infecting *Prochlorococcus*, other photosynthetic genes have been identified including the *hli* gene that encodes a high-light-inducible protein which may confer improved fitness upon the host bacterium in surface waters, suggesting that phage may play a role in determining the ecological niche in which marine bacteria are able to exist (Lindell *et al.*, 2004). Whilst the role of phage in contributing to carbon release in marine environments following phage infection of cyanobacterial blooms is well recognized (see Suttle, 2005), the additional role of phage in promoting carbon fixation and primary production in the oceans stresses the importance of phage-mediated HGT as a driver for ecological adaptation.

In contrast to phage, plasmids are more typically recognized as the key mediators of HGT, and their role in enabling a diverse selection of phenotypes is firmly established, with leading examples of adaptive phenotypes including antibiotic resistances, xenobiotic degradation and symbiotic functions. In addition to carrying such accessory functions, which provide a selective pressure for maintenance of the plasmids, plasmids are themselves defined by their core features of replication, maintenance and transfer. Our understanding of such systems is that these can be considered as plasmid-selfish genes ensuring survival and enabling dissemination of the plasmids, yet recent research has demonstrated that some core plasmid genes can also confer important phenotypes on the host bacterium. Of particular interest is the discovery that a number of different conjugative plasmids belonging to a variety of plasmid incompatibility groups, including IncFI, IncFII, IncI1, IncN and IncW, can induce biofilm formation in their host bacteria. This phenomenon was first demonstrated in *E. coli* strains containing the F plasmid by Ghigo (2001), who further demonstrated by mutagenesis that plasmids in which the *traA* gene (encoding the pilin subunit of the conjugative pilus) was mutated no longer promoted biofilm formation and, additionally, that a *traA* gene supplied *in trans* would restore the promotion of biofilms. Further mutagenesis studies demonstrated that biofilm formation did not require conjugative transfer of DNA to occur, rather, that this phenotype was conferred by the presence of a functional conjugative pilus. Of those plasmids tested which conferred biofilm formation, a number of these plasmids have constitutive synthesis of the pilus. A number of other plasmids which have inducible pilus synthesis, including members of IncA/C, IncI1 and IncP groups and in addition the ICE R391, were shown not to promote biofilm when expression of their pilus synthesis genes was repressed. However, when cells containing these plasmids (or ICEs) were added to plasmid-free cells of *E. coli* MG1655, the subsequent initiation of plasmid transfer and the resultant derepression of the genes encoding transfer machinery then permitted biofilm formation to occur. Following transfer of plasmid DNA, Ghigo (2001) suggests that the conjugative pili would recede, but that the biofilm could be maintained by secretion of other host-encoded adhesion factors, e.g. exopolysaccharides. Subsequent microscopy-based studies of a series of mutants (Reisner *et al.*, 2003) have confirmed that the key requirement for plasmid-induced biofilm formation by the F plasmid is the production of the *traA*-encoded pilus and additionally that the absence of chromosomally encoded flagella or fimbriae did not affect biofilm development.

The above example demonstrates how MGEs can confer unusual responses upon their host and affect the environment in which the host inhabits, i.e. changing from a planktonic to a biofilm habitat. Conversely, the environment of a bacterium carrying an MGE can also exert an influence on the activity of the MGE. Such systems include the induction of transfer of the conjugative transposon Tn*916*, which carries tetracycline-resistance genes, by the presence of tetracycline, demonstrated initially in culture (Showsh & Andrews, 1992) and subsequently in the intestines of gnotobiotic rats (Bahl *et al.*, 2004). Induction of transfer genes in the presence of tetracycline enables Tn*916* to disseminate rapidly through a population and raises obvious concerns in therapeutic treatment, whereby the application of tetracycline actively stimulates transfer of the resistance gene carried on the MGE. A second example of the external environment inducing transfer of an MGE has been highlighted in the SXT element. SXT (and the closely related R391) carries a conjugative transfer system related to that on the plasmid R27. In addition, the integration system consists of a phage lambda-related integrase controlling site-specific chromosomal integration. Elsewhere on SXT and R391, there are a number of repressor genes, including the *setR*-encoded repressor that shows similarity to the phage lambda CI repressor which regulates control of lysogeny in lambda. Beaber *et al.* (2004) have demonstrated recently that SetR represses expression of two other regulatory transcriptional activators (*setC* and *setD*) on SXT that control SXT excision and transfer (Beaber *et al.*, 2002). This regulatory cascade has additionally been shown to be susceptible to DNA damage in the SOS response, resulting in stimulation of chromosomally encoded RecA co-protease activity leading to autoproteolysis of the SetR protein and enabling expression of *setC* and *setD* and subsequently transcription of excision and transfer genes. SOS responses can be stimulated by damage to DNA caused by UV light or by some antibiotics, e.g. ciprofloxacin, with the latter shown to induce SXT transfer in *E. coli* $RecA^+$ strains (Beaber *et al.*, 2004). Such a system enables SXT transfer to be stimulated in the presence of UV damage and/or challenge by antibiotics and, as seen in the tetracycline-inducible transfer of Tn*916*, promotes wider dissemination of the MGE. More recently, the role of the SOS response in promotion of MGE transfer has been observed in a second ICE, namely the ICE*Bs1* element from *Bacillus subtilis* (Auchtung *et al.*, 2005), showing that such responses are found in both Gram-negative and Gram-positive bacteria.

CONCLUDING COMMENTS

In conclusion, this review has highlighted the mechanisms and agents by which HGT has contributed to the tremendous physiological, metabolic and genetic diversity observed within the prokaryotes. Our understanding of the extent to which HGT facilitates evolutionary adaptations in the prokaryotes is being enriched by genomic and metagenomic comparisons, yet it is clear from such studies that, whilst HGT is indeed widespread, there is considerable variation in the relative importance of HGT between different species and indeed organisms inhabiting particular habitats. In marine environments, where prokaryotes exist in planktonic forms, phage-mediated transduction represents a major mechanism for HGT whilst in soils (and sediments), and indeed in clinical environments such as the intestinal tract, the formation of biofilms and the transfer of DNA via transformation and conjugation may be more important mechanisms for HGT. In the coming years it is expected that the discovery of important phenotypes that are disseminated by HGT will continue apace.

REFERENCES

Ackermann, H.-W. & DuBow, M. S. (1987). *Viruses of Prokaryotes*, vol. 1, *General Properties of Bacteriophages*. Boca Raton, FL: CRC Press.

Alm, E. J., Huang, K. H., Price, M. N., Koche, R. P., Keller, K., Dubchak, I. L. & Arkin, A. P. (2005). The MicrobesOnline web site for comparative genomics. *Genome Res* **15**, 1015–1022.

Arber, W. (1991). Elements in microbial evolution. *J Mol Evol* **33**, 4–12.

Auchtung, J. M., Lee, C. A., Monson, R. E., Lehman, A. P. & Grossman, A. D. (2005). Regulation of a *Bacillus subtilis* mobile genetic element by intercellular signaling and the global DNA damage response. *Proc Natl Acad Sci U S A* **102**, 12554–12599.

Avery, O. T., MacLeod, C. M. & McCarthy, M. (1944). Studies on the chemical nature of the substance inducing transformation of pneumococcal types. I. Induction of transformation by a deoxyribonucleic acid fraction isolated from pneumococcus type III. *J Exp Med* **79**, 137–158.

Bahl, M. I., Sørensen, S. J., Hansen, L. H. & Licht, T. R. (2004). Effect of tetracycline on transfer and establishment of the tetracycline-inducible conjugative transposon Tn*916* in the guts of gnotobiotic rats. *Appl Environ Microbiol* **70**, 758–764.

Beaber, J. W., Hochhut, B. & Waldor, M. K. (2002). Genomic and functional analyses of SXT, an integrating antibiotic resistance gene transfer element derived from *Vibrio cholerae*. *J Bacteriol* **184**, 4259–4269.

Beaber, J. W., Hochhut, B. & Waldor, M. K. (2004). SOS response promotes horizontal dissemination of antibiotic resistance genes. *Nature* **427**, 72–74.

Bernal, A., Ear, U. & Kyrpides, N. (2001). Genomes OnLine Database (GOLD): a monitor of genome projects world-wide. *Nucleic Acids Res* **29**, 126–127.

Böltner, D., MacMahon, C., Pembroke, J. T., Strike, P. & Osborn, A. M. (2002). R391: a conjugative integrating mosaic comprised of phage, plasmid, and transposon elements. *J Bacteriol* **184**, 5158–5169.

Breitbart, M., Salamon, P., Andresen, B., Mahaffy, J. M., Segall, A. M., Mead, D., Azam, F. & Rohwer, F. (2002). Genomic analysis of uncultured marine viral communities. *Proc Natl Acad Sci U S A* **99**, 14250–14255.

Broothaerts, W., Mitchell, H. J., Weir, B., Kaines, S., Smith, L. M. A., Yang, W., Mayer, J. E., Roa-Rodriguez, C. & Jefferson, R. A. (2005). Gene transfer to plants by diverse species of bacteria. *Nature* **433**, 629–633.

Burrus, V., Pavlovic, G., Decaris, B. & Guedon, G. (2002). Conjugative transposons: the tip of the iceberg. *Mol Microbiol* **46**, 601–610.

Cascales, E. & Christie, P. J. (2003). The versatile bacterial type IV secretion systems. *Nat Rev Microbiol* **1**, 137–149.

Cérémonie, H., Buret, F., Simonet, P. & Vogel, T. M. (2004). Isolation of lightning-competent soil bacteria. *Appl Environ Microbiol* **70**, 6342–6346.

Chen, I. & Dubnau, D. (2004). DNA uptake during bacterial transformation. *Nat Rev Microbiol* **2**, 241–249.

Cohan, F. M. (2002). What are bacterial species? *Annu Rev Microbiol* **56**, 457–487.

Dobrindt, U., Hochhut, B., Hentschel, U. & Hacker, J. (2004). Genomic islands in pathogenic and environmental microorganisms. *Nat Rev Microbiol* **2**, 414–424.

Farrow, K. A., Lyras, D. & Rood, J. I. (2001). Genomic analysis of the erythromycin resistance element Tn*5398* from *Clostridium difficile*. *Microbiology* **147**, 2717–2728.

Franke, A. E. & Clewell, D. B. (1981). Evidence for a chromosome-borne resistance transposon (Tn*916*) in *Streptococcus faecalis* that is capable of "conjugal" transfer in the absence of a conjugative plasmid. *J Bacteriol* **145**, 494–502.

Fuhrman, J. A. (1999). Marine viruses and their biogeochemical and ecological effects. *Nature* **399**, 541–548.

Ghigo, J.-M. (2001). Natural conjugative plasmids induce bacterial biofilm formation. *Nature* **412**, 442–445.

Griffith, F. (1928). Significance of pneumococcal types. *J Hyg* **27**, 113.

Grohmann, E., Muth, G. & Espinosa, M. (2003). Conjugative plasmid transfer in gram-positive bacteria. *Microbiol Mol Biol Rev* **67**, 277–301.

Hacker, J., Knapp, S. & Goebel, W. (1983). Spontaneous deletions and flanking regions of the chromosomally inherited hemolysin determinant of an *Escherichia coli* O6 strain. *J Bacteriol* **154**, 1145–1152.

Hall, R. M., Collis, C. M., Kim, M.-J., Partridge, S. R., Recchia, G. D. & Stokes, H. W. (1999). Mobile gene cassettes and integrons in evolution. *Ann N Y Acad Sci* **870**, 68–80.

Handelsman, J. (2004). Metagenomics: application of genomics to uncultured microorganisms. *Microbiol Mol Biol Rev* **68**, 669–685.

Hochhut, B. & Waldor, M. K. (1999). Site-specific integration of the conjugal *Vibrio cholerae* SXT element into *prfC*. *Mol Microbiol* **32**, 99–110.

Holmes, A. J., Gillings, M. R., Nield, B. S., Mabbutt, B. C., Nevalainen, K. M. H. & Stokes, H. W. (2003). The gene cassette metagenome is a basic resource for bacterial genome evolution. *Environ Microbiol* **5**, 383–394.

Ivanova, N., Sorokin, A., Anderson, I. & 20 other authors (2003). Genome sequence of *Bacillus cereus* and comparative analysis with *Bacillus anthracis*. *Nature* **423**, 87–91.

Karlin, S., Mrázek, J. & Campbell, A. (1997). Compositional biases of bacterial genomes and evolutionary implications. *J Bacteriol* **179**, 3899–3913.

Klockgether, J., Reva, O. N. & Tümmler, B. (2006). Spread of genomic islands between clinical and environmental strains. In *Prokaryotic Diversity: Mechanisms and Significance* (Society for General Microbiology Symposium no. 66), pp. 187–200. Edited by N. A. Logan, H. M. Lappin-Scott & P. C. F. Oyston. Cambridge: Cambridge University Press.

Kurland, C. G. (2000). Something for everyone. Horizontal gene transfer in evolution. *EMBO Rep* **1**, 92–95.

Lacour, S., Gaillard, M. & van der Meer, J. R. (2006). Genomic islands and evolution of catabolic pathways. In *Prokaryotic Diversity: Mechanisms and Significance* (Society for General Microbiology Symposium no. 66), pp. 255–273. Edited by N. A. Logan, H. M. Lappin-Scott & P. C. F. Oyston. Cambridge: Cambridge University Press.

Lawrence, J. G. & Ochman, H. (1998). Molecular archaeology of the *Escherichia coli* genome. *Proc Natl Acad Sci U S A* **95**, 9413–9417.

Lawrence, J. G. & Roth, J. R. (1996). Selfish operons: horizontal transfer may drive the evolution of gene clusters. *Genetics* **143**, 1843–1860.

Lederberg, J. & Tatum, E. L. (1946). Gene recombination in *Escherichia coli*. *Nature* **158**, 558.

Leplae, R., Hebrant, A., Wodak, S. J. & Toussaint, A. (2004). ACLAME: a CLAssification of Mobile genetic Elements. *Nucleic Acids Res* **32**, D45–D49.

Lindell, D., Sullivan, M. B., Johnson, Z. I., Tolonen, A. C., Rohwer, F. & Chisholm, S. W. (2004). Transfer of photosynthesis genes to and from *Prochlorococcus* viruses. *Proc Natl Acad Sci U S A* **101**, 11013–11018.

Lyras, D. & Rood, J. I. (2000). Transposition of Tn*4451* and Tn*4453* involves a circular intermediate that forms a promoter for the large resolvase, TnpX. *Mol Microbiol* **38**, 588–601.

MacLeod, D., Charlebois, R. L., Doolittle, F. & Bapteste, E. (2005). Deduction of probable events of lateral gene transfer through comparison of phylogenetic trees by recursive consolidation and rearrangement. *BMC Evol Biol* **5**, 27.

Mahillon, J. & Chandler, M. (1998). Insertion sequences. *Microbiol Mol Biol Rev* **62**, 725–774.

Mann, N. H., Cook, A., Millard, A., Bailey, S. & Clokie, M. (2003). Bacterial photosynthesis genes in a virus. *Nature* **424**, 741.

Mølbak, L., Tett, A., Ussery, D. W., Wall, K., Turner, S., Bailey, M. & Field, D. (2004). The plasmid genome database. *Microbiology* **149**, 3043–3045.

Nakamura, Y., Itoh, T., Matsuda, H. & Gojobori, T. (2004). Biased biological functions of horizontally transferred genes in prokaryotic genomes. *Nat Genet* **36**, 760–766.

Nesbø, C. L., Boucher, Y., Dlutek, M. & Doolittle, W. F. (2005). Lateral gene transfer and phylogenetic assignment of environmental fosmid clones. *Environ Microbiol* **7**, 2011–2026.

Neumann, E., Schaefer-Ridder, M., Wang, Y. & Hofschneider, P. H. (1982). Gene transfer into mouse lyoma cells by electroporation in high electric fields. *EMBO J* **1**, 841–845.

Novick, R. P. (1987). Plasmid incompatibility. *Microbiol Rev* **51**, 381–395.

Ochman, H., Lawrence, J. G. & Groisman, E. A. (2000). Lateral gene transfer and the nature of bacterial innovation. *Nature* **405**, 299–304.

Omelchenko, M. V., Makarova, K. S., Wolf, Y. I., Rogozin, I. B. & Koonin, E. V. (2003). Evolution of mosaic operons by horizontal gene transfer and gene displacement *in situ*. *Genome Biol* **4**, R55.

Osborn, A. M. & Böltner, D. (2002). When phage, plasmids and transposons collide: genomic islands, and conjugative- and mobilizable-transposons as a mosaic continuum. *Plasmid* **48**, 202–212.

Price, M. N., Huang, K. H., Arkin, A. P. & Alm, E. J. (2005). Operon formation is driven by co-regulation and not by horizontal gene transfer. *Genome Res* **15**, 809–819.

Ptashne, M. (2004). *A Genetic Switch: Phage Lambda Revisited*, 3rd edn. Cold Spring Harbor, NY: Cold Spring Harbor Laboratory.

Reisner, A., Haagensen, A. J., Schembri, M. A., Zechner, E. L. & Molin, S. (2003). Development and maturation of *Escherichia coli* K-12 biofilms. *Mol Microbiol* **48**, 933–946.

Salyers, A. A., Shoemaker, N. B., Stevens, A. M. & Li L.-Y. (1995). Conjugative transposons: an unusual and diverse set of integrated gene transfer elements. *Microbiol Rev* **59**, 579–590.

Schouls, L. M., Schot, C. S. & Jacobs, J. A. (2003). Horizontal transfer of segments of the 16S rRNA genes between species of the *Streptococcus anginosus* group. *J Bacteriol* **185**, 7241–7246.

Shoemaker, N. B., Wang, G.-R. & Salyers, A. A. (1996). The *Bacteroides* mobilizable insertion element, NBU1, integrates into the 3′ end of a Leu-tRNA gene and has an integrase that is a member of the lambda integrase family. *J Bacteriol* **178**, 3594–3600.

Showsh, S. A. & Andrews, R. E., Jr (1992). Tetracycline enhances Tn*916*-mediated conjugal transfer. *Plasmid* **28**, 213–224.

Spratt, B. G., Bowler, L. D., Zhang, Q. Y., Zhou, J. & Smith, J. M. (1992). Role of interspecies transfer of chromosomal genes in the evolution of penicillin resistance in pathogenic and commensal *Neisseria* species. *J Mol Evol* **34**, 115–125.

Suttle, C. A. (2005). Viruses in the sea. *Nature* **437**, 356–361.

Thomas, C. M. & Nielsen, K. M. (2005). Mechanisms of, and barriers to, horizontal gene transfer between bacteria. *Nat Rev Microbiol* **3**, 711–721.

Toussaint, A. & Merlin, C. (2002). Mobile elements as a combination of functional modules. *Plasmid* **47**, 26–35.

Tribble, G. D., Parker, A. C. & Smith, C. J. (1997). The *Bacteroides* mobilizable transposon Tn*4555* integrates by a site-specific recombination mechanism similar to that of the gram-positive bacterial element Tn*916*. *J Bacteriol* **179**, 2731–2739.

Tsirigos, A. & Rigoutsos, I. (2005). A new computational method for the detection of horizontal gene transfer events. *Nucleic Acids Res* **33**, 922–933.

Venter, J. C., Remington, K., Heidelberg, J. F. & 20 other authors (2004). Environmental genome shotgun sequencing of the Sargasso Sea. *Science* **304**, 66–74.

Yoon, S. H., Hur, C.-G., Kang, H.-Y., Kim, Y. H., Oh, T. K. & Kim, J. F. (2005). A computational approach for identifying pathogenicity islands in prokaryotic genomes. *BMC Bioinformatics* **6**, 184.

Zechner, E. L., de la Cruz, F., Eisenbrandt, R. & 8 other authors (2000). Conjugative-DNA transfer processes. In *The Horizontal Gene Pool*, pp. 87–174. Edited by C. M. Thomas. Amsterdam: Harwood Academic Press.

Zinder, N. D. & Lederberg, J. (1952). Genetic exchange in *Salmonella*. *J Bacteriol* **64**, 679–699.

INDEX

References to tables/figures are shown in italics

***A**cetohalobium arabaticum* 237
Acidianus filamentous virus 1 134
Acidobacteria *164*
Acinetobacter 207
ACLAME *282*
Actinobacteria *164*, 165, 168
Actinopolyspora halophila 237
adaptive evolution
    genetics 92–93
    micromutationism 93
adaptive radiation 91–92
Aerococcus 216
AF0070 *49*
AF1870 *49*
Akkermansia muciniphila *71*, *75*
Alphaproteobacteria 32, *35*, 115
Aminobacterium 170
Aminomonas 170
Anaerobaculum 170
Anaeroglobus 166
Anaerostipes caccae *74*, 76
Anaplasma marginale 115
antibiotic production
    encoded by gene clusters in streptomycetes
        205–208
antibiotic resistance and integrons 215
Archaea 9
archaeal viruses
    fuselloviruses
        biogeography *140*
        SSV-1 (*Sulfolobus spindle-shaped virus 1*)
            134, 135–137
        SSV-2 *134*, 135–137
        SSV-3 *134*, *136*, 137, 138
        SSV-K1 *134*, 135–137
        SSV-RH *134*, 135
    PSV (Pyrobaculum spherical virus) 133
    rudiviruses
        Acidianus filamentous virus 1 134
        ARV-1 (Acidianus rod-shaped virus 1) 133
        SIFV (*Sulfolobus islandicus filamentous*
            virus) 134
        Sulfolobus virus SIRV-1 133–134
        Sulfolobus virus SIRV-2 133–134
        TTV-1 (*Thermoproteus virus 1*) 133
        from Yellowstone National Park 133
ARV-1 (Acidianus rod-shaped virus 1) 133
atherosclerosis
    role of *Chlamydia pneumoniae* 145–155
        animal models 151–154
        antibiotic treatment 153–155
        clinical trials 154–155
        inflammatory processes 148–149
        in vitro studies 150–151
        in vivo studies 150–151
        presence in lesions 146–147
        serology 146–147
Atopobium 66, *69*
auxiliary genes 2
Azoarcus 264
Azotobacter vinelandii *194*

***B**acillus*
    B. anthracis, genomic evolution 182
    B. cereus 182
    B. subtilis 111, 203, 289
        sin operon 98
bacterial signal transduction cascades, modular
        structure 97
bacteriophage
    biogeography 133
    diversity 132–133, 286–287
    in marine environments 286–287
Bacteroides *74*, *77*, *164*, 283
    B. eggerthii *71*
    B. fragilis *71*
    B. ovatus *71*
    B. thetaiotaomicron *71*, *75*
        starch utilization 72, *73*
    B. vulgatus 71
    polysaccharide degradation 73
Bacteroidetes 66, 67, *69*, 165
Betaproteobacteria 32, *33*
BH0338 *48*
Bifidobacterium 66
    B. adolescentis *71*, *78*
    B. bifidum *71*, *75*
    B. breve *71*
    B. infantis *71*
    B. longum 72
    Bifidobacterium spp. *71*
biodiversity of micro-organisms 20–35
    co-occurrence 26
    distance-decay relationships *24*, 25–26, *28*
    fundamental patterns 22–23
    macro-organisms, comparison 20–28
    measurement 20–22

parasexuality 23–25
phylogenetic patterns 20–35, 26–28
see also taxonomic richness
biogeography 10–12, 131–132
animal viruses 132
archaeal viruses 133–140
bacteriophage 133
plant viruses 132
BLAST 2 analysis 265
'*Blochmannia pennsylvaticus*' *114*
bluensomycin *204*
Bordetella
B. bronchiseptica 183
B. parapertussis 183
B. pertussis 183
Borrelia 115
B. burgdorferi 284
Brevibacterium spp. 216
Buchnera aphidicola 108
Buchnera aphidicola BCc 106, *114*, 116
Burkholderia
B. cepacia 214
B. mallei
genome islands 189
genomic evolution 183
B. pseudomallei 183, *194*
genome islands 189
B. xenovorans 193, 264

C*andida* 164, 165
'*Candidatus* Blochmannia' 108
'*Candidatus* Phytoplasma asteris' 115
caries associated bacteria 165–166
cas gene systems 40–61
CASS (CRISPR-associated system) 41–42, 50–60
protein components *48–49*
related genes 59
CASS1 46, 50
CASS2 46
CASS4 46, 50
CASS5 46, 50
CASS6 46
CASS7 46, 50
catabolic pathway evolution 268
cellulolytic gut bacteria 72
Chlamydia
C. pneumoniae 115
pathogenesis *149*
role in atherosclerosis 145–155
see also atherosclerosis
C. psittaci 145–146
C. trachomatis 109, 115, 145–146, 148, 151
Chlamydiae *164*, 165
Chloroflexi *164*, 165
Cinaria cedri (aphid) 116
Citrobacter diversus 188
clc element 193, 195
BLAST 2 analysis of related elements 265
Burkholderia xenovorans 263–264
characteristics 260–261
conjugation *266*
DNA transfer 266–268
excision and integration 261–263
Pseudomonas sp. B13 *258*
Ralstonia sp. JS705 263–264
related pathogenicity islands 264
clostridial clusters
IV 81
IX *68*, *77*
XIVa 67, *68*, 78, 81
XIVb *68*
XVI *68*
Clostridium butyricum *71*
co-occurrence patterns in communities 26, *27*
COG1203 43, 46, *48*, 51, 53
COG1343 (CASS polymerase) 42, 43, 46, *48*
COG1353 46, *48*
COG1421 *48*
COG1468 46
COG1517 47, *49*, 50
COG1518 41–48
COG1688 43, 46
COG1857 43, 46, *48*, 50, 53
COG2462 *48*
COG3337 *49*
COG3513 *49*
Collinsella aerofaciens *71*
community composition
amplified fragment length polymorphism (AFLP) analysis *24*, 25
co-occurrence patterns 26, *27*
distance-decay relationships *24*, 25–26, 28
habitat filtering 27–28
phylogenetic clustering 27–28
phylogenetic patterns 25–28
comparative genomics
horizontal gene transfer (HGT) 284–286
minimal genomes 107–110
pathogens 175–183
Streptomyces 209, *210*
conjugation in horizontal gene transfer *278*, 280–281
core genes 2
evolution 123–128
Corynebacterium
C. diptheriae 206
C. glutamicum 114, 216

crenarchaeota 9
CRISPR elements (clustered regularly interspaced short palindrome repeats) 40
  RNAi defence system 50–58
Cupriaviridus
  C. campiniensis 195
  C. metallidurans 195
  C. oxalatica 267

'**D***einococcus–Thermus*' *164*, 165
demand theory in phenotype evolution 98
dentoalveolar abscesses
  associated bacteria 165
Desulfobulbus 166
Desulfovibrio sp. *74*
Dethiosulfovibrio 170
Dialister invisus 165
disease
  associated unculturable bacteria 170
  atherosclerosis 145–155
  caries 165–166
  dentoalveolar abscesses 165
  plague 175–176
distance-decay relationships *24*, 25–26, 28
diversity
  evolution 91–92
  marine bacteriophage 286–287
  oral micro-organisms 164–165
Dunaliella 242

Endosymbionts of insects
  minimal genomes 108–109
Entamoeba gingivalis 164
Enterobacteriaceae
  genomic islands 188–189
  integrons and antibiotic resistance 215
environmental sequencing 6–9
Erwinia carotovora atroseptica *194*
Escherichia coli 4, 78, 96, 114, 123, 126, 165, 203, 207, 210, 284, 289
  genomic islands 188, 189
  lac operon 98
  propanediol reductase origin 94
essential genes 111, *112*, *113*, *114*
Eubacterium
  E. cylindroides *68*
  E. eligens *71*
  E. hallii *68*, *74*, *78*
  E. ramulus *75*
  E. rectale *68*, *71*, *76*
evolution of diversity 91–92

Faecal bacteria *70*
Faecalibacterium prausnitzii *68*, *74*, *76*, *77*
FAS (fatty acid synthases) 210–211
Firmicutes 66, 67, *164*, 165, 168, 169
FISH probes 66
fosmids 285, 286
Francisella tularenis, genomic evolution 182
fuselloviruses
  biogeography 140
  sequence tree *139*
  SSV-1 (*Sulfolobus spindle-shaped virus 1*) 134, 135–137
  SSV-2 *134*, 135–137
  SSV-3 *134*, *136*, 137, 138
  SSV-K1 *134*, 135–137
  SSV-RH *134*, 135
Fusobacteria *164*

G*ammaproteobacteria* 108, 115, 126, 127, 193, 195, 265
GEIs (genomic islands)
  catabolic pathway evolution 268
  characteristics 187–188, 256–257
  Citrobacter diversus 188
  classification 259
  clc element 193, 195
    BLAST 2 analysis 265
    Burkholderia xenovorans 263–264
    characteristics 260–261
    conjugation *266*
    DNA transfer 266–268
    excision and integration 261–263
    Pseudomonas sp. B13 *258*
    Ralstonia sp. JS705 263–264
    related pathogenicity islands 264
  Enterobacteriaceae 188–189
  Escherichia coli 188, 189
  excision 259
  integration 190–191, 257–259
  Klebsiella 189
  Pseudomonas 188–195
  Salmonella 188
  spread among species 193, 195–196
  structure 257
gene clusters
  antibiotic production 205–208
  definition 202
  HGT 210–211
  integrons 214–216
  origin 203–204
  selection 208–209
  'selfish operon' 205
  statistical testing 202–203

transcriptional processes 207–208
xenobiotic compound degradation 214
gene duplication 203–204
genome evolution
gene replacement 124–125
HGT (horizontal gene transfer) 124–128
homologous genes 126–127
innovation 126
genome minimization
experimental methods 114–115
in nature 115–117
free-living organisms 116–117
intracellular organisms 115–116
LUCA (last universal common ancestor) 117
minimal gene set 117
Genomes Online Databases *282*
gut bacteria *see* intestinal bacteria

Habitat diversity 12–13
habitat filtering in microbial communities 27–28
Haemophilus
H. ducreyi 194
H. influenzae 107, 126, 193, *194*, 265
minimal genome 107–108
Halanaerobium lacusrosei 237
haloadaptation 239
haloarchaea 224–229
archaeal characters 225, 228
growth conditions 225, 228
haloalkalophiles 228
model organisms 228–229
square bacteria 228, 229–235
isolation and culture 231–233
phase-contrast microscopy *230*
Haloarcula 228
H. japonica 231
H. marismortui 229
H. quadrata 231
Halobacteriaceae
genera and species validly published *226–227*
Halobacterium 225, 226, 228
H. salinarum 229
Halobacteroides lacunaris 237
Halococcus 225, *226*
Haloferax 227, 228
Halogeometricum borinquense 232
Halomonas (*Flavobacterium*) *halmophila* 237
Halomonas (*Paracoccus*) *halodenitrificans* 237
halophiles
definition 223–224
Dunaliella 242
extreme halophile bacteria 235–237
fungi 243
moderate halophile bacteria 237–240
Gram-positive and negative genera 237, *238*
haloadaptation 239
industrial use 239–240
Salinibacter ruber 235–237
viruses 240–242
see also haloarchaea
Halorhodospira halophila 237
Halorubrum 228
haloviruses 240–242
Halovivax asiaticus 225, *226*
heart disease
role of *Chlamydia pneumoniae* (*see* atherosclerosis)
HGT (horizontal gene transfer) 124–128
assessed by comparative genomics 284–286
barriers 276
differential gene mobility 276–277
in gene cluster origin 210–211
marine phages 286–287
plasmids 288–289
'selfish operon' model 205
and species concept 276
tRNA synthetases 128
Histophilus somni 194
Horizstory and Lumbermill *282*
HPIs (high pathogenicity islands) of enterobacteria 188–189
hydrogen metabolism in intestine 77
hypersaline habitats 223–224

Integrons 214–216
resistance properties 215
intermediate disturbance hypothesis 32–34
intestinal bacteria
anaerobic metabolism 75–79
carbohydrate utilization 70–75
cross-feeding 72–73, 75
culturability 67
enzymatic activity 70–73
host-derived substrate metabolism 74–75
hydrogen metabolism 76–79
methane producers/non-producers 80
oxygen sensitivity 67
pH effect 78–79
plant compound metabolism 74–75
protein degradation 75
16S rRNA analysis of diversity
human 65–66
pig 66, 67
ruminants 67
short-chain fatty acid (SCFA) metabolism

acetate 74, 77
branched-chain compounds 78
butyrate 74, 76
lactate 77–78
pH 78–79
propionate 74, 77
syntrophy 67
zonation 70
intestinal microbiota
acquisition 80
in adults, individual composition 80
antibiotic-induced changes 81
development 80
host diet 80–81
in intestinal disease states 81
interaction with host 82
intestine
biofilms 70
faecal bacteria 70
microenvironments 70
intraspecies genomic variation 2–3
'island' patterns in taxonomic richness 31

Jaccard index *24*

Ketosynthase genes 211
Klebsiella 189

LA3191 *49*
Lachnospira 68, 166
Lactobacillus/Enterococcus 69
Legionella pneumophila 194
LUCA (last common universal ancestor) 117, 123

***M**arinococcus* (*Planococcus*) *halophilus* 237
marine phages 286–287
Megamonas hypermegale 74
Megasphaera 68, 166
Megasphaera elsdenii 74
Mesoplasma florum 115
Mesorhizobium loti 257
metagenomics 6–8
evaluation of horizontal gene transfer 285–286
marine bacteriophage diversity 286–287
Sargasso Sea library 285–286
Methanobrevibacter 164
Methanobrevibacter smithii 66, *74*
MGEs (mobile genetic elements) *278*, 281–284
integrons 282
ISs (insertion sequences) 281–282
plasmids 288–289
transposons 282–283
Microbes Online Database *282*
micromutationism 93
minimal cells 105–106
minimal gene set 117
minimal genomes 106–118, 124–125
comparative genomics 107–110
endosymbionts of insects 108–109
essential genes 111, *112*, *113*, *114*
experimental methods 110–111
gene loss 107
genetic information processing 111–112, 124
genome minimization 113–117
Haemophilus influenzae 107–108
minimal gene set concept 117
Mycoplasma genitalium 107–108, 109
proteins 107, 108
modular organization of organisms 97–98
Mycobacterium
M. leprae, genomic evolution 183
M. smegmatis 132
M. tuberculosis 206
Mycoplasma 115
M. genitalium 107, 109, 284
M. pneumoniae 108

***N**atronobacterium 227*, 228
Natronomonas 227, 228
N. pharaonis 229
neighbourhood analysis 39
Neisseria gonorrhoeae 194
Nesterenkonia (*Micrococcus*) *halobia* 237
network analysis 95
Nitrosomonas eutropha 194

OP11 *164*, 165
operons 202
'selfish operon' model 205, 285
oral bacteria 163–172
caries 165–166
culture-independent analysis (PCR) 165–166
biases 170–171
dentoalveolar abscesses 165
mouth as a habitat 164
oral diseases 163–164, 165, 166, 171
plaque 166
species diversity 164–165
spirochaetes 170
unculturable species, phylogeny 166–172
ORFans 126

PAGI1 189–191, 193, 195
PAGI2 190–191, 193, 195
PAGI3 190–191, 193
PAPI-1 191, 193
parasexuality, effect on diversity patterns 23–25
see also HGT
PCR analysis of unculturable bacteria 165–166
biases 170–171
'*Pelagibacter ubique*' (SAR-11) 12
Peptostreptococcus 166
PH0918 *49*
phage lambda 98, 280
phage-mediated transduction 280
phenotype origin 93–94
Escherichia coli, propanediol reductase 94
network analysis 95
power-law distribution 95–96
phenotypic evolution
demand theory 98
modularity 96, 97–98
networks 95–96
pleiotropy 94, 97
power laws 95
Pseudomonas fluorescens, WS genotype 99–100
robustness 96, 98
Photorhabdus luminescens subsp. *laumondii* *194*
phylogenetic clustering in communities 27–28
pKLC102 192–193, *194*
PKS (polyketide synthase pathways) 210–211
plague 175–176
plaque, dental associated bacteria 166
Plasmid Genome Database *282*
Potyviridae 132
power-law distributions 95–96
Prevotella 67
P. bryantii
xylanase activity 72
primary productivity, effect on taxonomic richness 31–32, *33*
Prochlorococcus 116, 280
P. marinus MED4 *114*
P. marinus SS120 *114*
Proteobacteria 69, 115, *164*, *194*, 237, 261
Providencia rettgeri 283
pseudogenes
Mycobacterium leprae 183
Yersinia 177–180
Pseudomonas 13, 165, 193
P. aeruginosa 193, 194, 265
chromosome regions 191
genomic islands 189–193
P. fluorescens 194, 265
phenotypic evolution, WS genotype 99–100
P. putida 190, *192*, 193
P. stutzeri 207
P. syringae 194
PSV (Pyrobaculum spherical virus) 133

***R**alstonia*
R. eutropha 264
R. metallidurans 264, *265*
Ralstonia sp. JS705 263–264
RAMPs (repair-associated mysterious proteins) 43, *45*, 46, 47, *48*
RecB-like nuclease *48*
recombination
Acinetobacter 207
Streptomyces 206–207
reticulate evolution 3
Rhodococcus sp. RHA1 106
Rickettsia 115
Rickettsia prowazekii 109
RISCs (RNA-induced silencing complexes) 51
RNA interference
in eukaryotes 51, *52*
in prokarytotes 50–60
robustness in phenotypic evolution 96, 98
Roseburia spp. *71*, *74*
R. intestinalis *71*, *76*, *77*
16S rRNA analysis of diversity
intestinal bacteria 65–67
oral bacteria 166, *167*, *168*, *169*
Rubrivivax gelatinosus *194*
rudiviruses 133–134
Acidianus filamentous virus 1 134
ARV-1 (Acidianus rod-shaped virus 1) 133
SIFV (*Sulfolobus islandicus filamentous virus*) 134
Sulfolobus virus SIRV-1 133–134
Sulfolobus virus SIRV-2 133–134
TTV-1 (*Thermoproteus virus 1*) 133
Ruminococcus 71
R. bromii *68*, *71*
R. flavefaciens *68*
cellulolytic enzymes 72, *73*
R. hydrogenotrophicus *74*, *76*
R. obeum *68*
R. torques *71*, *75*

***S**alinivibrio costicola* 237
Salmonella 265
genomic islands 188
S. enterica 193, *194*, 203
genomic evolution 183

Sargasso Sea 285–286
secondary ion mass spectrometry (SIMS) 8–9
secondary metabolites 205–210
  selective pressure 208–210
  substrate specificity 208
  survival value 208–209
Selenomonas *68*, 166
short-chain fatty acid (SCFA) production in intestine *74*, 75–78
  acetate *74*, 77
  branched-chain compounds 78
  butyrate *74*, 76–77
  lactate 77
  propionate *74*, 77
SIFV (*Sulfolobus islandicus filamentous virus*) 134
Silicibacter pomeroyi 12
soil as microbial habitat 201–202
soil prokaryotes 201–217
  streptomycetes 206
  taxon-richness 29
species–area relationships (SAR) 28
species concept 1–4
Spirochaeta halophila 237
Spirochaetes *164*
Spy1049 *49*
square bacteria 228–235
SSV-1 (*Sulfolobus spindle-shaped virus 1*) 134, 135–137
SSV-2 *134*, 135–137
SSV-3 *134*, *136*, 137, 138
SSV-K1 *134*, 135–137
SSV-RH *134*, 135
stable isotope probing (SIP) 8–9
Staphylococcus 216
Streptococcus agalactiae 3
Streptococcus/*Lactococcus* *69*
Streptomyces
  linear chromosome function 206–207, 209, *210*
  S. argillaceus 208
  S. avermitilis 209
  S. bluensis *204*
  S. coelicolor 206, 209
  S. glaucescens *204*
  S. griseus *204*, 208
streptomycetes
  antibiotic production 205–208
streptomycin biosynthesis 204, 211–213
Sulfolobus
  S. acidocaldarius 140
  S. islandicus 133, 140
  S. shibatae 133, 140
  S. solfataricus 140
Sulfolobus turreted icosahedral virus (STIV) *5*
  capsid protein *5*
Sulfolobus virus SIRV-1 133–134
Sulfolobus virus SIRV-2 133–134
'*Synergistes*' *164*
systems biology *95*

Taxa–area relationships 28–31
taxonomic richness
  distance-decay relationships 30
  intermediate disturbance hypothesis 32–34
  'island' patterns 31
  and primary productivity 31–32, *33*
  richness–area relationships 28–31
  richness–disturbance relationships *32*–34
  species–area relationships (SAR) 28
  taxa–area relationships (TAR) 28–31
    micro-organisms in water-filled holes *30*, 31
Thermanaerovibrio 170
Thermoproteus tenax 133
TM7 *164*
transcription
  coupling 207
  interference 207–208
transduction in horizontal gene transfer *278*, 279–280
transformation in horizontal gene transfer 278–279
transposons 282–283
Treponema 170
  T. pallidum 115
Trichomonas tenax 164
tRNA genes as integration sites 186, 188–190, 214, 257–261, 283
tRNA synthetases 128
Tropheryma whipplei 115
TTE2665 *49*
TTV-1 (*Thermoproteus virus 1*) 133
type IV secretion systems 281

Unculturable bacteria 163–172
  oral diseases 163–164, 165, 166, 171
  PCR analysis of oral microflora 165–166
    biases 170–171
  phylogenetic distribution of species 166–170
universal genes 124–125
  homologues 126–127
  phylogenetic core 128
  tRNA synthetases 128
Ureaplasma urealyticum 115

V*eillonella* *68*, *74*, 166
Vibrio
  V. cholerae 11, 126, 266, 283
  V. splendidus 11
viral evolution 93
viral lineages 5–6
viruses
  biogeography
    animal 132
    Archaea (*see* archaeal viruses)
    bacteriophage 133
    plant 132
  diversity measured by PCR 138–140

W*heat streak mosaic virus* 132
Wigglesworthia glossinidia 108
Wolbachia 115
Wolbachia wBm *114*

X*anthomonas*
  X. axonopodis *195*
  X. campestris *195*, 264

xenobiotic-degrading gene clusters 214
Xylella fastidiosa *195*, 264, 265

Y1724 *49*
Yersinia 265
  Y. enterocolitica *195*
  Y. pestis
    biovars 176, 177
    evolution from *Y. pseudotuberculosis* 180–182
    genes *178*
    genetic evolution *178*
    genome characteristics *176*
    genome comparison 176–180
    plasmids 177, 180
    pseudogenes 177–180
  Y. pseudotuberculosis *195*
    genome characteristics *176*
    genome comparison 176–180
    pseudogenes 177–180
YgcL *48*